Elementary Statistics

Bernard W. Lindgren
Donald A. Berry

University of Minnesota

Elementary Statistics

MACMILLAN PUBLISHING CO., INC.

New York

Collier Macmillan Publishers

London

Macmillan Publishing Co., Inc.
866 Third Avenue, New York, New York 10022

Collier Macmillan Canada, Ltd.

Library of Congress Cataloging in Publication Data

Lindgren, Bernard William, (date)
  Elementary statistics.

  Includes index.
  1. Statistics. I. Berry, Donald A., joint author.
II. Title.
QA276.12.L56   519.5   80-12920
ISBN 0-02-370790-9

Printing: 2 3 4 5 6 7 8   Year: 1 2 3 4 5 6 7 8

Preface

THIS BOOK IS INTENDED as a text for a one-quarter or a one-semester introductory course in statistics, for college freshmen or sophomores who have had some algebra. It could also serve as a text in a course for good high school seniors.

Originally conceived as a revision of the first author's *Basic Ideas of Statistics*, it is a totally different book, at an appreciably lower level of sophistication. Nevertheless, it is still our intention to convey ideas rather than to teach manipulation of formulas. Our goals for students using this book are modest, but we require that the student think and analyze while reading.

The ideas and methods of statistics are encountered as investigative tools in most disciplines and as operating tools in industry, business, and government, and we have drawn from many of these applications in our examples and problems—without emphasizing any particular one. But whatever the student's major interest, he or she will have encountered and will continue to encounter statistics in the "media," often relating to health or welfare, and we take advantage of this common experience in our use of many illustrations from newspapers, magazines, and television.

Coverage of statistical topics is more traditional than not, but we are not always constrained by tradition in discussing them. Our treatment of probability is deliberately less formal and less detailed than what has become common in elementary texts. Our approach is intuitive, without sacrificing mathematical correctness.

Not wanting students to lose track of ideas in the tedium of arithmetic, we have occasionally used artificial data involving convenient numbers and easy arithmetic. Moreover, we assume that each student has a hand calculator—one that does square roots as well as the elementary operations (+, –, ×, and ÷).

Most data sets used in our examples and problems are real, but some are more illustrative than useful. Indeed, some are included to point out that not all statistical analyses should be taken seriously, even when seriously presented. We do not hesitate to include grains of salt needed to swallow some research reports. In both problems and examples we often discuss points that may seem distracting to the student who is looking only for a tidy procedure to follow. However, we prefer that the student profit from such discussions rather than learn the mechanical procedure as a tool to be applied indiscriminately. Real situations are seldom so "clean" that results do not have to be hedged.

Sets of problems are provided at the end of each chapter, some identified as to the section to which they pertain, and some not. These are a most important part of any student's experience. Many problems should be attempted, although more problems are provided than most students will manage to work through. Answers are given in the back of the book for starred problems, often including comments, interpretations, warnings, and hints.

An instructor's manual is available from the publisher upon request. This includes answers for problems without answers in the text as well as solutions to the problems.

B. L.

D. B.

Contents

Elementary Statistics

1

Populations and Samples

STATISTICS AFFECT MANY FACETS of our lives. They play a central role in the search for a cure for cancer. The federal government uses them in making decisions. Even television programming is dominated by statistics, as seen in the following extract from an article in *TV Guide:*†

In Dunedin, Fla., a town of 20,000 near the Gulf of Mexico, there is a small computer with giant responsibilities. Each day it sends out telephonic signals to automatic printers all over the country, robots typing out reports that finally determine what will be available on television for you and me and every other American.

The computer is owned by the Media Research Division, A. C. Nielsen Co.—the same people who from the first years of television broadcasting have been giving us "The Ratings."

Ratings act as judge and jury for all network shows, deciding which ones stay on the air and which are canceled. Ratings are the basis for celebration and mourning, hirings, firings and the risking of billions of advertising dollars every year.

Behind the scenes where the money changes hands, the game of who pays how much for what is based on one scorekeeper—the Nielsen numbers. Networks base their charges for commercial time on ratings. A commercial minute on a top-rated show like Three's Company costs an advertiser $147,000, compared with a charge of $60,000 or even less for shows low in the rankings. That is why

†David Chagall, "Can You Believe the Ratings?," *TV Guide* **26**, Nos. 25 and 26 (1978). This and subsequent quotes are with permission of *TV Guide.*

new shows die so quickly—often within six weeks—and why old shows with respectable numbers hang on so long.

Tough, big-money decisions are based on those numbers. Harried advertising executives must account for expensive buys of television time with rising sales of their client's products. When the products don't move, there is a lot of explaining to do.

Television ratings are "statistics," and analyzing or interpreting them is a statistical activity.

You may share the common notion that "statistics" means facts and figures about the way things are, or the way things happened.† This is, to be sure, one of the meanings of the word:

- Labor statistics describe various aspects of the labor force.
- Vital statistics tell about births and deaths.
- Economic statistics describe various features of a nation's economy.
- Football statistics summarize what happened in a game or series of games.
- Accident statistics inform us about numbers, types, and locations of accidents.
- Census statistics give a multitude of facts about people in a country or other geographical unit.
- Medical statistics summarize the results of medical treatment and the incidence of diseases.
- Agricultural statistics report how much of various crops were grown or harvested.

Why all the interest in such "statistics?" Knowing what has happened, or the way things are, helps in making decisions, in understanding "laws" of nature, in making predictions, and in taking steps to change things for the better.

However, and this is a big "however," it is often impossible to know how things really are—to know the whole story, in any case. We must then deal with *partial* information, that is, with *samples*. Quoting further from *TV Guide:*

The ratings typically are based on a secret sample of 1170 TV homes. . . . A ratings number is the percentage . . . tuned to the same show. Nielsen then assumes the same proportion of all American TV households are watching that show.

†Indeed, the root of the word is "state," and it originally referred to facts and figures about the state as a political entity.

The essence of the modern subject of statistics is gathering and analyzing data that, while incomplete, permit drawing educated conclusions—with some element of uncertainty—about the way things really are.

This first chapter is intended as an introduction to some of the most basic ideas and language encountered in the design and analysis of statistical experiments. It should not be dwelt on at length at first reading (or at the beginning of a course), but you may find it useful to come back to after experiencing the material of Chapters 7 to 9. The one topic included here that should be absorbed the first time through is that of random numbers and their use in sampling.

1.1
SAMPLING

Sampling—checking out a small part of a whole to determine something about the whole—is a common procedure. For instance:

- We taste a few grapes to see if the whole bunch is sweet enough to serve.
- A heroin buyer (at least in TV crime shows) samples what he is about to buy to see if the quality of the lot is up to his standard.
- A new diet for young pigs is tried out on a sample of pigs before being marketed, to see how it will work for young pigs generally.

No doubt you can add to this list from your own experience.

In each case mentioned, the real interest is in the "whole" from which the sample is drawn—in the whole bunch of grapes, in the whole lot of heroin, and in the whole collection of pigs that might be given the new diet. The term *population* is used to denote the whole set or class of individuals to which any conclusions based on characteristics of a sample will be extended.

Sometimes, it is possible to survey a population completely, that is, to take a "100 percent sample." But surveys are costly, and the larger the group surveyed, the more costly the survey. Other reasons for settling for only a partial look or proper "sample" are apparent in the examples cited above. In the case of the grapes, testing the whole bunch would leave nothing to serve, and testing the quality of the heroin alters it chemically so that it is no longer usable. And the complete population of pigs that might be given a new diet is simply inaccessible—they have not all been born yet.

There is a basic assumption in using sample results to draw conclusions about a population. It is that the sample is "representative"—that the sample characteristics are just about the same as the population characteristics.

Example 1-1 *The Nielsen Family*

The complete population of TV viewers is there—it is real. But like many other populations of interest, it is practically impossible to get at. Information about TV viewing is vital to advertisers and to network vice-presidents, but they cannot afford complete surveys. What Nielsen does is described in the *TV Guide* article cited earlier, as follows:

Nielsen statisticians pick random locations from a Census Bureau Master list and make sure they are spread geographically across the country. Field representatives drive to the designated neighborhoods and try to persuade families to join their ratings panel. To help convince them, they offer prospects $25 to sign a contract, another $25 for each of the five years they stay on the panel and half of all TV-set repair costs during that period.

Once the homemaker says "Yes," the hardware moves in. Her house gets an audimeter wired from a closet or basement to every TV set in the house. The meter records when the sets are off or on and which channels are being viewed. Special telephone lines feed off the meter, allowing Nielsen's Florida computer to poll the device twice a day, collecting in a few seconds all the TV watching in that house during the past 24 hours. The family gets regularly mailed literature, chatty newsletters, notices and periodic calls from their field rep. The home has become a "Nielsen Family."

During any biweekly rating period, not all Nielsen families are counted. Some have agreed to allow the hardware into their homes but have not yet been visited by a rep and connected up. Others are lost because of malfunctions in the telephone lines or in the Nielsen meter. Nielsen has no way of knowing if these homes are watching TV. Thus, on a typical ratings day last winter, Nielsen reported viewing information from about 993 homes. Included are sets that burn out or go dark because of power failures, families out for the evening or even away on vacation—all are counted as "no viewing" for that day.

Concerning the use of sample information to draw conclusions about the population of TV viewers, the article says:

Nielsen has the responsibility to prove its sample truly represents all the United States. ▲

How can anyone be sure a sample is representative? The answer is that one can never be absolutely sure without a survey of the whole population to check the sample. (If this were done, of course, there would be no point to using the sample.) A sample that is not representative, favoring some segment of the population or other, is said to be *biased*.

In Example 1–1 it was stated that "Nielsen statisticians pick random locations. . . ." We shall see that the notion of a random selection is the key to having a good chance of obtaining a representative sample. But incorporating an element of randomness somewhere in the sampling process is not enough if it is defective in other ways. The next few examples illustrate how biases are apt to be introduced in certain nonrandom or only partially random sampling schemes.

Example 1-2 *The Nielsen Family (continued)*

Although Nielsen statisticians pick random locations, within constraints to assure geographical spread, the field representatives must then "try to persuade families to join their ratings panel," and they offer monetary inducements as part of their persuasion. Some families, perhaps those with lower incomes, would jump at the offer, perhaps for the opportunity of being a "Nielsen family" as well as for the financial considerations. And just as surely, some families would find it unattractive, families whose viewing habits would likely differ markedly from those who are eager to take part. A "top-10" ad agency executive is quoted in *TV Guide* as saying, "I would not let them put a meter and a telephone hookup in my house even though I have plenty of professional interest in television."

A rather disturbing feature of the sample of "Nielsen families" is that if it is biased, it would *stay* biased week after week—or, at best, change slowly as some families drop out and others are added to replace them. Another questionable aspect of the ratings is the awareness of those being surveyed that their choices are being used for crucial decisions. Can they forget this as they turn on their sets and select their programs? Referring to the diary system used by Nielsen and others to get detailed information about viewers and viewing habits, a research firm representative said (again in the *TV Guide* article): "People change their behavior when they know

they're being watched. Anything socially questionable or incon-
sistent with the image they want to project tends to be omitted." ▲

Example 1-3 *The Literary Digest Poll*

In 1936, the *Literary Digest* magazine published results of a poll of
2,400,000 individuals, showing 57 percent in favor of Alfred Landon
for president over Franklin Roosevelt. Of course, Roosevelt won
the election in one of the biggest landslides in history, winning 62
percent of the popular vote and losing only two states. The magazine
went bankrupt not long after this.

The poll was conducted by mailing questionnaires to 10,000,000
persons listed in telephone books and other sources that tended to
omit the poorer voters. It has been argued that, in 1936, political
preference tended to be according to economic level, so that neglecting the poor would bias the sample, no matter how large it was.
Others have given figures suggesting that this "selection bias" was
small because the segment omitted either did not vote or voted in
the same way as those included. But whatever this effect may have
been, relying on the sample of those who took the trouble to return
the questionnaire would have been enough to invalidate the poll.
(Such mailings typically experience a response rate of about 1 in 4.)
Those who respond are usually *not* representative of the population
in regard to the topic of the survey. ▲

Example 1-4 *A Marriage Study*

A campus newspaper carried the following classified ad under the
"Wanted" heading:

> Couples married 1 day to 10 years wanted for study on marital/
> sexual health. Conducted by U FSoS PhD candidate. Will receive
> free assessment of your relationship.

Such a study does not have much value if its results apply only to
the particular couples who take part in it. But, to what larger group
or population of married couples *would* this study apply? That
is, what population would the sample of participating couples
represent?

The sample is drawn from the campus community, and there
would surely be differences between campus couples and couples
from the population at large. Perhaps more serious is the fact that
only those willing to *volunteer* will be used. The circumstances and

personal characteristics of those willing to take part would certainly be very special and could introduce significant bias. ▲

Example 1-5 *Left-Handedness*

A high school teacher decides to estimate the proportion of people who write left-handed by counting the number of left-handed students in a class of 30 students as they are writing an examination. Does this sample have any chance of being representative?

Generally speaking, samples that are conveniently at hand are apt to be biased, in ways and degrees that are hard to determine. One reason that this particular sample may not reflect the general population is that left-handed children are now less apt to be trained to write right-handed than a few decades ago. But if the inference about left-handedness will only be extended to high school students, it may be that the class of 30 is nearly representative. The problem with such conveniently available samples is that there may be hidden biases. For instance, it could be that this is a chemistry class, which tends to be the choice of "bright" students, and that these students are more often left-handed than students of that age in general. ▲

Example 1-6 *Quota Sampling*

A college newspaper reporter interested in student opinion on a proposal to increase tuition decides to interview 40 students, 25 men and 15 women—reflecting the 5:3 ratio of men and women on that campus. Going out onto the campus walks, he interviews the first 25 men and 15 women he can collar.

But, despite his effort to obtain a representative sample with regard to sex of the student, the sample will likely be biased in various other ways. For instance, the students he can get hold of in a catch-as-catch-can approach are likely to have different characteristics than students in general. And the female students he approaches are apt to be those he finds personally attractive. ▲

In summary, bias is almost assured when:

- Samples are constituted of those who respond voluntarily.
- Samples are chosen by methods that omit segments of the population.
- Samples are used because they are handy.

- Samples are selected by someone with a vested interest in the outcome.
- Samples include individuals whose knowledge that they are being used as experimental subjects might influence their response.
- Samples are taken according to quotas (in attempting to obtain representativeness), including so many of each sex, each age, each race, each economic status—but without other restrictions on the interviewers.

It is possible for one of the various unsatisfactory ways of obtaining a sample to produce a sample that *is* representative of the target population, at least in regard to the particular characteristic being studied. But you would never know it, and the odds are against it. The safest rule is never to use voluntary response samples, convenient samples—or, in general, samples that neglect some segments of a population, as a basis for conclusions about the population.

Sometimes, however, economics or expediency makes it desirable to try to salvage something from a sample obtained by one of these unsatisfactory methods. Thus a voluntary response sample might be adjusted to have some validity if a follow-up study is made of the characteristics of those who do and those who do not respond. And if some segment of the population is neglected, conclusions may be usefully limited to the portion of the population actually represented, or possibly adjusted if the relation between the included and neglected portions is understood.

The bias in samples of convenience is sometimes obvious—as, for instance, in the case of a class of students in a physics course that (even today, perhaps) would enroll a larger proportion of males than in the campus community generally and in the population as a whole. Sometimes it is not obvious that there is any bias, and an unnoticed bias can invalidate any inferences. A qualified inference might be worth the risk, depending on the seriousness of the consequences and the cost of using a more satisfactory sampling method.

1.2
THE U.S. CENSUS—A 100 PERCENT SAMPLE?

In the Constitution of the United States we find in Article 1, Section 2, that representation of the states in the federal government shall be "according to their respective numbers." It goes on to say

The actual enumeration shall be made within three years after the first meeting of the Congress of the United States, and within every subsequent term of 10 years, in such manner as they shall by law direct.

The first enumeration, authorized in 1790, took 18 months to complete and showed a population of 3,929,326.

It is clear that because of births, deaths, and migration, the population is continually changing. Even now, the enumeration takes many weeks, which means that there are certain to be errors. Moreover, people do not all line up to be counted, and some are missed. George Washington was certain that the count in 1790 was too low, and spot checks after the 1970 census indicated that they may have missed 2.5 percent or more of the people. The task of counting illegal aliens was viewed as a special challenge of the 1980 census.

Before 1970 the count was done by census takers, with (beginning in the 1930s) some sampling to check on their footwork. In 1970 came the use of mailed census forms for about three-fifths of the population. In the first census, only five questions were asked of a household:

Name of the head of household.
Number of persons.
Number of males, females.
Number of slave, free.
Number 16 and over.

Now, in addition, much personal information is solicited, some at the behest of special interests outside of the government.

The census information is made available to the public, down to the point where individual confidentiality might be violated. Libraries have numerous publications of the Bureau of the Census containing selected results of the census, and computer tapes with census data are available from the Bureau.

The Bureau of the Census is a large, expensive department of the government. Even so, it cannot make accurate, 100 percent surveys of the population. One might wonder if it should try.

1.3
RANDOM SAMPLING

The primitive notion of "selecting at random" is a common one. Everyone is familiar with this in some form or other—to mention a few instances:

- Drawing names from a hat.
- Drawing winning lottery tickets from a drum.
- Drawing cards from a deck ("pick a card, any card").
- Drawing numbered chips from a box, for bingo.

The pervading idea is that the tickets or chips or cards are indistinguishable from each other, so that there is no reason to prefer one over the other—each has the "same chance" at a given draw. People seem to understand, intuitively, what this means, even though they may not be able to explain it clearly.

The concept of random selection is an ideal, and it can never be established whether a given selection process is truly random.† What is generally done, to try to achieve the ideal in practice, is to mix thoroughly and draw blindly—from a hat or drum or other container in which are placed slips of paper or chips or capsules, one representing each member of the population from which the selection is to be made.

Example 1-7 *The 1969 Draft Lottery*

Thirteen men sued the Selective Service System after the 1969 draft lottery, contending that the lottery was biased against birthdays near the end of the year. Pentagon manpower officials had said that men with numbers in the highest third (mid-200s to 366) would probably escape the draft entirely, and the challenge was based on the observation that birth dates early in the year tended to be drawn late in the lottery and have large numbers.

The lottery was conducted by selecting capsules containing slips labeled with the days of the year, one at a time and each at random from those remaining. The sequence of days in the year as drawn would provide the order in which, according to their birthdays, 18-year-olds would be called up in the draft. It is a fact, and intuitively clear, that when the numbers are drawn individually at random to form the sequence, every possible sequence has the same chance of occurring.

The U.S. District Court judge, although denying a temporary restraining order, said: "I find that the selection made in the December 1 drawing was not a perfectly random selection." He defined "ran-

---

†It is hard enough to determine when two amounts of tangible things are precisely equal, let alone when chances are equal.

domness'' as "that quality which makes any one sequence of [birth-days] as likely as the selection of any other sequence." So far, so good. But he went on to conclude that "there is a substantial dis-crepancy between a perfect selection, on the one hand, and the selection which resulted from the December 1 drawing, on the other." This is nonsense. If all sequences are equally likely, the one obtained is a random sequence—whatever quirks it may appear to have. Moreover, *any* sequence has some kind of quirk; if you look long and hard enough, you will find one.

The real question is this: What steps were taken to make the selection process a sequence of random selections? *The New York Times* printed a description by the man in charge, the chief of public information for the Selective Service System, Captain W. S. Pascoe:[†]

Over the weekend before the Dec. 1 drawing, Pascoe and Col. Charles R. Fox, under the scrutiny of John H. Adams, an editor of *U.S. News and World Report*, set up the lottery.

They started out with 366 cylindrical capsules, one and a half inches long and one inch in diameter. The caps at the ends were round.

The men counted out 31 capsules and inserted in them slips of paper with the January dates. The January capsules were then placed into a large, square wooden box and were pushed to one side of the box with a cardboard divider, leaving part of the box empty.

The 29 February capsules were then poured into the empty por-tion of the box, counted again, and then pushed with the divider into the January capsules. Thus, according to Pascoe, the January and February capsules were thoroughly mixed.

Then the same process was followed with each subsequent month, counting the capsules into the empty side of the box and then pushing them with the divider into the capsules of the pre-vious months.

The box was then shut, and Fox shook it several times. He then carried it up three flights of stairs, a process that Pascoe says further mixed the capsules.

The box was carried down the three flights shortly before the drawing began. In public view, the capsules were poured from the box into the two-foot-deep bowl.

Once in the bowl, the capsules were not stirred. The last draft lottery, in 1940, was conducted differently from the recent one. But officials remembered that when the capsules were stirred then some of them broke.

The persons who drew the capsules last month generally picked

[†]© 1970 by The New York Times Company. Reprinted by permission.

ones from the top, although once in a while they reached into the middle or the bottom of the bowl.

(It has been observed that in lotteries like this, conducted by having someone reach in and draw a capsule or a slip of paper, the person will usually pick from the top. If the contents of the drum or bowl are thoroughly mixed, this should not matter.)

Was the selection random? You be the judge. Was it fair? Why not (assuming no vested interest on the part of Pascoe and Co.)? No one knew the result far enough in advance to plan his birthday accordingly. Nevertheless, in the next draft lottery, the final order was determined by a double selection—one capsule drawn at random from the 366 dates matched with one capsule drawn at random from capsules numbered 1 to 366. Randomness in either selection would have sufficed, but perhaps the double randomness lent an additional air of fairness to the proceedings. ▲

The randomness of a selection process can be neither verified nor disproved by looking at a sequence of results and trying to spot any "nontypical" aspects or inequities. This is because there is always something peculiar about a sequence of random draws. Very "unlikely" things can and do happen.

Example 1-8 *Dealing Cards*

Dealing a card from the top of a deck is usually considered a random selection, provided that the deck has been thoroughly shuffled. (Incidentally, studies have suggested that cards are not usually shuffled well between deals.)

If you are in a poker game and someone is dealt a royal flush— a most unusual combination of cards—do you protest that the deal was not random? Probably not. Indeed, if the dealing is done by random selections, then it turns out that every combination of cards has the same chance of being dealt. And it is the occasional appearance of hands of special value that make the game what it is. ▲

When it comes to sampling human populations, one finds that it is not easy to pick a person at random without resorting to some such device as putting a slip of paper in a hat for each person. If it is some characteristic of the individual (such as political preference or age) that is of interest, it would suffice to note that characteristic, for each person, on the slip—the name would not be necessary. Picking a slip at random from the hat takes the place of selecting an individual and recording the characteristic of interest.

Chips are easier to manipulate than are slips of paper, and we shall usually speak of selecting chips from a bowl. We think of the selection at random of a chip from the bowl, which contains a chip for each population member, as a *model* for the random selection of an individual from the population. It can also be a means for accomplishing that selection, although what is usually done in practice involves random-number tables, to be taken up in the next section.

In selecting a chip from a bowl representing a population, one might wonder whether a chip that is selected should be replaced, so that the population being sampled is the same at each draw. In dealing cards, of course, each card is picked at random from those not yet dealt, and there is no replacement. In sampling a shipment of heroin or a bunch of grapes, anything tested would not be put back because it has been altered. And in surveying human populations, it would seem awkward if a person could be drawn a second time and interviewed again. Nevertheless, as it turns out, the analysis of a sampling process is easier if items selected *are* replaced and the population then mixed or shuffled to restore the symmetry of equal likelihood for the next draw. Moreover, we shall be using chip-from-bowl models for experiments in which selection with replacement is required for a correct representation of the experiment.

Random sampling that is carried out without ever replacing individuals already drawn is sometimes referred to as *simple random sampling*. The term *simple* refers to a comparison with more complicated or sophisticated schemes, such as cluster sampling and stratified sampling (see Example 6–19).

Random sampling

1. *With* replacement: Individuals are drawn at random, one at a time, each one selected being put back and thoroughly mixed with the rest of the population prior to the next selection.
2. *Without* replacement (simple random sampling): Individuals are drawn, one at a time, each time in a random selection from those not previously drawn.

The reason that random sampling with replacement is mathematically simpler is that each selection is precisely the same experiment, whereas when items are not replaced, the available population changes from one selection to the next. However, intuition suggests that sampling without replacement is more efficient. For in the extreme case in which sampling continues until the sample size equals the population size, the whole population would then be known if there were no replacement. But in sampling with replacement, there would always be some uncertainty about the population.

If the pool from which selections are made is huge compared to the size of the sample to be drawn, then it is practically immaterial whether individuals selected are replaced or not. Moreover, whether there is replacement or not, as long as the individual selections are random, *all individuals in the population have the same chance of being included* in a sample of given size.

A sampling method that gives each individual in a population the same chance of being included in the sample not only seems "fair" but has a good chance of producing a sample that represents the population in *every* characteristic. Moreover, as we shall see in later chapters, the mathematical structure of such a method permits us to assess the chances of representativeness. These are the two reasons for taking random sampling as the ideal to strive for in obtaining a sample.

1.4
RANDOM NUMBERS

In tossing an ordinary die you are, in effect, selecting one of the set of numbers {1,2,3,4,5,6} at random. If you do not have a die, you can accomplish the same thing by rolling a pencil with hexagonal cross section, the sides numbered from 1 to 6. Similarly, an integer from 1 to 50 can be selected at random by selecting a chip at random from a bowl containing 50 chips numbered from 1 to 50. And you can accomplish the same thing by spinning a "wheel of fortune" to select one of 50 equal sectors numbered from 1 to 50.

There is a method of selecting integers at random, however, that does not depend on the availability of dice, wheels of fortune, or numbered chips in a bowl—one that is not biased by inadequate mixing or insufficient spinning. This method employs what are called random-number tables, or tables of *random digits*.

A set of *random digits* is a random sample (*with* replacement) from the population of the 10 digits 0, 1, 2, . . . , 9.

Table VIII of Appendix B is an excerpt from a table of one million random digits prepared by the Rand Corporation in Santa Monica, California. This huge sample was constructed by computer simulation of random sampling with replacement from the 10 digits.

A table of random digits can be used for random sampling from any list of integers, 1, 2, . . . , N—with or without replacement. Taking random digits in pairs yields random integers in the range 00 to 99; in threes, random integers from 000 to 999; and so on. To restrict the range of integers to any specified list, simply ignore those drawn that are not on the list. For sampling without replacement, an integer that has been drawn once is just ignored if it occurs again.

Random digits can be used to form random decimals, to any prescribed number of decimal places. Putting a decimal point in front of a five-digit integer, for example, results in a five-place decimal number that can be regarded as chosen at random from all the numbers (rounded to five decimal places) between 0 and 1.

To use a table of random digits in producing a random sample of integers, one should use a *randomly chosen starting point* in the table. For informal purposes, starting wherever your finger lands when placed casually on any page is good enough—but do not always use the same page! (Some feel that it is safer to get a random starting point by first using the random drop of a finger to select a preliminary sequence of digits for selection of a page number, column number, and row number for the beginning of the sample sequence.)

To use a random-digit table for sampling a population of individuals, the individuals must be numbered—say, from 1 to N if there are N individuals in all. Selecting a number at random from 1 to N by means of the random-digit table then selects the individual who has been assigned that number.

Example 1-9

Suppose that a random sample is needed from a list of 4100 students. Students on the list are numbered from 0001 to 4100, and the table of random digits is then used to select integers at random on this range. Starting at an arbitrary or random point in the table and taking digits in successive groups of four yields a sequence of random integers in the range 0000 to 9999. Ignoring 0000 and anything larger than 4100 results in a sequence of integers in the desired range. For sampling without replacement, one would ignore any integer that has already occurred in the sequence.

Because the random-digit table is arranged in groups of five, horizontally and vertically, it may be convenient (and it is proper) to use just the first four out of five, ignoring the fifth digit in a group. One can proceed either across the table in rows, or down in columns (or through it in any systematic fashion). ▲

Example 1-10 *White Shoots Black*

A black man was shot and killed by a white policeman in Eagan, Minnesota, a predominantly white suburban community of about 20,000, under questionable circumstances. A college newspaper† conducted a survey to determine whether the Eagan townspeople believed the shooting to be justified. The poll was conducted by phone, using four-digit random numbers, alternating between the two exchanges that serve the community.

Random-digit dialing does have the advantage, over using directories, of not neglecting households with unlisted numbers. Moreover, it can be easily automated, with a tape recorder to ask the questions and record the replies.

Restricting the poll to those with telephones would not be apt to introduce a bias in this case, since in this particular community there are few households without a telephone. But there may be other problems, as the published report recognized:

> More than 300 . . . numbers were dialed to obtain 120 respondents. Some effort was made to call back those who were not originally at home. This was not completely successful. Although the *Daily* tried to alternate between men and women, this was not always possible, because there were more women than men at home. ▲

We mentioned earlier that tables of random digits are ordinarily constructed by means of a computer. Numerical analysts have developed schemes which, although really nonrandom, produce num-

†*Minnesota Daily*, February 6, 1976.

bers that are about as "random" as any drawn from a hat. Some hand calculators have a "RAND" key that will produce, as it is pressed repeatedly, a string of two-digit random numbers. The RAND key on some desk digital calculators will produce eight-digit decimals in the interval 0 to 1, the sequence depending on a starting "seed" number that is chosen arbitrarily and entered into the calculator's storage.†

1.5
CONCEPTUAL POPULATIONS

So far, in discussing sampling, we have assumed that there is an identifiable group of individuals—the population—in which the real interest lies. In sampling at random from this population, all members of the population are equally likely. Variation in a particular characteristic from one observation to another is due to the differences in the population from one member to another.

We want to be able to use the language of "population" and "sampling at random from the population" in situations in which there is not an identifiable group of tangible subjects constituting an actual population but in which there is variation from one observation to another that is usefully conceived of as "random." In the sense that we can at least think about such populations, they are *conceptual*.

Sometimes data are gathered from samples of actual persons or things, but the collection of all such individuals to which we should like to extend an inference is not available to be listed or identified because the individuals are not yet all produced. For instance, one might observe:

- The weight of a package of cereal.
- The length of a fish of given species in a certain region.
- The time it takes for a rat to run a maze.

Here the interest lies in a class of individuals—packages of cereal, fish, rats—that are being produced indefinitely, with perhaps no end in sight. That class of individuals can be thought of as a population.

†One calculator, starting with the seed, multiplies each number by 29 and drops the integer portion of the product, using the decimal part as the next number in the sequence.

An important type of experiment involving populations that are usually not identifiable deals with the effect of *treatment*. For instance, one might observe:

- The degree of relief from pain when a prescribed dosage is administered to a patient.
- The amount harvested from a specified amount of seed planted under given conditions.
- The score in an achievement test following a prescribed type of training.
- The failure or success of a juvenile offender in keeping out of further trouble following a certain correctional program.
- The yield from a chemical reaction using a certain catalyst.

Here the interest is in the treatment, and different individuals and different samples of experimental materials respond differently to the treatment. The population is a population of treated individuals, but the individuals who will actually and ultimately be treated are not known and may not even be around yet. Indeed, if the experiment turns out unfavorably for the treatment, the treatment may be discarded without ever being used on other individuals. But we can think about the "population" of experimental subjects that might have been used.

Selecting an individual at random from a population that is partly available and in large part unavailable is not an easy matter. In the case of weights of packages produced by a certain machine, one might assume that there is a kind of time invariance that permits simply taking successive packages as they come off the line and regarding them as randomly drawn from the population of packages that could be produced. In the case of fish, we might select one at random from those already produced and presently swimming in a lake, and take this to be a random selection from the conceptual population of all fish of the given species that might ever be produced. In the case of individuals who are to be treated for the purpose of learning the effectiveness of the treatment on patients as yet unknown, one would have to try to sample from a pool of individuals as much like those who will need treatment as possible. Samples of convenience are often used in such studies, and we have already pointed out the possibilities of bias with such samples. There is no easy answer.

Some statistical problems involve data that are obtained by doing an "experiment" repeatedly, but the experiment may not involve selecting individual subjects from a population of tangible persons or things, available or not. Some examples:

- Making a measurement (e.g., of time, dimension, chemical property, or IQ).
- Keeping track of the total rainfall in a month at a particular place.
- Determining the radiation at a certain distance from a nuclear phenomenon.
- Driving over a certain course in traffic to see how long it takes.
- Counting the number of incoming telephone calls to a given exchange in a given period of time.

Repeating such experiments will produce variable results. The phrase "random variation" is used to mean that there is variation introduced by causes that are just too complex to be accounted for, or even understood. The object of collecting data is to understand the nature of the randomness and to extract from the randomness the essential component that is really of interest.

An experiment of this type, but simpler than those listed above, is the toss of a coin. This can be thought of as picking, at random, one of the two sides of the coin; and in a sense those two sides can be thought of as constituting a "population." If there is no coin handy, on the other hand, one could get along with a bowl containing just two chips, one marked "Heads" and the other marked "Tails." Selecting a chip at random from the bowl is like tossing a coin, in that the "chances" are the same; and people would usually accept one experiment in lieu of the other. Moreover, a bowl containing any even number of chips marked "Heads" and "Tails" in equal numbers would serve just as well. (Note that the symmetry of the coin makes it possible actually to construct an equivalent chip-from-bowl experiment.)

The toss of a thumbtack is similar to the toss of a coin, but there is a complication. Tossing a thumbtack involves "chance," for we cannot know whether the tack will come to rest with its point either straight up (U) or tipped over (D). The toss "selects" one of these two outcomes, but this selection is not a "random selection" in that the outcomes are not equally likely. However, one can imagine an *equivalent* chip-from-bowl experiment, with some chips marked U and some D in numbers whose proportion makes the random selection of a chip equivalent to a toss of the thumbtack. For instance, if D is twice as likely as U, the bowl would contain twice as many D-chips as U-chips. The chips in the bowl can be thought of as a population, even though we do not know the ratio of U's to D's.

Example 1-11 *Particle Counts*

A certain substance emits α-particles at a steady rate, overall, but at haphazard points in time. Suppose that one counts the number of particles emitted in a 10-second interval. This is an experiment that "selects" one of the integers 0, 1, 2, . . .—but *not at random*, which is to say, not with equal chances. However, one can imagine a bowl of chips—some numbered 0, some numbered 1, some numbered 2, and so on, with numbers of 0-chips, 1-chips, and so on, in proportion to their chances of occurring as a count of α-particles. Drawing a chip at random from this bowl is equivalent to counting the particles emitted in a 10-second interval of time. ▲

Example 1-12 *Measuring a Distance*

Suppose that it is desired to measure the length of a marathon course. However this is done, there are numerous sources of error that make the measured distance come out differently every time. The measurement can be any one of the infinitely many values in some interval of values (near 26 miles), if one uses—as is customary— a "continuous" scale of measurement. (With scales whose reading involves rounding off, the number of values may be finite, but enormous.) Making a measurement selects, in effect, one of these possible values—but not as a random selection. Again, however, we can imagine a bowl with infinitely many chips, marked with possible values in such proportions that the experiment of drawing a chip at random from the bowl is equivalent to making the measurement. ▲

Since the definition given earlier for "random sampling" assumed a finite, actual population, a definition of "random sample" that will cover situations involving conceptual populations is in order:

When an experiment of chance is repeated in n independent trials, the results are said to constitute a *random sample of size n.*

Obtaining such a sample is the same as random sampling with replacement from the equivalent chip-from-bowl model.

(For the moment, the qualifier "independent" may be interpreted in an intuitive sense: The experiment, at each trial, is the same experiment no matter what the outcomes of any other trials may be. The notion will be taken up again in Chapters 4 and 5.)

What was termed "simple random sampling" earlier does not yield a random sample in this special sense, for drawing a population member and then *not* replacing it leaves an altered population for the next draw. However, sampling one at a time from a finite population at random and *with* replacement does produce a random sample in the sense defined here. As pointed out before, if the sample size is small compared to the population size, the difference between sampling with replacement and sampling without replacement is negligible.

1.6
SOME PROBLEMS IN COLLECTING DATA

Random sampling produces data that have an excellent chance of being representative. But it is seldom easy to get a truly random sample from the target population. The following examples illustrate some of the problems involved.

Example 1-13 *The Pinto Gasoline Tank*

In 1977 to 1978 safety authorities were prodded to investigate allegations that the Ford Pinto gasoline tank was vulnerable to rupture in a rear-end crash. A series of newspaper articles presented arguments involving consumer advocates, the manufacturer, and the National Highway Traffic Safety Administration. The manufacturer quoted figures from NHTSA showing that 12 of 848 people who died in 1975 in accidents involving fires were in Pintos, in contrast with a claim that 70 or more persons burned to death each year because of fuel-tank fires in Pintos.

The data sources are unreliable, for most states do not code the type of vehicle (e.g., Pinto, Maverick, Mustang) on police accident reports and, further, have no coding for "fire." So it is necessary to hand-sort all accident records, hoping that the investigating officers at least noted when a fire occurred.

This situation is typical of "retrospective" studies—those that hunt up data on things that have happened in the past. In this case, engineers did conduct "laboratory" tests to determine the amount of leakage upon impact at various impact speeds, but these results are not easy to extend to actual crashes.

A 1980 criminal trial of Ford Motor Company in an Indiana court resulted in a verdict of "not guilty" on three counts of reckless homicide. ▲

Example 1-14 *Crime Statistics*

A basic question in the allocation of police resources is how the amount of crime in an area is related to the amount of police manpower and equipment in the area. Good data are hard to get, for several reasons. (1) Continual shifting of crime patterns makes any controlled study difficult. In particular, sociologic or economic changes may produce changes in crime patterns that are independent of policing. (2) The true amount of crime is never known because only a fraction of it is reported. The proportion reported may increase if police manpower increases. (3) More intensive policing in an area may just reapportion crime to other areas, in amounts hard to detect because of random fluctuations.

[An account of "Police Manpower versus Crime" by J. Press appears in Tanur et al., eds., *Statistics: A Guide to the Unknown* (San Francisco: Holden-Day, Inc., 1972).] ▲

It is sometimes necessary, perhaps because of the length of time that would be required to carry out an investigation in a manner that is more at the control of the investigator, to try to learn something from what has already happened. A *retrospective* study uses existing records for making statistical inferences. This is in contrast to a *prospective* study, in which the experimental design and the experimental subjects are at the choice of the investigator. In particular, he can use random samples, rather than data whose mode of collection is not under his control and, indeed, sometimes not known.

When carried out by experienced and knowledgeable investigators, a retrospective study can be useful, as in the following example.

Example 1-15 *The National Halothane Study*

A few years after achieving rather wide acceptance as a general anesthetic, *halothane* came under suspicion when reports appeared of massive and fatal liver damage in patients who had recently been given this anesthetic. A national committee was appointed to gather evidence that might help answer the questions that had been raised. Laboratory experiments with mice would not help, since there was no conclusive reason to believe that mice would behave like human beings in this regard.

Because the death rate was small (around 2 percent), a large number of trials would be needed to compare the various anesthetics. A prospective study involving medical treatment of human subjects is always controversial on ethical grounds, and might take several years in this case, so a retrospective study was conducted. Fortunately, a particular group of 34 hospitals that kept good records had information on 850,000 operations, including facts about deaths, surgical procedures, anesthetics, sex, age, and status prior to surgery for each patient.

Because the experiment was not planned so that patients were comparable in age, type of operation, and so on, there were substantial differences in the kinds of patients receiving the various anesthetics. For instance, cyclopropane had been given several times more often than halothane to patients who were older or in poor physical condition. Complicated adjustments were necessary to sort out the "true" rates from the various confounding variables. After adjustment, death rates were as follows:

Halothane	2.1 percent
Pentothal	2.0 percent
Cyclopropane	2.6 percent
Ether	2.0 percent
Other	2.5 percent

As to liver damage, the total deaths from this source were few and autopsies were not always conducted, so firm conclusions were not possible.

The study was reported in 1969 and published by the U.S. Government Printing Office. A brief account is given in Tanur et al., eds., *Statistics: A Guide to the Unknown* (San Francisco: Holden-Day, Inc., 1972) in the chapter "Safety of Anesthetics," by Moses and Mosteller (two statisticians who participated in the study).

(In 1978, in response to a question by one of the authors, an anesthetist said: "Oh, we don't use halothane anymore—they found it was causing damage to the liver.") ▲

It is especially difficult to get reliable data when people are surveyed about very personal practices or opinions. Suppose that a man is asked: "Do you beat your wife?" He may realize that he does but not want to admit it to the interviewer—or even in a written questionnaire, despite assurances of anonymity. Indeed, he may have himself convinced that he does not beat his wife, even if he does.

One way around an individual's reluctance to respond to personal questions is the method of "randomized response." The idea of the method is to introduce an alternative, innocuous question at random, in such a way that only the person responding knows which question is being answered. Each respondent does a random experiment of known structure to determine whether he or she then answers the sensitive question or the harmless one. With this procedure the interviewer does not know which question has been answered. The respondent may or may not understand this, but there would appear to be a better chance of an honest answer with this randomized response method than with direct questioning.

Example 1-16 *A Private Affairs Survey*

To find out what proportion of married people have "affairs," given a suitable sample, the interviewers question can be posed as follows:

Take a coin and toss it, but do not show me how it lands. If it falls heads, answer Question 1. If it falls tails, answer Question 2:

Question 1. Have you had an extramarital affair?
Question 2. Is the next to the last digit of your phone number an even number?

Given a sample of responses, and assuming that the digit in Question 2 is equally likely to be even or odd, it is possible to estimate the desired proportion, without knowing whether an individual response is to Question 1 or to Question 2. (This will be discussed in a later chapter, but you may be able to figure out how to do it on your own, even now.) ▲

1.7
SAMPLES FOR EVALUATING A TREATMENT

A common application of statistical inference—drawing conclusions about a population on the basis of sample data—is in the evaluation of a *treatment*. A "treatment" is some particular regimen, an environment, or the application of an agent, which may be intended to improve a response, but not necessarily. For example, one may wish to determine whether

- Different catalysts have the same effect on yield of a production process.
- A new type of teaching increases student comprehension.
- A new drug offers more relief from pain than do standard drugs.
- Smoking affects health adversely.
- An advertising campaign increases sales.
- The time of conception affects the sex of an infant.

Variability in response to treatment is the result of individual differences or of variation in the experimental material. The group of individuals or the particular specimens that are to be treated for the purpose of evaluating the treatment can be thought of as a sample from the population of all who are or might be treated. As discussed before, one is rarely in a position to know that a sample is random in such situations. The usual procedure is to analyze the sample one *can* get as though it were random, stating the *assumption* of randomness clearly in reporting conclusions, while privately worrying that it is not really satisfied.

An aspect of a study of treatment effect that is sometimes overlooked is the need for *comparison*. Without knowing how individuals respond who are *not* given the contemplated treatment (or who are given some standard treatment) there is not much that can be said about its relative virtues.

Example 1-17   *"Scared Straight"*

A program with this title, the subject of a 1979 TV documentary, uses scare tactics by hardened prison inmates in an attempt to keep youthful offenders from entering a career of crime. A success rate of about 80 percent was claimed as evidence of the success of the treatment. However, this figure does not mean much by itself, for one would want to know what the success rate is when the offenders are *not* given this treatment. Indeed, another study concluded that those who are not given such a treatment fare even better—among both offenders and nonoffenders. ▲

When a comparison is to be made between responses of treated and untreated individuals, the treated individuals constitute the *treatment group.* This is a sample from the population of all those who might be given the treatment. Individuals who are not given the treatment are referred to as *controls.* It may be that one has had lots of experience with untreated individuals and knows essentially all about the control population. If not, a sample of individuals is drawn—the *control group*—whose responses will be compared with those of the individuals in the treatment group. (For various reasons not appropriate for discussion at this point, applied statisticians recommend always using a control group.) Ideally, both samples would be random samples.

Two kinds of studies are common. *Observational* studies are those in which the investigator observes what happens or has happened, without having had the opportunity to select the subjects used in the study or to determine which ones get the treatment. *Controlled* experiments are those in which the investigator plays a role in the design of the experiment, and in particular, *does* select the subjects who will be given the treatment. And if the treatment is assigned to subjects by a random scheme (such as tossing a coin to decide if a subject gets the treatment), there is some basis for drawing a conclusion about the treatment. The conclusion may be weaker than one might like, depending on the representativeness of the subjects in the experiment. For example, if they are all white, or all male, or all young, extending conclusions to a wider class of individuals may not be warranted and should only be done with caution.

About the most that can be gained from observational studies is a suggestion that further studies are warranted. For, it may be that when the choice of treatment is up to the individual, the factors that determine his choice also determine his response. Those factors are then "confounded" with any treatment effect. A controlled experiment, on the other hand, can give statistical evidence of an effect. But even more convincing would be an analysis (physical, chemical, physiological, or whatever is appropriate) that explains a mechanism of causation.

Example 1–18 *Aortocoronary-Artery-Bypass*

The need for control seems so obvious that it is disconcerting to read† "the insistence on the use of prospective randomized studies . . . reflects a naive obsession with this research tool." The need

†M. E. DeBackey and G. M. Lawrie, "Aortocoronary-Artery-Bypass: Assessment After 13 Years," *J. Am. Med. Assoc.* **239**, 837–839 (1978).

for controlled studies is well illustrated in the field of coronary surgery:[†]

More than 30 different surgical procedures have been employed in past years to relieve angina and improve myocardial blood supply. In only two instances were appropriate control studies performed. Internal mammary ligation was shown to be of no value in the treatment of angina by two randomized studies. Internal mammary implantation was shown to result in no improvement in survival by a randomized study. Two other well-conducted follow-up studies demonstrated no improvement in exercise performance or ischemic responses to hypoxia. Unfortunately, before these studies were carried out, thousands of needless operations had been performed. There is a clear and continuing need for further randomized studies to evaluate other surgical procedures employed in coronary artery disease. ▲

Example 1-19 *Smoking and Health*

An early study of the effect of smoking on health[‡] showed a 65 percent survival rate for nonsmokers and a 43 percent survival rate for smokers. The "treatment" is smoking, and its application was at the choice of the subject—it was not controlled. Since that time many studies have been reported with similar results, but since it is not possible to assign the treatment to the subjects in a free society, a controlled experiment is not feasible. A major criticism of the existing data is that they continue to confound the treatment with factors that cause the subject to choose the treatment. ▲

In clinical trials for the evaluation of a medical treatment, besides randomizing in the assignment of subjects to treatment and control, one should take steps to avoid confounding results with a "placebo effect." A placebo is a nontherapeutic "treatment" resembling the actual treatment as much as possible—for example, a sugar pill that looks like the real thing. It has been found that people respond differently when they know or think they are being treated than if they do not. When possible, therefore, in what are called *blind* trials, the control subjects are given a placebo; but neither control nor treatment patients know what they are getting.

[†]H. N. Hultgren et al., "Aortocoronary-Artery-Bypass Assessment After 13 Years," *J. Am. Med. Assoc.* **240**, 1353-1354 (1978). (References are given there for the studies mentioned.)

[‡]Raymond Pearl, "Tobacco Smoking and Longevity," *Science* **87**, 216-217 (1938).

There are also reasons to keep from the patient's physician the knowledge of whether the patient is getting the treatment or is a control. The physician can affect the outcome of the trial, for example, by advising control patients to modify their diet or by giving some other supplementary treatment. And there is evidence that if the physician thinks his patient is getting the treatment, this can affect the outcome. Thus, trials designed to test the effectiveness of drugs in treating disease are often *double-blind*. That is, neither patient nor physician knows whether the patient is in the treatment group or in the control group. (Clearly, a treatment such as heart-bypass surgery cannot be double-blind!)

1.8
FROM SAMPLE TO POPULATION—A CAVEAT

Drawing conclusions about a population from sample results always has an element of risk, even when straightforward and based on a random sample from the population. When inferences are in the hands of those who have a point to make, however, one must be skeptical. Conclusions are often drawn that are unwarranted, possibly having little to do with the experiment carried out. A skeptical attitude seems especially needed when one encounters advertising—particularly in viewing TV commercials, which are often hit-and-run assaults on credibility.

Example 1–20 *Beer Tasting*

A TV commercial for Blatz beer showed a skeptical drinker choosing Blatz, from two cans with labels covered, as the better-tasting light beer. He was then properly astounded to learn that the one he did not choose was his favorite brand, Pabst. The one conducting the test asked him: "What beer are you going to buy now?" The drinker (still reeling from the blow) replied: "I'm going to buy both of them and do the test again!"

The Blatz people were content with his reply, presumably thinking that viewers would be convinced by the one test shown. The drinker, on the other hand, realized the possibility of a random element in the test and the need for more data—*he* had not been convinced by the single trial. (And, of course, we'll never know how many times they tried the test before finding someone who chose Blatz!) ▲

Example 1-21 *Pepsi Versus Coke*

In 1977, Pepsi decided to challenge Coke's 3-to-1 sales lead in Dallas. (Nationally, their shares of the market were 17.4 percent for Pepsi, 26.2 percent for Coke.) A promotion was concocted which "showed" that more than half the Coke drinkers tested preferred Pepsi's flavor. During the test, Coke was served in a glass marked Q and Pepsi in a glass marked M. (The subjects did not know this code.) Within a year the Coke lead was cut to 2 to 1.

Naturally, Coke officials were perturbed and conducted their own test. They put Coke in all glasses, some marked M and some marked Q. Most people tested expressed a preference for M, a result that was headlined in Coke ads: "Coke beats Coke." Pepsi countered by doing the test again with Pepsi in an L-glass, Coke in one marked S. People preferred Pepsi (or L!). Coke executives again cried "foul," claiming that people prefer the letter L to the letter S. ▲

Example 1-22 *Colgate's MFP and Cavities*

Table 1-1 is taken from a glossy brochure sent out by the Colgate people to those who wrote in for details of test results referred to in a TV commercial. As might be expected (they surely would not publicize results that were in the wrong direction), children using Colgate with MFP fluoride developed fewer cavities, in each test. Whether the noted differences are explainable as chance variation or must be attributed to the effectiveness of MFP will be taken up in Chapter 9. One might wonder, though, whether comparisons with a "nontherapeutic" toothpaste are helpful to consumers. when nearly all brands of toothpaste on drug store shelves claim to be therapeutic.

The most remarkable feature of the results, evident to anyone (except Colgate?) without a statistician's help, is that by far the fewest cavities were found among children whose brushing was supervised at least twice a day or who lived in an area with naturally occurring fluoride in the water supply. ▲

Table 1-1 Results of Dental Studies

Study	Location	Period	Ages	Brushing
I	Southeastern United States	2 yr	6–17	Supervised
II	New England	3 yr	8–14	Home
III	Texas	3 yr	9–13	Home
IV	Southeastern United States	2 yr	9–15	2×, supervised
V	Texas	22 months	9–19	3×, supervised
VI	Denmark	2.5 yr	10–21	1×, supervised and at home
VII	Australia	2 yr	11–13	Home

Example 1-23 *Bikes Versus Autos*

A newspaper column of questions and answers included the following:

Q. My wife and I are thinking of taking up bicycling as a hobby but, in view of the number of cars on the road, we want to know how risky it is?—M. L.

Meals	Water	Toothpaste	Number of Children	New Cavities
Institution		MFP	153	2.99
		Leading SF	140	4.11
		Gardol	135	4.18
Home		MFP	208	19.98
		Leading SF	201	22.39
		Gardol	193	21.64
Home	Fluoride	MFP	172	4.60
		Leading SF	193	4.83
		Non-therapeutic	190	5.57
		Gardol	163	5.37
Orphanage		MFP	149	2.58
		Modified MFP	141	2.60
		Non-therapeutic	156	3.94
Institution		MFP	228	3.89
		Stable SF	229	4.49
		Non-therapeutic	229	4.91
Home		MFP	294	10.88
		Non-therapeutic	284	13.41
Home		MFP	422	9.76
		Stable SF	422	9.67
		Non-therapeutic	422	12.23

A. Statistics show you're safer on two wheels than on four—some 800 bicyclists are killed in accidents yearly, while 56,000 motorists die during the same time.

The "statistics" quoted may or may not be accurate. But they have little to do with the question. The same "logic" would lead to the conclusion that climbing Mount Everest is safer than riding in an automobile because only a few climbers are killed per year. (The phrase "statistics show" seems invariably to be followed by something not substantiated by statistics.) ▲

KEY VOCABULARY

Sample
Population
"Representative" sample
Selection bias
Response bias
Sample of convenience
Random sampling
Sampling with and without replacement
Simple random sampling
Random digits
Conceptual population
Retrospective and prospective studies
Control group
Treatment
Observational study
Controlled experiment
Placebo effect
Double-blind experiment

QUESTIONS

1. Is there any way of guaranteeing that a sample will be representative? How, or why not?
2. Would a poll of adults in your state that is based on telephone numbers (nowadays, in contrast with 1936) have a "selection bias"?
3. A sign is posted by elevator doors in the University Hospital building asking for normal volunteers to fill out a questionnaire for a psychiatric study. Is this a good way to get a "control" sample?
4. Suppose that you want a sample of adults. To get one, you take a random sample of telephone numbers and talk to whoever answers the telephone when you call. Does everyone (in tele-

phone households) have the same chance of being included in the sample?

5. Why do we like to use *random sampling* in gathering data for an inference about a population?

6. Explain how a table of random digits is used in obtaining a random sample.

7. When a new cold remedy is to be tested on a sample of individuals, what do we mean by the "treatment population?"

8. When you toss a coin, what "population" are you sampling?

9. Why is it difficult to use retrospective studies as a basis for a statistical inference?

10. What is a "controlled study" in the evaluation of a treatment?

11. Why is a controlled experiment better for inference than an observational study?

12. Are observational studies of any use at all?

13. What is the purpose of "double-blinding" in a clinical trial?

14. Is it important to know who conducted and who sponsored a given study?

PROBLEMS

Sections 1.1-1.4

*1. To assess public support for a certain tax measure, a school board sent letters to 3000 randomly selected members of the community, asking their opinions. Of 1192 who responded, 714 indicated support. Is the board justified in accepting this result as indicating public support? If not, how might they attempt to validate their conclusion?

2. Suppose that you are a university student planning to take an introductory physics course from a certain professor. In a survey taken among the 12 students awarded Bachelor's degrees in physics from the university in the previous year, this professor was unanimously regarded as being an "excellent" teacher. Does this give you reason to believe that you will also find her to be "excellent"?

*3. A telephone survey is conducted to determine the number of persons per household. A first attempt obtains 313 responses out of calls to 500 residential telephones chosen at random. A second try results in 41 more responses from the 187 not at home at the time of the first call. Is the sample of 354 useful in learning about the number of persons per household? Explain.

4. To get an idea as to college student opinion about drugs, is it a good plan to send five interviewers out on the central mall of the campus during a noon hour to interview as many students as they can? Explain.

*5. A department of education in a state wants to conduct a survey of teacher opinion, and obtains a sample by first selecting three schools at random from the state and then either (a) surveying each teacher in the schools selected, or (b) taking a random sample of teachers from each school selected. Does either approach yield a random sample of teachers in the state?

6. An educator wants to find out information about the families represented in a school system and decides to pick a random sample of children from those registered in the schools in the system for a survey—to ask such questions as: How large is your family? How much education do your parents have? And so on. Is there anything wrong with this sampling plan?

*7. To obtain a sample of 200 university students to learn about student opinions on abortion, which of the following methods would have a good chance of getting representative responses? (And why are the other methods not as good?)

   (1) Station interviewers at the door of a building who would question whatever students they could until they got 200 interviews.

   (2) Distribute questionnaires in a mathematics lecture section of 200 students, wait for students to fill them out, and collect them.

   (3) Use a table of random numbers to pick 200 pages of the student directory and again to pick a name on each page; the person named would be interviewed by phone or in person, the interviewer persisting until he got the interview or established that the person named is no longer a student.

   (4) Using the student directory, take every 100th name, starting at the top of a randomly chosen page. Interview the persons selected as in method (3).

   (5) Send a team of five interviewers out to question students lounging on the campus mall during the noon hour.

8. At a large university, which method of obtaining a sample

would be best for learning about the feelings of fraternity and sorority members toward their organizations? Explain why the other methods are not as good.

(a) Pick three houses at random from the list of fraternity and sorority houses, and solicit interviews with each member of these groups.

(b) Visit *every* fraternity and sorority house and interview two members of each organization.

(c) Stop people on campus and ask if they are members of a fraternity or sorority; if so, interview them. Keep this up until 50 interviews are obtained.

(d) Interview 50 workers during preparations for the annual "Campus Carnival," sponsored by the "Greeks."

(e) Obtain membership lists from each organization, number the members on all lists, and use a table of random numbers to select 50 individuals for a personal interview.

*9. Starting at a random point in Table VIII of Appendix B, obtain a sample of 50 digits.

(a) Are these equally divided among the 10 digits? Should they be?

(b) Are they equally divided between even and odd digits? Should they be?

10. Explain how you could use the table of random digits to obtain a random sample of 20 pages in a book of 482 pages.

*11. How could you use the table of random digits (Table VIII of Appendix B) to simulate the throw of two dice?

12. The caller in a bingo game picks chips (without replacement) from a bowl of chips numbered from 1 to 75. How could you use a random-number table to do the calling if you misplaced your box of numbered chips?

*13. Pick a row of random digits in Table VIII of Appendix B and for each digit write down O if the digit is odd and E if it is even.

(a) Is there any noticeable pattern in your sequence of O's and E's?

(b) Is there anything peculiar about it?

(c) Would you be apt to encounter this sequence?

> OEOEO   EOEOE   OEOEO   EOEOE

Or this?

> OEEOO   EOOEE   EEOOO   EOEEO

Sections 1.5–1.8

*14. One of two drugs is to be administered to each patient partici-
   pating in a certain clinical trial.
   (a) The clinician assigns the drugs to patients as they arrive in
       the trial, trying to keep the number of patients assigned to
       the two drugs roughly equal. If he favors drug A over drug
       B (for whatever reason), state how he might bias the trial,
       thereby invalidating any conclusions that might be drawn.
   (b) Suppose, instead, that the clinician is required to assign
       patients to drugs by tossing a coin—a patient gets drug A
       if it falls heads and drug B if it falls tails. Further, he must
       follow each patient's progress and administer care as he
       deems necessary. If, again, he favors drug A, how can he
       bias the trial?
   (c) Suppose that the sequence in which drugs are assigned to
       patients as they come in is dictated in advance (for in-
       stance, A, B, A, B, etc.) but that the clinician is to decide,
       in each case, whether the patient is to be included in the
       study at all. How can he bias the results in favor of drug A?

*15. In 1960, there were 30,435,000 persons 12 years and older
   who fished and/or hunted. In 1965, the figure was 32,881,000
   (according to an issue of the U.S. Census *Pocket Data Book*).
   Does this show that hunting and fishing were becoming more
   popular?

*16. A campus newspaper published these enrollment figures, with
   the headline "*U Enrollment declines, CLA shows largest
   drop*":

	Year N	Year $N + 1$
General College	3,057	2,816
Liberal Arts (CLA)	17,501	16,687
Institute of Technology	3,938	3,559
Graduate School	8,025	7,588
College of Education	2,851	2,523

Is this headline justified?

17. According to a note flashed on the screen during a football
   game in Seattle, Fran Tarkenton had had 11 passes inter-
   cepted so far that year, 10 of them on third downs. Should

you conclude that he was more careless on third downs? (If you are not a football fan, talk to someone who is.)

*18. A Congressional subcommittee's report on surgical care included this statement: "It would seem that a person living in the greater Boston area needing open-heart surgery might like to be aware that within a 10-mile distance are three hospitals that have death rates of 49, 22, and 10 percent, respectively." Is it obvious that you would want to choose the hospital with the 10 percent death rate for your open-heart surgery? (What pertinent information, not given here, would you want to have in evaluating these rates?)

19. An AAA publication reported that 54 percent of the auto fatalities in Minnesota in 1968 had measurable amounts of alcohol in the blood. On the basis of this figure, the report asked this rhetorical question: "When will people learn that drinking affects their driving capability?" Is it fair to say that the given data show that drinking is bad for driving?

20. An experiment to test the effectiveness of a training program for graduate student instructors was conducted for the Economics Department of a large university. For a control group, students in a fall-quarter freshman class in the principles of economics were divided into 14 sections with average size 25. Student enrollment in each section was a self-selection process on a first come, first served basis, but the seven instructors (with two sections each) were not identified in the class schedule. Neither instructors nor students were aware of the experiment. The same seven instructors were used for the 14 sections (averaging 31 each) of the same course in the winter quarter. During the quarter, the instructors were given the training system being tested, so the winter-quarter students constituted the treatment group. The same instructional materials and syllabus were used in all sections, both quarters. At the beginning and end of each quarter, students were given a nationally normed and validated test of their understanding of economics, to determine how much they had gained in this understanding. Comment on the value of this experiment in the testing of the training program.

21. A one-time network vice-president proposed (in a 1980 book[†])

†Matthew Culligan, *How to Be a Billion Dollar Persuader*, New York: St. Martin's Press, 1979.

a scheme he called "Mass Participation Response," for improving the quality of TV ratings. The scheme involves computer cards that would contain questions about how much the viewer enjoyed the TV shows he watched. One card, he said, could contain room to answer up to 400 questions, with the viewer answering by punching the card with a pencil. He was quoted in news reports as follows: "Cards could be distributed totally at random. We could put them in boxes with new TV sets. We could ask people who want to have a voice in what is on the air to write in and request the cards. We could even work out an agreement with, say, General Mills, to put the cards on the back of breakfast cereal boxes." Comment on this scheme.

2
Summarizing Data

IN ORDER TO APPRECIATE the meaning and discern the implications of a set of data, one must ordinarily summarize it in some way. This chapter takes up some elementary methods of tabulation, graphical representation, and description, many of which you will have already encountered. They will be presented as applying to any list of data, whether considered as a population or a sample, and whether the sample be obtained honestly by some useful type of sampling or not.

Tables, graphs, and various descriptive measures are enormously valuable in uncovering and highlighting pertinent features of a set of data. Sometimes these features are made so clear, in the process, that no further statistical analysis is needed. And the situation is rare in which a graphical representation is not useful as a check and a guide for more sophisticated analyses.

Like all good things, tabular and graphical displays can be misused. Frequently, they are designed to make a particular point. There is always some arbitrariness in choosing the form of a display, so the displayer can choose (consciously or not) one that best makes a point, sometimes at the cost of distorting the meaning of the data. There is no foolproof way of detecting deceptions, but you should always view displays of data with the displayer's motives in mind.[†]

[†]The book by Darrell Huff and Irving Geis, *How to Lie with Statistics* (New York: W. W. Norton & Co., Inc., 1954), deals with a variety of such deceptions.

2.1
VARIABLES

When an individual is selected from an actual population, what is usually of interest is some quantity or attribute associated with the individual—age, pulse rate, test score, sex, political preference, or blood type, to name some common variables. These are "variables" in the sense that, usually, not every individual in the population has the same value of the variable.

Sometimes the term *variable* is used for, specifically, a *numerical* characteristic. Of the examples mentioned above, age, pulse rate, and test score are numerical-valued variables. But sex, political preference, and blood type are *attributes*, and the "values" of these variables are categories in some scheme of classification. We refer to such variables as *categorical*, and to observations on categorical variables as *categorical data*.

Numerical variables are often treated as categorical, by dividing the scale of its values into nonoverlapping intervals called *class intervals*. Thus, for annual income in dollars, one might use the intervals 0–4999, 5000–9999, 10,000–14,999, and so on. Class intervals are defined implicitly whenever data have been subjected to roundoff, as when body temperatures are recorded to the nearest two-tenths of a degree. All temperatures between 98.1 and 98.3, for instance, are rounded to 98.2, and this value—called the *class mark*—is used as a label for the class interval. Such class intervals are *ordered* categories.

Categories may be ordered, even if they are not assigned specific, numerical values. Thus, one might be asked to rate the taste of a brand of beer, a musical performance, or an instructor's teaching as excellent, good, fair, poor, or awful. And sometimes such ordered categories are assigned numbers (e.g., A = 4, B = 3, C = 2, D = 1), and treated as though the numbers meant what numbers usually mean. (Is an A and a C, for instance, really the same as two B's, on the average?)

Example 2-1 *Health Plan Subscribers*

Table 2-1 gives the values of several variables associated with the parents in each family of a simple random sample of 30 families

Table 2-1   Results of Smoking Study

Couple Number	Wife Education	Wife Smoking	Wife Age	Husband Education	Husband Smoking	Husband Age
1	2	N	41	0	S	43
2	3	Q	23	3	N	21
3	4	Q	29	4	Q	31
4	4	N	26	3	N	26
5	4	Q	37	4	N	35
6	3	N	37	4	N	41
7	2	N	41	0	S	43
8	2	N	28	2	S	25
9	3	N	23	4	N	24
10	4	N	24	4	N	29
11	2	S	24	3	S	24
12	2	S	27	4	S	33
13	2	S	29	2	Q	31
14	2	N	25	2	S	28
15	2	N	22	3	S	30
16	3	S	23	2	Q	28
17	2	S	26	1	S	31
18	3	Q	30	4	Q	34
19	1	Q	28	1	S	32
20	3	N	24	3	Q	27
21	3	N	31	4	N	35
22	4	N	38	2	S	39
23	2	S	18	3	S	23
24	4	N	33	4	N	32
25	2	S	18	2	N	23
26	2	N	21	4	S	33
27	2	Q	42	3	Q	42
28	2	S	23	1	S	29
29	4	N	34	4	N	38
30	2	N	33	3	Q	39

drawn from subscribers to a certain California health plan.[†] The code used in the tabulation is as follows:

Education	Smoking History
1 = less than high school graduate	S = currently smokes
2 = high school graduate	N = never smoked
3 = some college	Q = did smoke, but quit
4 = college graduate	

The ages given are those recorded at the birth of a particular child. The education and smoking histories were taken about 10 years later.

Age is a numerical variable, rounded off by most people to the age at their last birthday. Education and smoking history are categorical, although the categories of education used here are ordered, as the coding suggests. ▲

A *discrete* variable is one whose values can be listed. Categorical variables are discrete by their very nature, but numerical variables can also be discrete. The most commonly encountered discrete numerical variables are those that count something, such as the number of children in a family, the number of failures of a system over a given period, the number of seeds that have germinated, the number of bacterial colonies on a petri plate, and so on. The list of possible values, in each case, is usually finite, although sometimes it is convenient not to impose a finite limit to the list and permit *any* integer, say, as a possible value.

The notion of a *continuous* variable is a mathematical ideal, useful in representing physical quantities such as time, temperature, pressure, distance, area, volume, concentration, and so on. The possible values of such variables, as we idealize them, cannot be put into a list—even an infinite list. Thus, time, as we conceive it, does not jump ahead second by second—or even microsecond by microsecond. It advances continuously. And the time for an object to traverse a given path, or the time to solve a maze, can have any value in some interval of "real numbers."

In practice, however, quantities such as *recorded* time, *measured* distances, and *observed* blood pressures *are necessarily discrete*—rounded off to the nearest tenth of a second, the nearest centimeter,

[†]The complete list is published in J. L. Hodges, D. Krech, and R. S. Crutchfield, *Stat Lab* (New York: McGraw-Hill Book Company, 1975).

the nearest millimeter of mercury. The degree of roundoff is dictated by the limitations of the instruments used for measurement. Even though such data are actually discrete, we may and often do use continuous mathematical *models*, just because they are usually simpler—and in fact may be more like the phenomenon being represented. Moreover, continuous models are often used for such quantities as income, white blood cell count, yield of corn per acre, population, and number of emitted α-particles, even though they are conceptually discrete. In each case, the number of possible values is so large that discrete models may be very cumbersome.

2.2
CATEGORICAL DATA

To summarize data on a single categorical variable, one prepares a list of the possible categories and then, for each individual, makes a tally mark opposite the category observed. The number of tally marks for each category is its *frequency*, and the list of categories and frequencies is termed a *frequency distribution.* It indicates how the individuals involved are distributed among the categories.

Example 2-2

The 30 males included in the list of Table 2-1 are summarized with regard to the variable "exposure to smoking" in the following tabulation:

Exposure to Smoking	Tally
S = smokes now	TTHL TTHL / / /
N = never smoked	TTHL TTHL
Q = did smoke, but quit	TTHL / /

The corresponding frequency distribution is as follows:

Exposure to Smoking	Frequency
S	13
N	10
Q	7
Total	30

The sum of the frequencies is the number of individuals whose smoking exposure is being summarized. ▲

A distribution is sometimes given in terms of *relative frequencies*, or proportions of the whole data set:

$$\text{relative frequency} = \frac{\text{frequency}}{\text{size of the data set}}.$$

Using relative frequencies is especially important when summaries for different groups are compared. A comparison of frequencies 13 from one data set and 20 from another, for instance, is meaningless without reference to the whole. However, a comparison of relative frequencies, such as 13/30 and 20/50, is proper and can be useful.

Example 2-3

The Ag Campus students listed in Data Set B (Appendix C) are classified there according to class. The frequency distributions for males and females are as follows:

	Males		Females	
	Frequency	Relative Frequency	Frequency	Relative Frequency
Freshmen	3	.057	3	.067
Sophomores	29	.547	20	.444
Juniors	13	.245	14	.311
Seniors	7	.132	6	.133
Adult special	1	.019	2	.044
Totals	53	1.000	45	.999

The *frequency* of juniors is about the same for males as for females, but a comparison of *relative* frequencies is more meaningful: .245 as compared to .311. Thus the proportion of juniors is appreciably greater among the females than among the males.

Observe that the relative frequencies add up to 1 (or to nearly 1, which is all that a given roundoff scheme may permit), since they are proportions of the whole. ▲

> In a frequency distribution, showing how the n observations in a data set are distributed among the possible categories, *the frequencies add up to n.* Equivalently, the relative frequencies add up to 1.

The content of a frequency distribution is often more vivid in a *graphical* representation. Such representations are familiar to you through their continual use in newspapers and magazines. To represent several different numbers—frequencies, in the present instance— one uses vertical bars, all of the same width, resting on a common baseline, with heights proportional to the numbers. (Sometimes the bars are drawn horizontally, in which case the picture is the same as the one we describe but placed on its side.) The bars must be placed in some order along the baseline, whether or not the categories they represent are ordered in a natural way. A scale along a vertical axis may be used to convey frequency information, or the class frequencies may simply be written within or at the top of the bars.

Example 2-4   *Ethnic Origin*

Figure 2-1 shows the distribution of the U.S. population according to ethnic origin, in a graph prepared by the Bureau of the Census. The bars are arranged according to frequency.

Reference to the table from which the graph was constructed shows that certain categories were omitted; "other," comprising 47.3 percent of the population and including more than 20 million blacks; and "not reported," comprising 22.9 million or 11.1 percent of the total. Although graphs are a useful visual aid, checking the tabular summaries on which they are based can prove enlightening. (In this example, the reason the authors checked the table was that, living in Minnesota, we found it strange that no Scandinavian countries were listed.) ▲

Sometimes small symbols are used to represent units. For instance, to represent numbers of births and deaths one could draw columns of storks and coffins, as is done in Figure 2-2. Sometimes a single symbol of height proportional to frequency is used—but with an un-

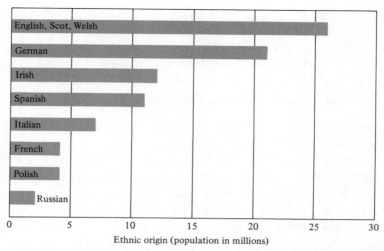

Figure 2-1.   Population by ethnic origin, March 1973. (From Bureau
of the Census, *1976 Pocket Data Book.*)

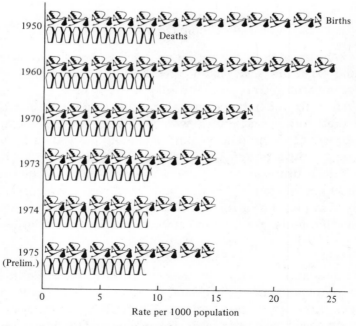

Figure 2-2.   Numbers of births and deaths per year. (From Bureau of the Census, *1976 Pocket Data Book.*)

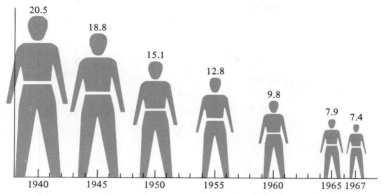

Figure 2-3. Man-hours in agriculture (in billions) represented by heights of farmers. (From Bureau of the Census, *1969 Pocket Data Book.*)

fortunate effect, as demonstrated in Figure 2-3. The 1940 farmer appears to be over five times as big as the 1967 farmer; but it is the ratio of *heights*, 2.77 : 1, that gives the correct ratio of man-hours. The eye tends to pick up an *area* relationship, and the area of the 1940 farmer in the picture is $(2.77)^2$ or about 7.67 times that of the 1967 farmer. A later version represents the information more fairly, as shown in the graph of Figure 2-4. There, in place of either bars or farmers, we find the points that would be bar tops connected with a line graph (and extended to years past 1967).

The type of distortion seen in Figure 2-3 is not uncommon. Another example is the figure from a Merrill Lynch ad reproduced in Figure 2-5. In this case, not only does the eye pick up an area relationship, but the mind extends this to one of *volume*—or to the

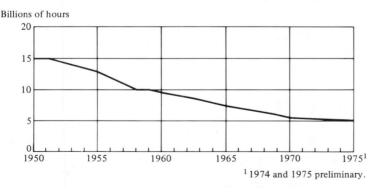

Figure 2-4. Man-hours in agriculture, 1976 version. (From Bureau of the Census, *1976 Pocket Data Book.*)

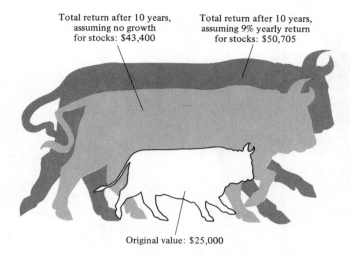

Total return after 10 years, assuming no growth for stocks: $43,400

Total return after 10 years, assuming 9% yearly return for stocks: $50,705

Original value: $25,000

Assumed results of investment strategy No. 1 using discount bonds and stocks.

Figure 2-5. Advertisement with misleading use of figures. (Redrawn from *Newsweek*, January 10, 1977.)

mass of the bulls. A bull that is twice as high as another would be expected to weigh on the order of *eight* times as much—a significant reality when you meet a bull! [Other examples of this type are found in Darrell Huff and Irving Geis, *How to Lie with Statistics* (New York: W. W. Norton & Co., Inc., 1954).] Again, the point is that the eye is trained to compare the *area* of plane figures. The ratio of their areas *is* the ratio of their heights if the figures have the same width, but not otherwise. We encounter this point again when dealing with bar graphs in which the bars are not all of the same width.

2.3
CROSS-CLASSIFICATIONS

Exploring relationships between variables is an important scientific activity, in which methods of statistics are an invaluable aid. A relationship between two variables can only be studied if the two variables for each individual are observed and recorded as a *pair*. In the case of categorical variables, this means that each individual must be *cross*-classified—classified into a category of one variable and simultaneously into a category of the other. Each combination of a category of one variable with a category of the other is a *cell* of a two-way array of such combinations. To record data on the two variables, one makes a tally mark in the cell corresponding to the observed combination of categories of the two variables.

Example 2-5 *Smoking and Education*

The 30 males listed in Table 2-1 (page 41) are classified there according to education and also according to smoking history. There are four categories of education and three for smoking history. Putting categories of education along the left margin and those of smoking across the top we have a two-way array of 12 cells (12 = 3 × 4). The 30 males are tallied in these cells as follows:

		Smoking History		
		S	*N*	*Q*
Education	1	TTHL		
	2	///	/	//
	3	///	//	///
	4	//	THLL //	//

(To be continued.) ▲

The number of individuals in each cell of a cross-classification is its *frequency*, and the array of cells with their corresponding frequencies for a given data set is called a *bivariate frequency distribution*. It shows how the individuals represented in the data set are distributed according to the two variables considered jointly. (The term *bivariate* is in contrast with the term *univariate*, which would be used in describing the distribution of a set of individuals according to a single variable.)

The sum of the frequencies in a row of a bivariate frequency table is the frequency of that row category, without regard to the column variable. These row sums constitute a univariate frequency distribution, the distribution of the data according to just the row variable. Because these univariate frequencies appear in a margin of the table, the distribution is called a *marginal distribution*. The column sums, in like fashion, are frequencies in the marginal distribution of the column variable—that is, the variable whose categories define the columns.

Example 2-6 *Education and Smoking (continued)*

The frequency distribution of the 30 males in Table 2-1 is obtained
by counting the tallies in the two-way array given in Example 2-5.
It is as follows, shown together with column and row totals:

		Smoking History			
		S	N	Q	Total
	1	5	0	0	5
Education	2	3	1	2	6
	3	3	2	3	8
	4	2	7	2	11
Total		13	10	7	30

Observe that the column totals, which give the frequencies of the
categories of smoking history without regard to education, constitute the frequency distribution for the single variable "smoking
history"—its marginal distribution. Similarly, the row totals constitute the marginal distribution of education, considered by itself.

The two-way table also gives a frequency distribution for education, for instance, for the 13 individuals who smoke—and a distribution according to smoking history for the 11 individuals in
educational category 4, and so on. ▲

The distributions of one variable *within* a category of another
variable, such as those mentioned at the end of Example 2-6, are
often of interest. Indeed, sometimes bivariate data are given in terms
of such distributions, as in the next example.

Example 2-7 *Government Employees*

The following table[†] gives distributions of employees in state and in
local governments according to type of activity:

[†]Taken from information in Bureau of the Census, *1976 Pocket Data Book.*

	Local (percent)	State (percent)
Education	52.6	52.3
Health	10.0	9.4
Police and fire	7.5	8.6
Highways	5.1	5.1
Sanitation	1.8	1.8
Natural resources	1.7	1.6
Financial administration	2.3	2.2
Other	19.0	18.9

These percentages are relative frequencies of the types of activity, for state employees—numbering 8,832,000—in one column, and for local employees of government—numbering 11,794,000—in the other. We can obtain the numbers involved by multiplying percentages by the total numbers in state and in local governments, and create therewith a two-way table of bivariate frequencies:

	Number (thousands)		Total
	Local	State	
Education	6,202	4,620	10,822
Health	1,183	833	2,016
Police and fire	887	756	1,643
Highways	596	450	1,046
Sanitation	216	162	378
Natural resources	197	143	340
Financial administration	274	198	472
Other	2,239	1,670	3,909
Total	11,794	8,832	20,626

From this array we can read out the proportion of individuals in education who are state employees: 4620/10,822, or 42.7 percent; the proportion of those in health activities who are employed at the local level: 1183/2016, or 58.7 percent; and so on. As before, we have in the column and row totals the marginal distributions of type of activity (without regard to level) and of level—state or local (without regard to the type of activity). ▲

A bivariate frequency distribution could be represented graphically by something like a bar graph, but the bars would require a third dimension. Such pictures are awkward on two-dimensional paper, and we are usually content with giving cell frequencies in tabular form.

It bears repeating that we need a two-way distribution of "joint" frequencies for investigating a possible relationship between two variables. That is, it is necessary to have the entries in the cells of the two-way table, not just the frequencies in the margins. Knowing the marginal frequencies does not mean that we know the joint frequencies, and, indeed, different joint frequency distributions could have the same marginal totals.

Example 2–8   *Religion and Marital Status*

A feature article in a newspaper's Sunday magazine section gave data on the marital status and religion of patients in an abortion clinic. The tallies for a given period of time were as follows:

Marital Status	Frequency
Never married	75
Married	12
Other	11
Total	98

Religion	Frequency
Protestant	41
Catholic	31
Jewish	0
Other	13
None or no answer	11
Total	96

(The totals do not agree, even though the same individuals are supposed to be included in the two tabulations; but we let that pass. Real data are seldom "clean"—free from errors in gathering and recording.) These frequency distributions tell about the variable *marital status* and the variable *religion*, considered separately, but tell nothing about any possible relationship. A two-way table would have these univariate distributions as marginal entries, but we have no information for the interior cells (because none was included in the article):

		Religion					
		P	C	J	O	N	
Marital Status	N						75
	M						12
	O						11
		41	31	0	13	11	

This lack is understandable, as there was apparently no plan to study the relationship between religion and marital status for the women who had abortions. But the point is that if you *do* want to study relationships between two variables, you have to gather data in a bivariate array—a cross-classification. ▲

Cross-classifications are useful in studying relationships or associations among categorical variables. Some ways of measuring the degree or extent of an association between two variables have been proposed, but we shall not take these up. Instead, we look at just two examples—one in which there *is* an assocation, and one in which there is *not* an association.

Example 2-9 *Education and Smoking (continued)*

The 30 individuals in the tabulation of Example 2-6 show an association between the amount of education and an individual's smoking history. For instance, the columns headed S and N are opposite in tendency; that is, those who smoke tend to have less education, and those who never smoked, more. This is evident from the relative frequencies in the columns—the relative frequencies of the categories of education among the smokers, among the "never-smoked," and among the "quit." These relative frequencies are obtained by dividing each entry in a column by the column total and are as follows (in percent):

		S	N	Q	All
Education	1	38.5	0	0	16.7
	2	23.1	10	28.6	20
	3	23.1	20	42.9	26.7
	4	15.4	70	28.6	36.7
		100	100	100	100

Knowing that an individual had never smoked, then, one could bet that he is *not* in education category 1; but if an individual smokes, education category 1 is the *best* bet. There *is* an association—at least, in this group of 30 individuals. (Whether this association has anything to say about an association among adults generally is a matter of inference that we encounter in Chapters 9 and 10.) ▲

Example 2-10 *Suits and Honors*

Consider an ordinary deck of 52 playing cards. Each card can be classified according to suit and also according to whether or not it is an honor card (10, jack, queen, king, ace). The result of this cross-classification of the 52 cards is the following frequency table:

	Suit				
	Spades	Hearts	Diamonds	Clubs	
Honor	5	5	5	5	20
Nonhonor	8	8	8	8	32
	13	13	13	13	52

The distribution according to suit is exactly the same (with equal relative frequencies) among the honor cards as among the nonhonor cards—and, indeed, the same as in the whole deck of cards. Similarly, 5/13 of the Spades are honor cards, as are 5/13 of the Hearts, 5/13 of the Diamonds—and, to be sure, 5/13 of all of the cards. The suit of a card tells nothing about whether or not it is an honor card. The variables here are *not* associated. ▲

2.4
MORE THAN TWO CATEGORICAL VARIABLES

In studying the relationships among more than two variables, both the tabulation and the graphical representation of data become more complicated. Thus, in the case of three variables, the rectangular array of a two-way classification—a two-dimensional "box"—expands into a three-dimensional box. The cells (cubicles) of the box correspond to the various combinations of a category of variable 1 with a category of variable 2 with a category of variable 3. Because of the difficulty of picturing such a three-dimensional array of frequencies on a two-dimensional page, such classifications are usually summarized as a set of two-way cross-classifications, one for each category of the third variable. (The "third" variable, of course, can be any one of the three.) The next example illustrates a three-way classification.

Example 2-11

The 30 couples classified in Example 2-5 according to education level and smoking history of the husband are classified further as to the smoking history of the wife, as follows:

		Wife's Smoking History								
		Smokes			Never Did			Quit		
		Husband			Husband			Husband		
		S	N	Q	S	N	Q	S	N	Q
	1	2	0	0	2	0	0	1	0	0
Husband's	2	0	1	2	3	0	0	0	0	0
Education	3	2	0	0	1	1	2	0	1	1
	4	1	0	0	1	6	0	0	1	2

These three two-way arrays can be thought of as "layers" of data in a box. They are shown in Figure 2-6 in a way that suggests how they could be stacked to form the box of the three-way array. Marginal totals are also shown, as is a summary layer which gives totals across the layers.

The wife's smoking history is the "layer variable," but we could have equally well chosen either of the other two variables to serve in this role. That is, the box can be layered in two other ways, by slicing it along the other two dimensions. For each such layering, the frequencies can easily be read from those shown in the layering of Figure 2-6. The best way to understand the construction of the box is to write down the layers corresponding to one of the other variables—education or smoking history of the husband. (See Problem 8.) ▲

In a three-way classification, associations can be more complicated than with just two variables. There may be associations in the marginal layers, and different associations in the interior layers, for instance. We shall not explore such things in detail, but present one example to illustrate some possibilities.

Example 2-12   *Education, Race, and Sex*

The Bureau of the Census publishes data on the educational level, race, and sex of adults. In one year the figures were something like this (modified from the actual data to achieve simplicity), in millions:

		Education Level			
		1 At Most Grade School	2 At Most High School	3 Beyond High School	Total
White	Male	13	20.5	10.5	44
	Female	13	26.5	8.5	48
	Total	26	57	19	92
Black	Male	2	1.7	.3	4
	Female	2	2.5	.5	5
	Total	4	4.2	.8	9
Both	Male	15	22.2	10.8	48
	Female	15	29	9	53
	Total	30	51.2	19.8	101

Some relationships are apparent here. For instance, in the white layer, more females than males are in educational category 2, whereas there are more males than females in category 3. Layering on the other variables may point up other features of interest. For example, using educational category as the layer variable, we have these tables:

	At Most Grade School		At Most High School		Beyond High School		Total	
	White	Black	White	Black	White	Black	White	Black
Male	13	2	20.5	1.7	10.5	.3	44	4
Female	13	2	26.5	2.5	8.5	.5	48	5

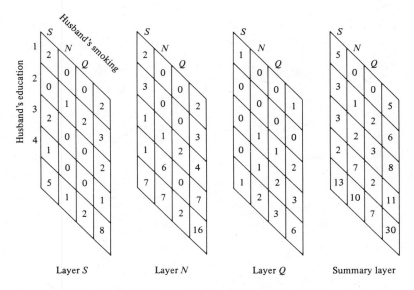

Figure 2-6. Three-way cross-classification (Example 2-11). Layers: wife's smoking history.

In the lowest educational level, race and sex appear to be unassociated—the proportion of males and females is the same for both races. But in the "beyond high school" category, the proportion of males is more than one-half for whites but less than one-half for blacks; so in that layer, race and sex are associated. ▲

The next example illustrates the important fact that, as in the case of two-way tables, the marginal frequencies cannot be used to reconstruct what is inside a multidimensional box array of frequencies.

Example 2-13   *Happiness is . . .*

A 1971 Gallup Poll of 1517 adults in more than 300 "scientifically selected localities" summarized the answers to a question as to how happy they are as percentages, as follows:

		Very	Fairly	Not
Race	White	46	46	5
	Nonwhite	20	63	12
Age	21–29	55	39	5
	30–49	42	51	4
	Over 50	38	50	8
Marital Status	Married	47	46	4
	Single	37	55	6
Income	More than $15,000	56	37	4
	$10,000–15,000	49	46	3
	$ 7,000–10,000	47	46	5
	$ 5,000–7,000	38	52	7
	$ 3,000–5,000	33	54	7
	Less than $3,000	29	55	13
Education	College	51	42	4
	High School	44	49	4
	Grade School	35	50	11

(The percentages unaccounted for in each row are for "no answer.")
Gallup states: "The happiest people among the U.S. adult population
are likely to be white, in their twenties, married, with a high income
and a college background." This statement, picking out the characteristic with highest proportion of "very happy" in each classification, is not justified if based only on these marginal breakdowns.
The cross-classification cell defined by "very happy, white, in their
twenties, married, with a high income, *and* with a college background," all stated characteristics holding *simultaneously*, need not
be the cell of highest frequency. It might even be empty! A simple
example illustrates this in the case of two variables:

	B_1	B_2	
A_1	2	4	6
A_2	4	0	4
	6	4	10

The combination $A_1 B_1$ is *not* the most frequent, even though A_1 and B_1, separately, are the most frequent categories of the marginal variables.

The raw data for the poll no doubt include cross-classification information, but the published results—the marginal totals—do not permit conclusions about the variables considered jointly. ▲

2.5
DISCRETE NUMERICAL DATA

A numerical variable that is discrete is also necessarily categorical—the various possible numbers in the list of values are categories. Frequency distributions and bar graphs again provide useful summaries of data on such variables, but now the list of values and the corresponding bars or rods are naturally *ordered*.

Example 2-14 *Family Size*

In preparing to tally data on the number of children in a family, one will certainly need categories 0, 1, 2, 3, and 4—but up to what number? There may be an occasional family with 12 or even 20 children, but perhaps most have 6 or fewer. Situations like this are sometimes handled by providing an "open-ended" category of the type "7 or more." Thus, the frequency distribution for 75 families might look like this:

Number of Children in a Family	Frequency
0	8
1	7
2	29
3	15
4	11
5	3
6	0
7 or more	2
Total	75

A graphical representation of this distribution is given in Figure

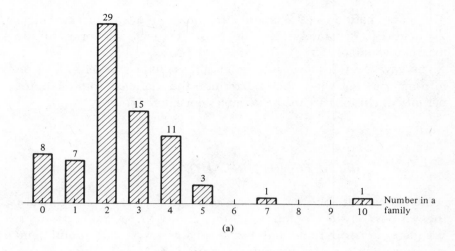

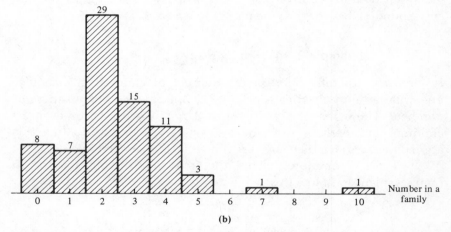

Figure 2-7. (a) Bar diagram for Example 2-14; (b) diagram for Example 2-14
with contiguous bars.

2-7(a) as a bar diagram. Notice the problem with the open-ended
category—it is not clear where to put the bar! One solution, possible
when the original (or "raw") data are available, is to see what went
into "7 or more." In this case it was one family with 7 children and
one with 10.

(Some difficulties of types encountered in actual surveys are
evident here. For example: What is a "family"? Would divorced parents constitute a family? If a child spends half his time with one
parent, is this "1/2 of a child?" One might consider children in a
"household" rather than in a "family," but then suppose that one

household is a small commune? Such questions must be resolved and the resolution described in presenting the data.)

The width of the bars in Figure 2–7(a) has no significance. It only serves to make them visible. Understanding that there are no fractional numbers of children, one would not be confused by the slopping over of bars into the space on the scale that would ordinarily represent numbers between successive integers. Sometimes bars are widened to the extent that they meet, and the data given above are shown in this kind of picture in Figure 2–7(b). (Observe that in this widening, the integrity of the bars with height 0 is preserved, even though they cannot be seen.) ▲

Sometimes it is desirable to *combine* values for purposes of a more convenient or more revealing frequency table. In such cases it seems unavoidable to use bars over the possible values in a group. This is not a problem if it is kept in mind just which values are possible. The next example illustrates this.

Example 2–15   *Highway Traffic*

To record the number of cars passing a certain point on a highway, each minute in a sequence of 60 consecutive minutes, one would need to prepare a sheet with categories 0, 1, 2, . . .—up to what? Perhaps a glance at the traffic suggests that 80 might be as high as the count would go. This involves 81 categories, and the data from 60 minutes would be rather sparsely distributed among them, which makes for a relatively uninformative picture. A grouping of neighboring values is helpful in this regard, such as in the following distribution, where any number of cars per minute from 0 to 9 is tallied in the category "0–9" and so on.

Number per Minute	Frequency
0–9	7
10–19	14
20–29	20
30–39	12
40–49	4
50–59	2
60–69	0
70–79	1
Total	60

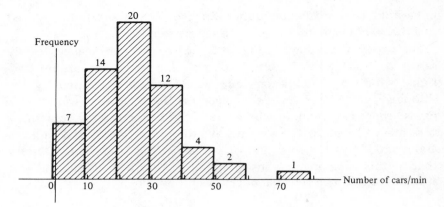

Figure 2-8. Histogram for Example 2-15.

Thus, in 7 of the 60 one-minute periods, the count was 9 cars or fewer, in 14 of the periods there were from 10 to 19 cars per minute, and so on. The distribution is shown in Figure 2-8 by means of a bar graph. The first bar covers the values 0, 1, . . . , 9. In the process, it also covers all the nonintegral numbers *between* 0 and 1, between 1 and 2, and so on. Observe that the bar over 0, 1, . . . , 9 extends from -.5 to 9.5, the next bar from 9.5 to 19.5, and so on; this makes the bars contiguous while treating all of the integers the same. (Think of each integer k as covered by the part of the bar from k - .5 to $k + .5$.) ▲

Representing a frequency by a bar of height proportional to frequency gives a fair picture as long as the *width* of the bar is the same for each category. The *areas* of the bars, which is what the eye tends to pick up, would be proportional to frequency. However, in grouping categories (as was done in Example 2-15) one must either use the same number of categories in each group or adjust the heights so that the areas are proportional to frequencies.

A diagram in which bars for adjacent groups of categories meet and in which the *area* of a bar over a group is proportional to the frequency of observations in the group is called a *histogram*. If the bars are *not* all of equal width, there is not a common vertical scale for frequency, in which case a frequency can be indicated for each bar.

Example 2-16   *Highway Traffic (continued)*

In the frequency distribution of the number of cars per minute given in Example 2-15, one might (because of the small frequencies) regroup the last four groups into one:

Number per Minute	Frequency
0–9	7
10–19	14
20–29	20
30–39	12
40–79	7

Using these frequencies as heights of bars would be a mistake, producing the picture shown in Figure 2-9(a). From that picture it

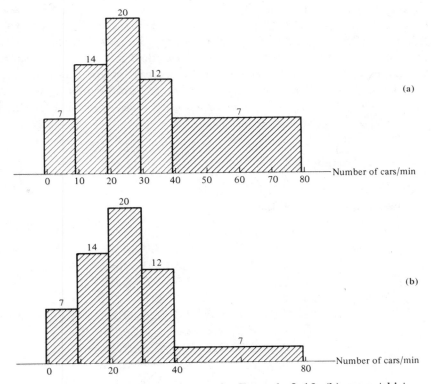

Figure 2-9.   (a) Incorrect histogram for Example 2-16; (b) correct histogram for Example 2-16.

would seem as though there were more observations in the range 40 to 79 than there are between 30 and 39. Adjusting the height of the last bar so that the *areas* are proportional to frequencies results in the graph of Figure 2-9(b)—a histogram. In that graph, the height of the last bar is 7/4, in terms of the unit used for the other bars, because the bar is four times as wide as the others. Another way of thinking of this adjustment is that the single bar for the four groups is given a height which is the average of the heights of the four bars for those groups in Figure 2-8: $(4 + 2 + 0 + 1)/4$. ▲

2.6
CONTINUOUS VARIABLES

Data from a continuous variable are necessarily read and recorded as though the variable were discrete, rounded off to a convenient or an imposed degree of accuracy—to the nearest tenth, or the nearest hundredth, for example. The degree of roundoff determines a corresponding set of categories. Thus, rounding to 27.1, 27.2, 27.3, and so on, amounts to classifying a measurement into one of the intervals 27.05 to 27.15, 27.15 to 27.25, and so on. Such categories are *class intervals*, and for any given scheme of roundoff in a set of data there is a frequency distribution based on the corresponding class intervals. This frequency distribution can be represented graphically by a histogram, in which there is a bar covering each class interval. (The bars are contiguous if the class intervals are.)

Example 2-17

Data Set B (Appendix C) lists the following grade-point averages for 71 male students in a statistics class:

3.0	3.1	3.0	3.5	2.2	2.3	2.2
2.7	3.2	3.0	2.5	3.0	3.0	2.6
2.4	3.8	2.5	3.0	2.5	3.0	3.3
2.5	2.9	2.0	2.1	2.5	3.6	3.3
2.7	2.2	2.9	3.7	3.0	2.9	3.0
2.1	2.5	2.7	2.8	2.6	2.8	3.2
2.7	2.8	2.5	2.4	2.8	2.1	3.0
2.7	3.6	3.6	2.7	3.0	2.6	2.7
2.7	3.0	2.7	2.7	3.9	2.7	3.5
3.6	2.3	2.3	2.8	3.5	2.6	3.0
						3.1

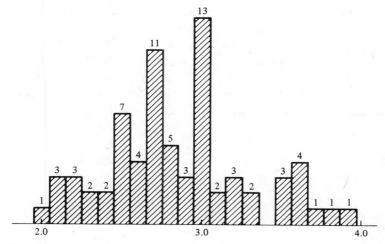

Figure 2-10.   Histogram of GPA's, Example 2-17.

These are summarized in a histogram in Figure 2-10. The graph is
quite rough, adjacent bars differing by as much as a factor of 6 (or
more, if we count the bar of zero height). But this roughness is
undoubtedly a manifestation of sampling fluctuations rather than
anything real in the population of student GPA's. For instance, in
the other section of the statistics class listed in Data Set B, the ratio
of male students with 2.7 to those with 2.8 averages is $7:7$, quite
different from the ratio $11:5$ in the data above. ▲

The irregularity of the top of the histogram in Figure 2-10 is
typical. To obtain a smoother or more regular picture, one more
suggestive of the true character of the population, it is often better
to use a scheme of categories or class intervals that is not as refined
as that implied by the roundoff scheme used in recording the data.
This can be accomplished by regrouping—combining groups of
adjacent class intervals into a smaller number of wider class intervals.
However, the regrouping must not be carried too far (using *one* large
class interval is the extreme), for the interesting and essential features
of the population would be smoothed out along with the sampling
roughness. A rule of thumb for more than 50 observations is to use
from 8 to 20 class intervals.

The choice of class intervals, both as to width and location, is arbitrary. So one should not be surprised to find the same data set
summarized by different people in histograms that look different
from one another.

Given the number of class intervals to be used, the class boundaries should be chosen as numbers that will not be encountered as data values. For instance, with data recorded to the nearest tenth, class boundaries should be chosen halfway between the tenths.

Also, to avoid the impression of accuracy that is not warranted, the boundaries should be chosen so that the center points are given to the same accuracy as the data. Using 26.75 to 27.85 as a class interval, for example, gives 27.3 as the center value for representing the interval.

The following steps might be followed in preparing to tally a set of numerical data:

1. Scan the list to get an idea of the range of values that will be encountered.
2. Choose a number of class intervals, from 8 to 20. This number is arbitrary. For 50 or so observations use about 8, and for 500 or more use 15 to 20 intervals. Using too few will tend to smooth out roughness that might be meaningful, and using too many introduces roughness in a picture that should be smooth.
3. Divide the range of values into class intervals of equal width,† choosing boundaries that are *between* values that will be encountered.

Example 2-18   *Students' Ages*

Table 2-2 gives ages in months of 113 students in an elementary statistics course. Glancing over the list shows that they run from around 210 to 572, the latter being an isolated value on the high end. The next highest is 384, so for deciding what class intervals to use in a frequency tabulation, consider the range from 210 to 384, which is 174 months wide. This range would be covered using 14 intervals if each is 13 months wide. The responses are given as a whole number of months, and we could use the intervals 210 to 222 (which includes 13 months), 223 to 235, 236, to 238, and so on; with this choice the tally is as follows:

---

†Occasionally, there is reason to use unequal interval widths, but the convenience and simplicity of equal widths are such as to make this the norm.

Interval	Tally
210–222	THL /
223–235	THL THL THL ///
236–248	THL THL THL THL THL THL THL ///
249–261	THL THL THL ///
262–274	THL ////
275–287	THL /
288–300	THL //
301–313	///
314–326	/
327–339	//
340–352	/
353–365	/
366–378	
379–391	//
392 and over	/

Table 2–2   Ages of Statistics Students (in Months)

252	227	247	243	259
384	237	254	242	264
252	279	238	244	361
245	269	276	264	241
243	213	240	253	300
247	258	228	287	261
273	261	237	254	249
237	235	214	235	252
348	238	297	216	240
312	227	237	211	276
229	246	256	239	257
232	264	299	293	338
241	276	248	243	229
288	249	240	241	219
242	240	226	273	224
239	230	386	226	245
240	252	320	303	252
262	236	267	234	243
241	252	236	229	240
234	276	262	260	334
243	240	247	288	247
	216	235	293	240
	572	306	232	228

The choice of 14 class intervals, 13 months wide, has turned out to give rather sparse tallies at the high end; but using wider class intervals would have made the bunching at the lower end even more pronounced. So perhaps it was not a bad choice.

In drawing the histogram, with bars that touch, it must be realized that an age of 222.3 months would be rounded down to 222, but 222.6 would be rounded up to 223. So for purposes of the histogram, the class intervals would be widened to meet each other, like this:

Class Interval	Frequency
209.5–222.5	6
222.5–235.5	18
235.5–248.5	38
248.5–261.5	18
261.5–274.5	9
etc.	

(The earlier scheme was perhaps a little easier to follow in doing the tallying.) Instead of drawing the histogram here, we point out that, in essence, the histogram is evident in the table of tallies. (Hold the page sideways, if you prefer to have histogram bars vertical.) It is clear that the lower ages are more frequent, and that the older students' ages spread out over a wide range. The distribution is said to be *skewed*, in contrast with the more symmetrical distribution we encounter in the next example. ▲

Example 2-19 *Viscosity Measurements*

Table 2-3 gives 200 measurements of viscosity, recorded in groups of 5 across the page, to the nearest tenth of a unit. These data are plotted in Figure 2-11 using three different class interval schemes— one using a class interval width of 1.1, and two using a class interval width of .9, but with different class interval centers.

In view of the differences among the three histograms, all purporting to represent the data in Table 2-3, only the grosser features of a histogram can have any significance. For not only sampling fluctuations, but also the different choices of class interval and class mark, will affect the detailed structure of a histogram. ▲

Table 2-3   Viscosity as Measured by an Ostwald Viscosimeter

37.0	36.2	34.8	34.5	32.0
31.4	31.0	30.8	30.0	33.9
34.4	33.5	32.9	32.8	31.3
33.3	33.7	34.3	35.9	31.0
34.9	33.4	33.3	32.4	32.0
31.7	32.8	27.6	31.6	35.0
30.0	30.7	32.2	35.3	35.5
34.4	30.2	32.0	31.3	31.6
32.0	33.1	31.2	31.9	31.8
29.8	29.6	31.7	32.0	29.9
31.6	34.4	33.3	35.9	35.4
31.6	32.1	34.2	33.8	33.4
31.3	33.1	32.2	33.3	31.2
34.6	32.7	32.1	32.0	30.5
32.6	31.5	30.8	31.2	31.0
31.5	30.7	32.5	33.0	32.9
34.5	33.7	32.0	31.3	30.9
32.3	32.0	31.9	30.0	33.9
33.8	34.8	34.2	33.8	33.8
31.1	31.8	26.8	32.3	32.8
31.7	33.5	34.5	33.6	32.5
32.5	32.7	32.4	33.6	34.1
31.8	30.5	30.8	30.6	30.7
31.9	31.4	30.9	32.0	30.8
29.9	31.6	29.5	31.7	30.4
30.7	31.3	28.7	28.8	31.0
27.6	31.9	31.3	29.9	32.1
31.2	30.5	30.9	29.8	28.7
30.0	29.1	29.7	31.0	29.0
30.9	28.9	29.5	28.8	29.3
28.4	30.0	29.3	30.2	31.5
31.3	30.3	32.5	31.8	30.8
28.7	30.7	32.3	33.5	32.9
32.1	29.8	29.9	30.8	30.7
31.9	32.0	29.5	30.6	30.1
32.8	31.0	30.0	29.2	32.5
30.2	31.3	29.5	29.0	30.3
30.2	30.7	30.8	31.6	32.7
32.4	32.8	31.3	31.4	30.7
30.7	30.6	31.3	30.6	31.7

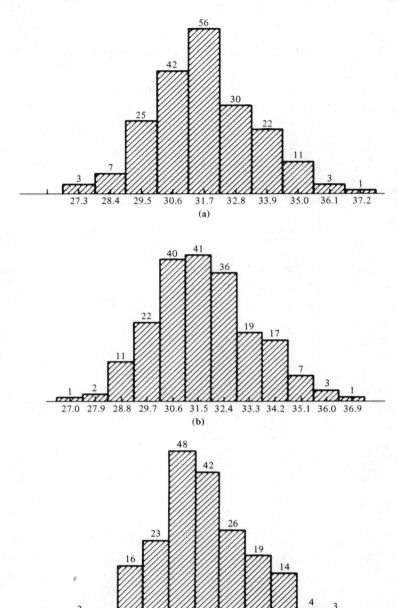

Figure 2-11.   Histograms for the data of Table 2-3, using various class-interval schemes.

Although the analysis of a frequency distribution is simplest when the class intervals are chosen to be of the same length, one occasionally encounters data summarized with unequal class intervals. And sometimes open-ended intervals are used when it is hard to anticipate just how large or small a value might come along.

Example 2-20 *Abortion Patients' Ages*

The data in Table 2-4 were published in a Minneapolis newspaper, in the form shown.

Table 2-4   Ages of Abortion Clinic Patients

Age	Tally
15 and under	///
16 and 17	7HL 7HL 7HL //
18 and 19	7HL 7HL 7HL 7HL 7HL
20 and 21	7HL 7HL 7HL 7HL
22 and 23	7HL ///
24 and 25	7HL /
26–30	7HL 7HL //
31–35	7HL
36+ (specify)	37, 38, 41, 36

A person's age is a continuous variable, but people usually give their ages as of their last birthday. For instance, the category 16 and 17 would include any one between the ages of 16 and 18. A frequency distribution, with class intervals designated accordingly, is given in Table 2-5.

To construct a histogram for this frequency distribution requires care, because not only are the class intervals of different lengths, two of them are of indefinite length! The highest age group (>36) can be handled because we know the actual ages (37, 38, 41, 36) of the four women in that group. But the ages in the lowest age group are not given. Perhaps it is safe to assume that the class interval 11–16 includes all of them.

Because the bases of the histogram bars are of different lengths, their heights should be adjusted so that the areas of the bars are proportional to frequencies. This is so that every category with the same frequency will appear to the viewer as having the same size

Table 2-5   Frequency Distribution

Age	Frequency
<16	3
16–18	17
18–20	25
20–22	20
22–24	8
24–26	6
26–31	12
31–36	4
>36	4
Total	99

(see also Figures 2–3 to 2–5). Thus, for each class interval of length 5, the bar height will have to be reduced to 2/5 of what it would be with base 2. The 12 for the interval 26 to 31 becomes $12 \times 2/5 = 4.8$, to give the correct impression of its relation to the frequency of 6 for the interval 24 to 26. The histogram is shown in Figure 2–12. In drawing it, we have taken the lower limit of the leftmost

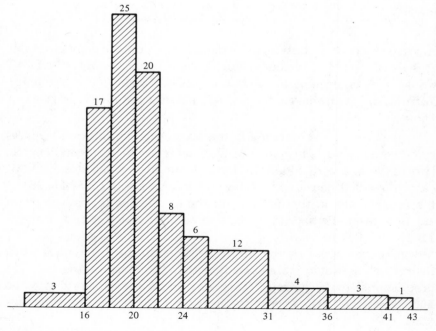

Figure 2-12. Histogram for Example 2-20.

class as 11, but have omitted the label "11" from the graph, admitting the possibility that an age might have been under 11. ▲

2.7
THE MEDIAN AND OTHER FRACTILES

Just as chest measurement, shoe size, weight, and center of gravity serve to characterize various aspects of a person's body distribution, each for a particular purpose, so it is useful to have ways of characterizing various aspects of a frequency distribution—or more basically, of a set of data.

The notion of the *center* of a set of numerical data, or of a corresponding frequency distribution, is useful in *locating* the distribution on the scale of values. A value in the center is what might be thought of as typical, or in a loose sense, "average." But "center" is a vague word for a vague concept; it can be made specific in a number of different ways. One way will be taken up here; others, in the next chapter.

Numerical data can be put in *numerical order*. For the moment we neglect the possibility of a "tie." With the data in numerical order, one can count halfway from either end to the middle of the sequence. When there are an *odd* number of observations, there is a middle value; when the number of observations is *even*, the middle falls between two values. A number that has the same number of observations on either side is called a *median*, or a median value. The customary definition is as follows.

Given a set of n numbers in numerical order,

$$\text{median} = \begin{cases} \text{middle number,} & \text{if } n \text{ is odd} \\ \text{average of middle} \\ \text{two numbers,} & \text{if } n \text{ is even} \end{cases}$$

When there are ties—and ties frequently occur in recorded data because of discreteness—we simply include the tied observations in the sequence, one after another, and count to the middle as before. There will again be as many numbers on one side of the median as on the other; but some of the numbers to the right of the median, or to the left, may be equal to the median.

Example 2-21

The median of this list of nine numbers:

$$2, \quad 19, \quad 12, \quad 30, \quad 19, \quad 23, \quad 22, \quad 26, \quad 15$$

is found by first putting the numbers in numerical order:

$$2, \quad 12, \quad 15, \quad 19, \quad 19, \quad 22, \quad 23, \quad 26, \quad 30,$$

and then counting over to the fifth one, the one in the middle:

$$\text{median} = 19.$$

If the list had contained a tenth number, say another 26:

$$2, \quad 12, \quad 15, \quad 19, \quad 19, \quad 22, \quad 23, \quad 26, \quad 26, \quad 30,$$

the median would be defined to be 20.5, the average of the two numbers (19 and 22) in the middle.

If the largest number had been mistakenly recorded as 300 instead of 30, the median would be the same. Indeed, changing any of the numbers to one side of the median, as long as they stay on that side, would not change the median. ▲

As seen in Example 2-21, the median does not depend on how big the biggest observations are, nor on how small the smallest ones are. This property of the median may be desirable in some situations and undesirable in others. (We touch on this point again in Section 3.1.)

Example 2-22 *A Senatorial Profundity*

A U.S. Senator from New York remarked on network television that "half the people in New York City earn less than the median income." He was trying to make the point that the city is economically disadvantaged. What he actually said is either a definition of "median" (if "median income" referred to New York City folks) or a statement that the New York City median is the same as the overall median (if "median income" referred to the people of the state or of the country as a whole). He might have said, and just as accurately, that half the people in New York City earn *more* than the median income, but the effect would not have been quite what he wanted.

Income distributions are often described in terms of the median because the median is not influenced by how high the highest incomes are. Thus, if a particular millionaire earned $2 million in a year instead of $1 million, the median income would be the same. On the other hand, the Internal Revenue Service would be more interested in total income, including the exact incomes of the richest people. ▲

In the case of large masses of data, it may be useful to divide the set of data into fractional parts other than one-half, to define what are called *fractiles* or *quantiles*. In particular, any three values that divide an ordered set of numbers into quarters are termed *quartiles*.† (Unambiguous definitions are not worth the trouble at this point.) Quartiles are usually numbered from left to right, the first quartile, Q_1, being a value with one-fourth of the data to its left (i.e., smaller) and three-fourths to the right (or larger). The second quartile, Q_2, is the median, because $2/4 = 1/2$. The third quartile, Q_3, has three-fourths of the data to its left and one-fourth to its right. As before, such statements about a fraction of the data to one side of a quartile are approximate if the number of observations is not a multiple of 4, and because of ties.

Percentiles are defined in like fashion. The 63rd percentile, for example, is a value such that 63 percent of the ordered data lie to the left and 37 percent to the right of that value.

Example 2-23 *Instructors' Salaries*

The American Association of University Professors published information on salaries (in a particular year) of those at the rank of "Instructor." Its summary is shown in Table 2-6. According to that table, 15.3 percent of the instructors earned $14,000 or more, which means that 84.7 percent earned less than $14,000. The median is not given, but it is clearly between $11,500 and $12,000 and perhaps closer to $11,500. The first quartile—the salary figure such that one-fourth of the instructors earn less and three-fourths more— is between $10,000 (the 17.3 percentile) and $10,500 (the 27.2 percentile). ▲

There are other measures of "center" that can be defined in terms

---

†Tukey calls the first and third quartiles *hinges*. [See John Tukey, *Exploratory Data Analysis* (Reading, Mass.: Addison-Wesley Publishing Company, Inc., 1977).] They include between them the middle 50 percent of the data.

Table 2-6   Salaries of College Instructors

Salary Interval	Percentage (Cumulative Downward)
21,500 and above	.5
21,000 and above	.7
20,500 and above	.9
20,000 and above	1.0
19,500 and above	1.1
19,000 and above	1.4
18,500 and above	1.5
18,000 and above	2.0
17,500 and above	2.4
17,000 and above	3.0
16,500 and above	3.9
16,000 and above	4.8
15,500 and above	6.1
15,000 and above	8.0
14,500 and above	11.2
14,000 and above	15.3
13,500 and above	19.9
13,000 and above	26.2
12,500 and above	34.0
12,000 and above	44.0
11,500 and above	53.1
11,000 and above	63.2
10,500 and above	72.8
10,000 and above	82.7
9,500 and above	88.1
9,000 and above	92.9
8,500 and above	95.1
8,000 and above	96.9
7,500 and above	98.1
7,000 and above	100.0

of the ordered observations in a data set. For instance, the "center" can be given as the average of the largest and smallest observations—the point on the scale of values that is halfway between the largest and smallest:

$$\text{midrange} = \frac{\text{smallest} + \text{largest}}{2}$$

Another approach is to take the value halfway from the first to the third quartile:

$$\text{midhinge} = \frac{1}{2}(Q_1 + Q_3).$$

Still another possibility is the average of the midrange and midhinge:

$$\text{trimean} = \frac{1}{4}(Q_1 + Q_3 + 2 \times \text{midrange}).$$

These various measures—all purporting to locate the center of the distribution of values in a data set—are usually different, although they could agree.

Another characteristic of a set of data that is often useful and important is the extent to which they spread, or are dispersed about the middle. Two measures of this characteristic, defined in terms of the ordered observations, are these:

Measures of dispersion:

range = largest - smallest
interquartile range = $Q_3 - Q_1$

Several of these various measures of location and spread can be calculated from five quantities:

Five-number summary for a set of numerical data:

Smallest observation
First quartile
Median
Third quartile
Largest observation

Ordering a set of observations, although rather trivial for fewer than 10 observations, is not at all trivial for a data set of 1000 observations, say. Computers can be programmed to do the ordering, but it takes more computer time than one might expect.[†]

To order a set of data by hand it is usually easiest to prepare a scale of values extending over a range including all values that one

[†]The order of the number of operations required for a data set of size n is n^2, using naïve algorithms, and at least $n(\log n)$ even with the cleverest schemes.

expects to encounter, and then make a mark at the place on the scale corresponding to each observation, as it is read. When data are tallied in a set of class intervals, they are, by that process, ordered to a degree; only the order of observations within a class interval is lost. The "stem-leaf" method of tallying to be described shortly avoids even this loss.

The median cannot be determined exactly when data are grouped into class intervals. However, it can be approximated rather well from a frequency distribution, by first locating the class interval containing the middle of the data and then interpolating within that interval. An example will illustrate the method.

Example 2-24

In Example 2-20 ages of women having abortions were given by means of a frequency table. The distribution is repeated in Table 2-7(a), and Table 2-7(b) gives *cumulative* relative frequencies as percentages.

Table 2-7(a)

Age	Frequency
<16	3
16–18	17
18–20	25
20–22	20
22–24	8
24–26	6
26–31	12
31–36	4
>36	4

Table 2-7(b)

Age	Percent Less
16	3.0
18	20.2
20	45.4
22	65.7
24	73.7
26	79.8
31	91.9
36	96.0
42	100.0

The median is the 50th observation from either end, or the fifth one from the smallest of the 20 observations in the interval 20–22. (There are 45 observations to the left of that interval.) The location of this fifth one within the interval is not known from what is given in the table. Although it may not be necessary to pinpoint its location, this can be done in various ways. One way is to assume that the 20 observations between 20 and 22 are more or less uniformly spread, one in each of the 20 subintervals of width 2/20. The midpoint of the fifth subinterval, 20.45, can be taken as an estimate of the median.

Another way would be to interpolate linearly between the 45.4th percentile, 20, and the 65.7th percentile, 22:

$$20 + \frac{50 - 45.4}{65.7 - 45.4}(22 - 20) = 20 + \frac{4.6}{20.3} \times 2 = 20.45.$$

Using this same type of interpolation to find the approximate location of the quartiles yields $Q_1 = 18.38$ and $Q_3 = 24.43$.

The five-number summary would then be as follows:

$$\begin{aligned}
\text{min} &= 12 \text{ (approx.)}\\
Q_1 &= 18.4\\
\text{median} &= 20.5\\
Q_3 &= 24.4\\
\text{max} &= 41
\end{aligned}$$

The maximum value of 41 was available in the original summary of Example 2-20. The minimum was not given, and the 12 shown is only a guess. From these five numbers we see that the range is about $41 - 12$, or 29; the midhinge is 21.4; and the trimean is 20.95. The interquartile range is $24.4 - 18.4 = 6.0$. ▲

2.8
STEM-LEAF DIAGRAMS AND BOX PLOTS

A frequency distribution and its representation by means of a histogram are useful ways of summarizing data, but the process of rounding off or grouping data to obtain them is a *reduction* of the data. The location of data points within the class intervals is lost in the process. A *stem-leaf diagram*† is a device for producing, in effect, a histogram without losing the original data.

Data are often given by two or three significant digits, and even data given with more than three significant digits often will have their variation occurring in the last two or three digits. The last digit will be thought of as a "leaf," growing on the "stems" of the earlier digits. Consider, as a first simple example, these numbers:

$$16, \quad 17, \quad 24, \quad 30, \quad 19, \quad 27, \quad 26, \quad 14,$$
$$32, \quad 23, \quad 28, \quad 33, \quad 19, \quad 22, \quad 26.$$

---

†Due to John Tukey, whose book cited earlier presents the method in greater detail and with variations (in its Chapter 1).

The first digit is either 1, 2, or 3, and these provide the stems, written down in a column to form the "trunk":

$$
\begin{array}{c|}
1 \\
2 \\
3 \\
\end{array}
$$

Data are then recorded by writing the second digit as a "leaf" on the appropriate stem—6 on the stem of 1 (for 16), 7 on the stem of 1 (for 17), 4 on the stem of 2 (for 24), and so on:

$$
\begin{array}{c|l}
1 & 67949 \\
2 & 4763826 \\
3 & 023 \\
\end{array}
$$

Although not really necessary, a dotted histogram has been sketched in just to point out that—if leaves are spaced evenly—the stem-leaf diagram is as good as a histogram. But the original numbers are still evident in this diagram; all that is lost is the original order of observation. The median is readily found to be 24, the third smallest leaf on the middle stem.

Example 2–25

Table 2–3 (page 69) gave 200 measurements of viscosity. A quick scan of the data shows a large value (37.0) right at the outset, but nothing close to it near the bottom. This raises the possibility that something might have been changing as the data were being recorded. To look into this, the data were divided into four segments (from top to bottom). Each segment of 50 observations was recorded in a stem-leaf diagram, the results given in Table 2.8. Also to look into this, the data were divided into four segments given are the median and the first and third quartiles, in each case. Even this quickly prepared summary shows that something was going on during the gathering of these data. The experiment being performed appears to have been changing. (Perhaps the person doing the measuring was learning as more and more measurements were made. The variability seems to be decreasing.) ▲

Sometimes the class intervals defined by the second-to-last digits are not the most convenient. Stems can be doubled or halved without much extra work. To make twice as many stems, use each

Table 2.8   Stem-Leaf Diagrams for Viscosity Data

26		8		
27	6		6	
28			97887	47
29	869		951759803	8395520
30	80072	85709	589678470959	0372868712277608637
31	403072763968	663522051893	789467239300	39850333647
32	9840008200	6172103050398	574051	1053984857
33	3574391	13834870898	566	5
34	489354	642582	51	
35	9305	94		
36	2			
37	0			

Quartiles:

Q_1:	31.3	31.5	29.8	30.2
Q_2:	32.0	32.3	30.9	30.7
Q_3:	33.9	33.8	31.9	31.8

Interquartile range:

	2.6	2.3	2.1	1.6

second-to-last digit twice, first with last digits 0 to 4, and then with last digits 5 to 9. This will be illustrated in Example 2–26.

For calculation of the quartiles, it is convenient to record the "depth" of each stem, alongside the trunk. Counting from the top and from the bottom, we accumulate the frequencies toward the middle, stopping short of half the data. Alongside the middle stem, the one that contains the median observation, we record just the number of leaves on that stem; the *depth* of any other stem is the number of observations on that stem plus the number on stems farther from the middle in the same direction. (Observe that the sum of the number written beside the middle stem and the depths of the stems just above and just below the middle stem is the size of the data set.)

Example 2–26

The 200 measurements of viscosity in Table 2–3 are entered in a stem-leaf diagram in Table 2–9, with 26, 27, 28 and so on, each serving double duty as stems. Depths are recorded at the left. The first quartile, obtained by counting from the top, will lie in the second 30-stem, about 7 observations in. (We have shown it midway between the 50th and 51st observations.) The third quartile is obtained

Table 2-9   Stem-Leaf Diagram for Viscosity Data

Depth	Stem	Leaves	
1	26	8	
	27		
3	27	66	
4	28	4	
10	28	978877	
16	29	103320	
29	29	8699575988955	
43	30	00204003212203	
70	30	87857958967879597868777 6867	
94	31	4030233220134233003033 34	
23	31	776968665589789 67998567	
83	32	4000200121030034011034	
61	32	98867598575598857	
44	33	343113340	
35	33	5798878985665	
22	34	4344221	Five-number
15	34	9856585	summary:
8	35	304	Smallest = 26.8
5	35	959	Q_1 = 30.65
2	36	2	Median = 31.6
	36		Q_3 = 32.8
1	37	0	Largest = 37.0

Sum = 200 { (bracketing stems 31, 31, 32 with depths 94, 23, 83)

similarly, counting up from the bottom, and keeping in mind that one enters the second 32-stem at the 9-leaves, then 8, and so on. (Incidentally, this summary of the 200 observations still exhibits the original numbers, but not in the original order of observation. The trend that became evident when we segmented the data in Example 2–25 is now hidden!) ▲

A quick visual summary of a set of numerical data is given by a *box plot*,[†] a graphical representation of the five-number summary. The plot consists of a rectangular box located between the quartiles, together with lines extending from the box, on either end, to the maximum and minimum values in the data. The median is marked as a line across the box, at its location on the scale of values. (The width of the box has no significance in the simplest type of plot, although variations of the box-plot idea do make use of that dimension.)

[†]Again due to Tukey; see *Exploratory Data Anlaysis*, pp. 39–41. His full name for this is "box and whisker plot."

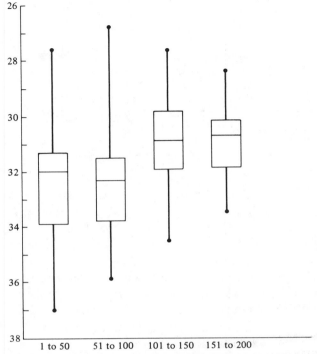

Figure 2-13. Box plots for Example 2-27.

Example 2-27

Box plots for each of the four segments of the viscosity data of Table 2-8 are shown in Figure 2-13. These plots show vividly, perhaps more so than the stem-leaf diagrams, that the situation was changing as the data were gathered. ▲

2.9
CONTINUOUS BIVARIATE DATA

Continuous variables, like categorical variables, are often studied in *pairs* in order to investigate the possible existence, the strength, and the nature of any relationship between them. The possibility of a relationship exists, in particular, when the variables are different numerical aspects of a single individual, family, epoch in time, place, happening, or experiment. That common element links or pairs the variables. Some examples:

X	Y	
Height of father	Height of son	(in a family)
Height	Weight	(of a person)
Score on pretest	Score on posttest	(of a student)
Amount of advertising	Sales	(of a product)
Date	Dow Jones average	(at a given time)
Soil acidity	Yield of a crop	(in a field trial)
Score by judge A	Score by judge B	(for a contestant)
IQ of husband	IQ of wife	(in a family)

The idea of "relationship" has to do with variation—a relationship between the variation of one variable and the variation in the other variable, from one individual to another or from one time or place to another—that is, from one trial of an "experiment" to another. To investigate a relationship, one observes the two variables for each of a number of individual persons or times—or trials of an experiment. The data consist of (say) *n pairs* of numbers: (X_1, Y_1), (X_2, Y_2), . . . , (X_n, Y_n). Each data pair can be represented geometrically as a *point* in a rectangular coordinate system, X being the horizontal coordinate and Y the vertical coordinate of the point (X, Y). The plot showing the n data points is called a *scatter diagram*.

We use the generic names X and Y in the general discussion of a pair of variables. If one variable can be thought of as "independent" or subject to control or choice, and the other as a "dependent" or response variable, the first will be called X and the second Y. Sometimes one variable will precede the other in time; in this case the earlier variable is the "independent" variable and is called X. (If the variables are not distinguished in any of these ways, the labeling will not matter.)

Example 2-28   Ages of Husband and Wife

Table 2-1 gave data on 30 couples in a certain study, including the ages of husband and wife. We repeat the age pairs here in the form (X, Y), where X is the wife's and Y the husband's age.

$$(41,43) \quad (24,24) \quad (31,35)$$
$$(23,21) \quad (27,33) \quad (38,39)$$

(29,31)	(29,31)	(18,23)
(26,26)	(25,28)	(33,32)
(37,35)	(22,30)	(18,23)
(37,41)	(23,28)	(21,33)
(41,43)	(26,31)	(42,42)
(28,25)	(30,34)	(23,28)
(23,24)	(28,32)	(34,38)
(24,29)	(24,27)	(33,39)

These 30 data points are plotted in Figure 2–14—the scatter diagram for these data. One might expect a tendency for younger women to be married to younger men, and this does show up in the scatter diagram. Points high on the graph tend to be far to the right, and the lower points tend to be to the left.

Although the relatively small number of data points may not warrant it here, a grouping of X-values into class intervals and Y-values into class intervals results in a bivariate frequency distribution. The roundoff used in recording an age as a number of years constitutes a grouping, but a regrouping with wider class intervals will be a little more manageable. The tally is as follows, using 5-year intervals:

H \\ W	20–24	25–29	30–34	35–39	40–44
15–19	//				
20–24	///	////	//		
25–29		///	////		
30–34			//	///	
35–39				//	/
40–44					///

This two-way array of the class cells, or one with cell frequencies in place of tallies, shows the distribution to some extent. A histogram would require a third dimension to represent frequency; a picture of the age data is shown in Figure 2–15, and it shows some of the problems involved in constructing such representations. Not only is it difficult to draw, but some bars are actually hidden. ▲

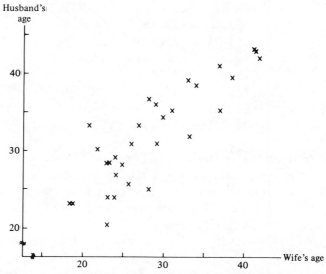

Figure 2-14. Scatter diagram for Example 2-28.

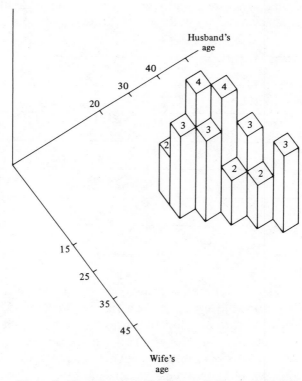

Figure 2-15. Bivariate "histogram" (Example 2-28).

Example 2-29   *A Study in Ophthalmology*

In studying the reaction rate of a certain synthetase of a bovine lens it is found that the rate is related to the substrate concentration. Two observations of reaction rate were obtained for each of several substrate concentrations and the results were plotted as reciprocals. The data are as follows:

Reciprocal Substrate Concentration	Reciprocal Reaction Rate
24	.429, .444
20	.293, .293
16	.251, .268
12	.207, .218
8	.239, .218
6	.180, .199
2	.156, .167

The data are plotted in Figure 2-16 with the controlled variable (substrate concentration) along the x-axis. As plotted, in terms of reciprocal concentration and reciprocal reaction rate, the points came close to falling on a straight line. The simplicity of a linear relationship is appealing and useful in this and, indeed, in any application.   ▲

In investigating a population relationship, one may not have the luxury of a large number of data points; but, of course, the more

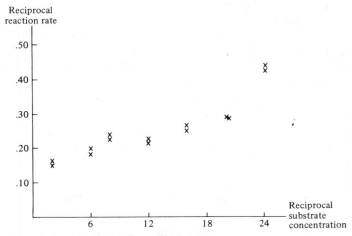

Figure 2-16.   Data for Example 2-29.

Figure 2-17. Data from an approximately related pair
(X, Y).

the data, the better the insight into any relationship in the population. Figure 2-17 shows a plot of a large number of data points,
which appear as a "cloud," somewhat indefinite, but yet with a
suggestion of shape or form that hints at a relationship. A large X-
value clearly does not guarantee a large Y-value, but there is an
evident tendency in this direction, a tendency that suggests a kind
of relationship between the variables. The looseness of this relationship may be the result of measurement errors or other chance components in the observations that are effectively masking what is
possibly a definite or "deterministic" connection.

A relationship, even though weak, may suggest the existence of
some underlying causal mechanism that might be investigated. But
whether or not there is a direct causal connection, one can exploit
a relationship—even an approximate one—to learn something about
one variable from an observation of the other. The term *correlation
analysis* refers to the study of data to determine the existence and
strength of a relationship of the simple type called "linear," seen
in Example 2-29. The term *regression analysis* refers to the use of
a relationship between the two variables in a bivariate pair for the
purpose of predicting one variable from the other, and also to the
uncovering of a deterministic or functional relationship of a response

to a controlled variable when that response is obscured by random error.

KEY VOCABULARY

Categorical variable

Numerical variable

Discrete variable

Continuous variable

Class interval

Class mark

Class frequency

Relative frequency

Frequency distribution

Bar graph

Cross-classification

Bivariate distribution

Marginal distribution

Histogram

Location (of a distribution)

Median

Quartile

Percentile

Dispersion

Range

Stem-leaf diagram

Box plot

Scatter diagram

QUESTIONS

1. In a "frequency distribution" what is it that is distributed, and among what?
2. A relative frequency is "relative" to what?
3. What kind of scheme must be used in collecting data on two variables in order to learn about their relationship to each other?
4. How does one get univarite distributions for the variables represented in a cross-classification from the frequencies in that cross-classification?
5. How can you reconstruct a cross-classification of two variables if you are given only the marginal totals?
6. In how many ways can data in a three-way cross-classification be layered?
7. In tallying data for a frequency distribution, what do you do with an observation that falls on the boundary between two class intervals?
8. What is the correct choice of class-interval scheme for recording data on a continuous variable?
9. In a histogram, what geometric quantity is used to represent frequency?

10. In constructing a histogram from a frequency table, what do you do with a frequency of 0—omit that class interval?

11. In constructing a histogram, what is the advantage of having used class intervals of the same size for summarizing the data?

12. If the class intervals in a frequency distribution are not all of the same size, how do you choose the bar heights in drawing the histogram?

13. In representing the same set of data should everyone's histogram look the same?

14. In what sense is the median of a set of numbers a measure of "middle?"

15. How do you put numbers in order when there are "ties?"

16. If (as one educator reputedly asserted) we need "more people above the median," how can this be accomplished?

17. Is a *percentile* of a distribution a percentage, or a value of the variable?

18. What advantage or advantages does a stem-leaf diagram have over a frequency distribution? (Any disadvantages?)

19. Explain the "depth" column in a stem-leaf diagram.

PROBLEMS

Sections 2.1–2.5

*1. Comment on the visual presentations of data on causes of death reproduced from a Bureau of the Census publication in Figures 2–18 and 2–19. (Is one better than the other? Does either have any flaw?)

2. Devise a graphical representation better than that in Figures 2–18 and 2–19, for the 1974 data on major causes of death.

*3. For the Main Campus students in Data Set B (Appendix C):
    (a) Make a frequency distribution according to class.
    (b) Make a frequency distribution according to major.

4. Make a frequency distribution according to class and a frequency distribution according to major for the Ag Campus students in Data Set B (Appendix C).

*5. Summarize the data on class and major of females in one cross-classification, and of males in another, for the Agriculture (Ag) Campus students in Data Set B (Appendix C). Then find:
    (a) The proportion of male sophomores with a business major.
    (b) The proportion of female home economics students who are seniors.

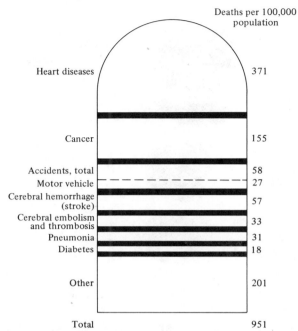

Figure 2–18. Major causes of death, 1966. (From Bureau of the Census, *1969 Pocket Data Book.*)

6. Students in a statistics class were classified according to sex and according to residence, as follows:

	Dorm	Commute	Walk
Male	10	33	15
Female	7	25	7

("Walk" means close enough to walk, but not in a dorm.)
(a) Find the proportion of males who commute.
(b) What proportion of dorm residents are female?

7. The cross-classifications obtained in Problem 5 are layers in a three-way classification (according to sex, class, major). Give the layer cross-classifications of the same data with class as the layer variable.

*8. Give a three-way classification of the data in Table 2–1 (page 41) on educational level of the husband, smoking history of the husband, and smoking history of the wife, using the husband's educational level as the layer variable. (This can be ob-

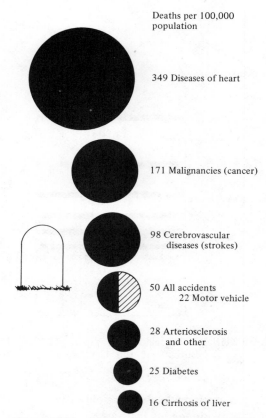

Deaths per 100,000 population

349 Diseases of heart

171 Malignancies (cancer)

98 Cerebrovascular diseases (strokes)

50 All accidents
   22 Motor vehicle

28 Arteriosclerosis and other

25 Diabetes

16 Cirrhosis of liver

Figure 2-19. Major causes of death, 1974.
(From Bureau of the Census,
1976 Pocket Data Book.)

tained from the summary of Figure 2-6, which uses the wife's smoking history as a layer variable.)

9. Fifty students in a dormitory gave their ages as of their last birthday as follows:

Age	Frequency
17	4
18	16
19	14
20	10
21	4
22	1
23	1

Represent this distribution by a histogram.

10. Make a frequency table and histogram for the number of mathematics courses taken by Main Campus males in Data Set B (Appendix C).

Sections 2.6 and 2.7

*11. Summarize the respiration data in Data Set A (Appendix C). Does the nature of the numbers reported give a clue as to how the respiration rate was obtained?

12. Summarize the temperature data in Data Set A (Appendix C) by means of a frequency table and histogram.

*13. Make frequency distributions summarizing the heights of females in Data Set B (Appendix C):
    (a) In the Main Campus section (numbers 73 to 102).
    (b) In the Ag Campus section (numbers 156 to 200).

14. A 1974 census report gave the income level of white and non-whites as follows (in percent):

Income	Whites	Nonwhites
0–$2,999	4	14
3,000–4,999	7	16
5,000–9,999	22	29
10,000–14,999	25	19
15,000–24,999	30	18
25,000–and over	12	4

Determine approximately the median income level for whites and for nonwhites.

*15. Scores on TOEFL (Test of English as a Foreign Language) for 112,241 foreign students in 1977 are summarized by percentile rank in Table 2–10.
    (a) Determine approximately the median.
    (b) What percentage of students scored between 460 and 480?
    (c) What percentage of students scored more than 580?
    (d) Determine the first quartile (approximately).
    (e) Draw a histogram for the distribution of scores.

16. Table 2–11 on page 95 gives a summary by percentile rank of Graduate Record Examination scores of 155,623 applicants in the United States in 1978.
    (a) Determine approximately the median.

Table 2-10

Score	Percentile Rank
660	99
640	98
620	94
600	90
580	84
560	76
540	67
520	58
500	48
480	38
460	29
440	21
420	14
400	9
380	5
360	2
340	1

(b) What percentage of students scored higher than 640?

(c) What percentage scored less than 440?

(d) What percentage scored between 500 and 600?

(e) What percentage scored more than 570? (Estimate this.)

(f) Determine approximately the upper quartile.

*17. Using the frequency distributions of Problem 13, determine the median height of females in the Main Campus section and the median height of females in the Ag Campus section of the students in Data Set B (Appendix C).

Section 2.8

*18. Hemoglobin levels (in grams per milliliter) of 50 lung cancer patients in a Veterans Administration hospital were recorded as follows:

13.5 15.6 16.3 12.3 13.1 14.2 12.4 11.3 14.0 14.6
13.6 14.8 12.7 10.9 11.0 11.4 15.0 10.1 15.4 11.3
10.7 14.6 13.5 15.1 12.1 12.0 14.2 11.4 15.0 13.3

13.2   9.1   16.9   14.2   15.0   13.6   14.8   11.4   14.8   15.7
13.5   13.5   12.9   13.8   13.8   13.7   16.3   11.6   14.2   10.7

   (a) Make a stem-leaf diagram for these data.

   (b) Give the frequency distribution suggested by the stem-leaf diagram, and sketch the corresponding histogram.

19. Given these heights (in inches):

72, 68, 67, 71, 73, 72, 68, 73, 66, 80, 69, 71, 76.

   (a) Make a stem-leaf diagram.

   (b) Determine the median height.

Table 2-11

Score	Percentile Rank
740	99
720	97
700	94
680	91
660	86
640	81
620	74
600	70
580	62
560	56
540	51
520	44
500	39
480	34
460	28
440	24
420	21
400	17
380	14
360	11
340	9
320	7
300	5
280	4
260	3
240	2
220	1

*20. Given the following stem-leaf diagram, find the median and the range:

Depth	Stem	Leaves
2	41	62
6	42	1738
$\overline{6}$	43	028693
$\overline{9}$	44	5147
5	45	70
3	46	317

Stems are 4.1, 4.2, and so on; leaves are hundredths.

21. Cars were counted passing a certain point on a roadway, over two 20-minute periods, the number recorded being the number of cars in each 1-minute period. In each case construct a stem-leaf diagram and box plot, noting the difference between the distributions and the possible explanation in terms of time. [Use two stems for $1x$ and two for $2x$ in part (b).]

(a) On a Monday, around 6 P.M.:

34, 31, 31, 35, 32, 23, 49, 35, 35, 83,
14, 51, 22, 51, 39, 45, 27, 23, 64, 54.

(b) On a Sunday, around 3 P.M.:

11, 29, 15, 21, 19, 21, 17, 14, 23, 17,
15, 28, 16, 20, 11, 14, 21, 21, 19, 13.

*22. Summarize the temperature data in Data Set A (Appendix C) in a stem-leaf diagram; from this determine the minimum, maximum, median, and quartiles, and make a box plot. (Can this stem-leaf diagram be obtained from the frequency distribution of Problem 12? How about the other way around?)

23. Summarize the life satisfaction indices of Main Campus males in Data Set B (Appendix C) in a stem-leaf diagram, using two stems for 10, two for 11, and so on. Determine the median.

Section 2.9

*24. Table 2–12 gives high and low November temperatures in Minneapolis in a certain year. Make a scatter plot of Hi versus Lo. Would you expect any relationship, and do you see any?

Table 2-12

Day	Hi	Lo	Day	Hi	Lo	Day	Hi	Lo
1	52	36	11	33	23	21	20	9
2	57	30	12	33	17	22	29	12
3	61	36	13	39	26	23	31	16
4	55	30	14	44	24	24	16	-4
5	56	34	15	43	33	25	-4	-12
6	51	47	16	45	30	26	12	-16
7	56	49	17	36	31	27	17	-1
8	58	53	18	36	28	28	24	-4
9	59	31	19	37	28	29	35	20
10	34	25	20	56	20	30	35	24

25. Make a scatter plot of confidence of doing well in statistics versus grade-point average, for the 72 males in the Main Campus section of the class listed in Data Set B (Appendix C). (Might you expect a relationship—and do you discern one?)

*26. Suppose that you have 10 pairs (F, C), where F is a temperature in degrees Fahrenheit and C is the same temperature expressed in degrees Celsius. What will the scatter plot look like?

27. Make a scatter diagram of radial versus apical pulse for the individuals listed in Data Set A (Appendix C). Do you discern a relationship? (Can you suggest why these two pulse measurements might differ? "Radial" means measured at the wrist, and "apical" means measured at the heart.)

*28. Given the systolic versus diastolic pressures in Data Set A (Appendix C):
(a) Make a scatter diagram for the first 50 individuals. Does there appear to be a relationship?
(b) Make a frequency table for systolic pressures from the scatter diagram of part (a), counting marks in "rows" corresponding to the values of pressure. (This gives the "marginal distribution" of systolic pressures.)

29. Can a football player avoid injuries by playing "on surfaces provided by Mother Nature rather than those manufactured by Monsanto?" This possibility was raised in a newspaper article (*St. Paul Pioneer Press*, November 13, 1977), which gave the data for 1976 shown in Table 2-13. Make a scatter plot for these data and comment on the evidence for or against a relationship between number of players on the injured reserve and the number of games played on artificial turf.

Table 2-13

Team	Number of Players on Injured Reserve	Number of Games on Artificial Turf
Atlanta	6	6
Baltimore	6	7
Buffalo	4	4
Chicago	4	10
Cincinnati	4	9
Cleveland	6	4
Dallas	4	12
Denver	6	5
Detroit	7	12
Green Bay	4	4
Houston	10	10
Kansas City	2	10
Los Angeles	8	4
Miami	7	4
Minnesota	1	2
New England	5	10
New Orleans	8	12
New York Giants	4	11
New York Jets	6	8
Oakland	6	4
Philadelphia	3	11
Pittsburgh	10	10
San Diego	5	5
San Francisco	5	10
Seattle	9	11
St. Louis	4	10
Tampa Bay	9	4
Washington	12	4

Review

30. In the histogram of Figure 2–20, the heights of the bars are in
the ratio $2:1:2$. If the sample size is $n = 28$, how many of the
observations lie between 20 and 30?

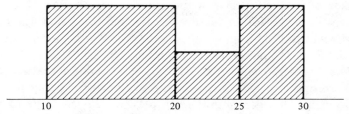

Figure 2-20. Histogram for Problem 30.

31. The following table gives the distribution of whites and blacks (in millions) in the United States in 1975, according to years of education:

		Less Than 8	8	Less Than 12	12	Less Than 16	16 or More
White	Male	5.13	5.18	6.90	16.32	6.70	9.07
	Female	5.21	5.81	8.71	22.52	6.63	6.03
Black	Male	1.20	.40	.99	1.23	.48	.31
	Female	1.47	.55	1.49	1.77	.53	.69

   (a) Find the proportion of males with 4 or more years of college.
   (b) Find the proportion of those with at most an eighth-grade education who are black.
   (c) Make a two-way table for males and a two-way table for females showing the cross-classification according to race and education.

32. Make a scatter plot of the females in the Main Campus section of Data Set B (Appendix C):
   (a) According to the variables X = height, Y = weight.
   (b) According to the variables X = weight, Y = satisfaction with life index.
   Would you expect, and do you discern, any relationship in either case?

33. Several thousand holes .018 inch in diameter must be drilled in a printed-circuit board to be used in a computer. None of these holes must contact the copper layer of the board, or it will short out. Instead, each hole must be completely within a circular opening in the copper layer, and each such opening is .036 inch in diameter. Owing to various kinds of error in drilling, the holes will not be perfectly centered. The following data represent the distance in thousandths of an inch from the center of the opening to the center of the hole, for 36 holes in eight different locations on a single board. A measurement over .009 inch means that the board is defective.

$$3.9 \quad 3.4 \quad 3.3 \quad 4.5$$
$$5.1 \quad 5.4 \quad 5.5 \quad 5.9 \quad 6.6$$
$$2.1 \quad 5.1 \quad 4.1 \quad 3.1 \quad 2.6$$

```
2.2  2.7  1.4  2.1
1.9  1.6  1.8  2.1
3.6  0.9  3.2  3.1  3.4
5.1  5.6  6.1  4.7
5.6  4.1  4.6  6.2  6.6
```

(a) Make a stem-leaf diagram for these data.

(b) Find the median and the range.

34. In the manufacture of a printed circuit for a computer, many holes are drilled through a plate. The holes are designed to have a diameter of .022 inch. A sample of 102 holes yielded the following measurements, in thousandths of an inch.

(a) Make a stem-leaf diagram for these data. (Use two stems for 20, two for 21, and two for 22.)

(b) Construct a histogram using about 12 class intervals.

(c) Determine the minimum, maximum, median, and quartiles, and construct a box plot.

21.4	21.3	21.8	21.2	21.5	21.3
22.1	21.7	21.6	21.9	22.0	22.9
21.5	21.3	21.5	20.5	20.7	20.4
21.5	20.8	21.0	21.3	22.2	21.2
22.3	21.9	22.7	22.2	22.7	22.2
22.1	21.8	22.2	22.1	21.5	22.2
21.3	22.0	21.2	21.8	22.5	22.2
21.6	22.1	21.9	22.4	21.7	21.6
22.5	22.0	22.9	22.6	22.9	23.2
22.3	21.7	22.2	21.8	21.2	21.6
21.1	21.6	21.2	21.8	20.0	21.4
22.2	22.6	21.8	22.2	22.0	21.6
22.3	21.2	22.2	22.7	22.6	21.6
21.9	20.9	21.5	21.5	21.0	21.4
21.5	22.6	21.2	21.8	21.2	22.0
21.6	21.5	21.8	21.8	22.7	21.1
21.7	22.7	21.2	21.8	21.5	21.5

35. If your SAT score were reported as being at the 73rd percentile, what would this tell you about how you stand with respect to the other students who took the test?

36. The percent butterfat of 30 Guernsey cows are as follows:

```
4.32  4.23  4.24  4.67  4.29  3.74
3.96  4.28  4.48  4.03  3.89  4.42
```

```
3.74  4.15  4.42  4.29  4.20  4.27
4.10  4.49  4.00  4.05  4.33  3.97
4.33  4.67  4.16  4.11  3.88  4.24
```

(a) Make a stem-leaf diagram.

(b) Find the median and range.

37. In one of its investigations, the U.S. Public Health Service obtained the following data about the number of children from a sample of women of ages 18 through 79:

Number	Percent
0	22
1	17
2	21
3	16
4	10
5	5
6	2
7	2
8	2
9 or more	3

(a) Construct a histogram for these data.

(b) Find the median.

(c) Do you see any use for this summary?

38. The table in Example 2–3 (page 44) gives, in effect, a cross-classification of the Ag Campus students in Data Set B according to class and sex. Deduce from what is given there the distribution of males and females in each class (i.e., determine relative frequencies that are "relative" to the number in a class). Do you see any association between class and sex?

39. A syndicated newspaper column (by Dr. Halberstan, November 21, 1979) reported on an English research study of patient attitudes toward breast cancer. Patients were given psychological tests shortly after diagnosis and classified as to attitude into one of four categories: denial, fighting spirit, acceptance, hopelessness. Sixty-seven (of 69) patients could be traced after five years, "of whom 33 were alive and had no sign of recurrence of cancer, 16 were alive but had some sign of spread of cancer, and 18 had died of breast cancer." The report continues: "Of the 20 patients who had shown earlier [denial or fighting spirit], 75 percent were still alive. On the other hand, only 35 percent

of those who had evinced acceptance or hopelessness were alive." Try to construct a cross-classification of the 67 patients, according to attitude (with categories "deny-fight", and "accept-hopeless") and according to result (with categories "lived" and "died").

3
Averages
and Deviations

IN CHAPTER 2 we took up some graphical means for presenting data and some measures used to describe aspects of data on numerical variables. In particular, the idea of a "center" or "typical" or "normal" value of a numerical variable was expressed in terms of the median and other measures calculated from an arrangement of the data in numerical order. In this chapter we consider some arithmetical procedures for distilling from a set of numerical data some other measures of what is typical and of how different from "typical" individual observations can be.

The word *average* is sometimes used in a rather general and vague sense, as in the phrase "the average man." Among all kinds of people, the "average" person is not extreme in any characteristic. He or she is somewhere near the center, or in the midst of the distribution of people.

An average of a set of numbers, however defined, must be such that it is in their midst. Some numbers are larger, and some are smaller than "average."

The averages we shall define have this property, and it is useful to have it in mind when calculating averages, as a rough but quick check on the calculation. Thus, if you come up with a calculated average that is outside the range of values of the data being averaged, you should realize that something has gone wrong in the calculation.

Averages of one kind or another are often used in comparing groups that are variable within the groups. We might say, for instance, that the average American family has more automobiles than the average Russian family. But this does not mean that every American family has more automobiles than every Russian family, for some have fewer—none. Rather, it means only that one "center" is to the left of the other. We describe this by saying that the one distribution is *located*, by its center, to the left of the other. The various "averages" we use are referred to as *measures of location*.

3.1
THE MEAN

The word *average*, even in its common usage, has a precise mathematical meaning in addition to its rather general meaning of typical, ordinary, or center.

Example 3-1

Four people eating dinner in a restaurant are told that, by management policy, they must be billed on a single check, which totals $66.80. They decide it would be too complicated to sort out who had which dinners, drinks, and desserts and agree to split the total bill four ways. Each then pays the *average:*

$$\text{average} = \frac{\text{total bill}}{\text{number of diners}} = \frac{\$66.80}{4} = \$16.70.$$

Some of them would pay more than the actual charges they incurred, and others less. But the average can be computed from the total bill, without knowing the actual charge for each person. ▲

The particular kind of average used by the diners in Example 3-1 is called the *mean* or, more precisely, the *arithmetic mean*.

The *arithmetic mean*, or *mean*, of a set of numbers
is their total or sum divided by the number of numbers
in the set.

In defining mathematical quantities, ordinary language becomes strained (as in "number of numbers"), so we find it helpful to use mathematical notation. In particular, we give the numbers in a sample literal names, to represent "whatever numbers." Thus, in speaking of a set of four numbers it is convenient to refer to them as X_1, X_2, X_3, X_4, which we can do without knowing their actual values. For a set of numbers of general or unspecified size, we often use the letter n for the size and refer to the n numbers as X_1, X_2, . . . , X_n. (Of course, we do not always have to use the letters X and n. We might have m numbers Y_1, Y_2, . . . , Y_m, or k numbers $a_1, a_2, . . . , a_k$.)

For a set of numbers referred to as X's, that is, $X_1, X_2, . . . , X_n$, the mean or average is commonly denoted by $\bar{X}$:

The mean of the numbers $X_1, X_2, . . . , X_n$ is

$$\bar{X} = \frac{X_1 + X_2 + \cdots + X_n}{n}$$

Similarly, the average of the numbers $a_1, a_2, . . . , a_k$ is denoted by $\bar{a}$, and the average of $Y_1, Y_2, . . . , Y_m$ by $\bar{Y}$.

It is worth noting that $\bar{X}$ is sensitive to each value in the list X_1, . . . , X_n. Changing any one of them changes the mean. This is in contrast with the median, which does not change if any numbers on one side of the median change, as long as they stay on that side. In particular, it does not matter for the median how extreme an extreme value is; but the mean is pulled toward an extreme value.

Example 3-2   *Dancers Short-Changed*

On the TV program "Dance Fever" four couples compete with a minute and a half program of disco-dancing. Three "celebrity" judges score each couple, and the *average* of their marks for any given couple is that couple's overall rating:

$$\frac{S_1 + S_2 + S_3}{3} = \text{rating}.$$

On one occasion, the scores were as follows:

	Judge 1	Judge 2	Judge 3	Average	Rank
Couple 1	91	92	93	92	2
Couple 2	90	88	77	85	4
Couple 3	95	89	100	95	1
Couple 4	100	97	69	89	3

Judge 3 (actor Hervé Villechaize) said, in giving his score for Couple 4, "I don't want to be mean, but I gave them a 69." But this score of 69 meant that Couple 4 could not possibly average more than 269/3. So, even though the other judges (and some viewers) thought Couple 4 was best, it came in third. If the overall rating were taken to be the median of the three scores, the ranking would have been different. By assigning scores that varied widely, Judge 3 almost succeeded in preempting the rating process. ▲

Example 3-3

Table 3-1 gives the combined state and local tax revenue in 1976 for each of the 50 states and the District of Columbia, relative to the total amount of personal income in each case. Thus, the average dollar earned in Alaska is taxed 21.8 cents:

$$\frac{\text{total state and local tax receipts in Alaska}}{\text{total of personal incomes in Alaska}} \times 1000 = \$218.$$

The U.S. mean of $125 shown in the table is *not* the mean of the 51 other means (which happens to be $122.57), as one naïvely might have thought. This is because both in the total receipts in the numerator and in the total income in the denominator of the U.S. average, the states are counted differently—according to dollars of personal income. ▲

Example 3-4 *Mean or Median?*

Since both the mean and median are intended as measures of "middle," or "typical," the obvious question is, which is better?

In a newspaper article,[†] the vice-president of the United States League of Savings Associations was quoted as saying that the "average" home price, currently about $50,000, can be "greatly distorted by a relatively few luxury homes in the $100,000-and-over

[†]*Minneapolis Tribune*, September 2, 1978.

Table 3-1   Annual State and Local Tax Collections per $1000 of Personal Income, 1976

Alaska	$218	New Jersey	118
New York	173	North Dakota	117
Vermont	151	District of Columbia	117
California	149	Idaho	116
Massachusetts	147	Pennsylvania	115
Hawaii	146	Iowa	115
Wyoming	144	Delaware	114
Minnesota	143	Illinois	114
Maine	142	Kentucky	114
Wisconsin	140	Connecticut	112
Arizona	139	Kansas	110
Montana	132	Georgia	109
New Mexico	128	Nebraska	109
Maryland	127	New Hampshire	108
Nevada	127	South Carolina	107
Louisiana	126	North Carolina	107
U.S. average	125	Virginia	107
Oregon	124	Texas	105
Colorado	124	Indiana	104
Utah	123	Missouri	104
Rhode Island	122	Oklahoma	103
South Dakota	122	Tennessee	101
Michigan	121	Florida	101
Mississippi	120	Ohio	100
West Virginia	120	Alabama	99
Washington	119	Arkansas	98

Source: Tax Foundation, Inc.

class." He further stated that a better indicator of the price of homes is the median price—around $44,000, which is "much less than the inflated average home price."

Which is better—mean or median—depends, of course, on the purpose one has in mind. A highway planner, for instance, would be more interested in the average value of a property along a certain proposed route, since muliplying that average by the number of homes to be taken in condemnation proceedings would yield the total cost. However, to give a flavor of price level for the ordinary homeowner, it may be that the median is more informative, as the savings association executive asserted. Given the median value, we then know that half of the homes sell for less than $44,000; but given only the mean value of $50,000, it is conceivable that half of the homes available might be worth $20,000 or less—if there

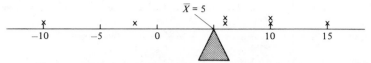

Figure 3-1. Sample mean as a center of gravity, or balance point.

happened to be enough million-dollar homes to raise the total value and, hence, the mean. ▲

An intuitive understanding of the mean is aided by interpreting it as a *balance point.* Imagine equal weights located on the scale of values at points corresponding to the numbers in a set of data; their balance point is computed by the same arithmetic that is used to compute the mean, $\bar{X}$.

Example 3-5

The following high temperatures were recorded for 7 days in January in Minneapolis:

$$10, 6, -10, -2, 10, 15, 6.$$

These are plotted in Figure 3-1 as x's along a scale of temperatures. An equilibrium of the corresponding weightless "teeter-totter" is achieved by putting a fulcrum at the mean, $\bar{X} = 5$. The gravitational force acting on the weights to the right of the mean tends to rotate the system clockwise; this tendency is exactly balanced or canceled out by the force acting on the weights to the left of the mean, which tends to rotate the system counterclockwise. ▲

A note on calculation by computer: It is not necessary that numbers be in numerical order to compute their mean, and the number of operations required is only of the order of n. So, less computer time is required than for determining the median. This contrasts with the fact that in hand calculations the median is usually easier to find than the mean.

3.2
THE MEAN FROM A FREQUENCY DISTRIBUTION

When numerical data are summarized in a frequency distribution, the mean can be calculated in a way that is especially adapted to such a distribution. We illustrate the method first in an instance in which the data are *counts*.

Example 3-6 *Particle Counts*

The following counts of α-particle emissions in 15 consecutive 1-second intervals were recorded:

$$3, \ 1, \ 0, \ 7, \ 3, \ 2, \ 2, \ 4, \ 3, \ 1, \ 2, \ 0, \ 2, \ 4, \ 1.$$

The sum of these 15 counts, rearranged so that equal values are adjacent, is

$$0 + 0 + 1 + 1 + 1 + 2 + 2 + 2 + 2 + 3 + 3 + 3 + 4 + 4 + 7 = 35,$$

and the mean is then 35/15. The part of the sum for each group of repeated values can be found by a multiplication; for instance, the sum of the four 2's is

$$2 + 2 + 2 + 2 = 4 \times 2.$$

With the sum of each such group written as a product, the sum of all the counts is

$$(0 + 0) + (1 + 1 + 1) + (2 + 2 + 2 + 2) + (3 + 3 + 3) + (4 + 4) + 7$$
$$= 2 \times 0 + 3 \times 1 + 4 \times 2 + 3 \times 3 + 2 \times 4 + 1 \times 7 = 35.$$

This computation is nicely organized by adding a column of products to the frequency distribution of counts:

Value (Count)	Frequency	Value Times Frequency
0	2	0
1	3	3
2	4	8
3	3	9
4	2	8
5	0	0
6	0	0
7	1	7
Sums	15	35

The graphical representation of Figure 3-2 shows the mean $\bar{X}$ = 35/15 = 7/3 to be quite reasonable as the balance point of the distribution. ▲

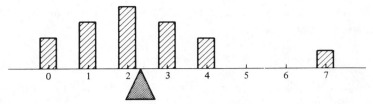

Figure 3-2. Mean as a balance point (Example 3-6).

To find the sum of the values in a set of numerical data
with repeated values, multiply each distinct value
by the number of times that value occurs, and add
these products.

There is a formula implicit in the foregoing statement, one that
needs a notation which means: "Add together all terms of the
following type." The capital Greek sigma is used for this purpose,
and we write

$$\bar{X} = \frac{\sum xf}{n} = \frac{1}{n} \sum xf,$$

where the typical term xf in the sum is the product of one of the
distinct values, referred to as x, and its frequency f, or the number
of times x occurs in the data set. In Example 3-6 these products
are shown in the last column.

We consider next an example in which continuous numerical data
are rounded off, recorded, and summarized in a frequency table.
Such a table ordinarily groups many possible values in one class
interval; nevertheless, the table can be used to calculate an approximation to the mean of the original observations. This is done by
picking some value on each class interval, usually the center point,
to represent all the observations that fall in that interval. This representative value is sometimes called the *class mark*.

Suppose, for example, there are six observations in the interval
from 30.45 to 30.75:

Class Interval	Frequency
30.45–30.75	6

The class mark is 30.6, the midpoint of the interval, and it is assumed that all six observations have that value. To find the sum of those six observations, then, we multiply the class mark by 6: 6 × 30.6 = 183.6. The six original observations would tend to be sprinkled throughout the class interval, some less than 30.6 and some greater than 30.6, so that their actual sum would be close to 183.6. (If the mean of the six numbers happened to be exactly 30.6, of course, then their sum is exactly 183.6.)

To calculate the mean from a frequency table summary of a set of continuous data, we use this same kind of approximation in each class interval. The required sample sum is then the sum of the products of distinct values and their frequencies. Dividing this sum of products by the sample size n, we obtain an approximation to the mean of the original data set. Naturally, there would be nothing wrong with computing the mean from the original or "raw" data set, if these data are available, but the approximation just described is often more convenient and usually adequate.

Example 3-7

In Problem 18 of Chapter 2, hemoglobin levels of 50 cancer patients were given by means of a stem-leaf diagram. The corresponding frequency distribution is given here, along with a column of products:

Class Mark, x	Frequency, f	Product, fx
9.45	1	9.45
10.45	4	41.8
11.45	7	80.15
12.45	6	74.7
13.45	12	161.4
14.45	10	144.5
15.45	7	108.15
16.45	3	49.35
Sums	50	669.5

The sum of the 50 observations is 669.5. (The sum of the three 16.45's is 49.35, the sum of the seven 15.45's is 108.15, and so on.) The mean is therefore

$$\bar{X} = \frac{\text{sum of observations}}{\text{number of observations}} = \frac{1}{n} \sum xf = \frac{669.5}{50} = 13.39.$$

(To the extent that the observations were rounded to class marks, this is an approximation to the mean of the original data.) The mean 13.39 is close to the middle of the fifth interval and seems reasonable as a balance point because there is more weight in the classes with the higher hemoglobin levels. ▲

We always hesitate to say how *not* to do something, but one type of error is made so often that a warning is in order: The number of class intervals, or the number of distinct values in a frequency table, *has no significance* in these calculations. *Never divide by it!* Better still, *do not even count* the class intervals!

The correct denominator in averaging is always the number of observations—the number of pieces of data. In looking at a frequency table, always ask: "How many observations are summarized here?" This number *is* evident in the table; it is the sum of the frequencies, $n = \sum f$, which may appear as a total of the column headed "frequency." In Example 3-7, for instance, there were eight class intervals —but this is irrelevant; there were *fifty* observations in the data set; and the sum of the frequencies gives that number, which is then the divisor to use in calculating the mean. Checking to see that the mean you calculate is within the range of values in the sample will usually tell you if you have divided by the correct number.

3.3
MODIFIED MEANS

It was pointed out in Section 3.1 that an arithmetic mean depends on the specific value of each observation. But extreme observations (the very large or very small ones) are sometimes in error,[†] resulting from unusual conditions (such as line-voltage drop when dealing with electrical equipment, or a gross misreading of a scale). These are referred to as *outliers*, although it is not always clear whether an extreme value is really an outlier or a bona fide observation. If it *is* an outlier, it should not be included in the data set when computing a mean.

[†]We do not mean to suggest that extremes are the *only* values that could be wrong. Other values are sometimes in error, too; it is just that when extreme values are wrong, they affect inferences in such a profound way that people worry about them more. (And, of course, it is also easier to guess that such values are incorrect!)

Example 3-8   *Judging Athletes*

In judging athletic competitions such as diving or figure skating, a panel of perhaps seven judges is used, each scoring individual contestants on a scale from 0 to 10. To lessen the influence of politically motivated favoritism, temporarily bad eyesight, or whatever, sometimes the highest and lowest scores are thrown out or "trimmed" and the middle five averaged to obtain an overall rating. For instance, if the judges score

$$9.9 \quad 9.1 \quad 9.2 \quad 9.2 \quad 9.3 \quad 9.2 \quad 9.2 \, ,$$

the mean rating before trimming is 9.3. The average of the five scores remaining after trimming 9.9 and 9.1 (the highest and lowest) is

$$\frac{1}{5} \, (9.2 + 9.2 + 9.3 + 9.2 + 9.2) = 9.22,$$

which is then reported as the overall rating. This is an example of a *trimmed mean.* Notice that if the 9.1 trimmed off the bottom had been 7.3, say, the trimmed mean would be the same.   ▲

Trimmed mean:
Average of the observations remaining after removal of the highest k and lowest k observations (for some integer k).

Midmean:
Average of the middle 50 percent of the data—a trimmed mean, with the top one-fourth and the bottom one-fourth of the data removed.

An advantage of trimming lies in minimizing the effect of possible outliers—incorrect extreme values that would affect the ordinary mean in a way that could mislead. A trimmed mean has characteristics of both the mean and the median.

Example 3-9

An extruding machine for polyethylene tubing is rated at 45 minutes per 6000 feet. On successive runs these times (in minutes) were recorded:

45, 93, 45, 60, 61, 49, 45, 48, 40, 45, 45, 65, 55, 40.

Their mean is 51.14, but after trimming the largest and smallest (93 and one 40), the new mean is 48.6. The 93, in particular, looks out of line—a possible outlier; and indeed, investigation showed that on that run something was different—a screen pack change was made during the run. ▲

3.4
ROOT-MEAN-SQUARE AVERAGING

Various kinds of "means" have been defined for use in special contexts. You may have heard of the *geometric* mean, for example. The geometric mean of three positive numbers *a*, *b*, and *c* is the length *d* of the side of a cube with volume equal to that of a rectangular box with dimensions *a*, *b*, and *c*. Like the ordinary mean, the geometric mean is no larger than the largest nor smaller than the smallest of the quantities being averaged; it is "typical," in a special way. And if the quantities being averaged are all equal, the geometric mean is that common value.

In statistics and probability, a kind of average that seems useful, second in importance only to the arithmetic mean, is the *root-mean-square* or r.m.s. average. To calculate it, one transforms a given set of numbers by squaring them, takes an arithmetic mean of the squares, and then goes back to the original scale by taking the square root. Thus, it is the square **root** of the **mean** of the **squares**. To express this in a formula, we again use "sigma notation," writing the sum of squares of a set of X's as $\sum X^2$:

R.m.s. average of $X_1, X_2, \ldots, X_n$:

$$\text{r.m.s.} = \sqrt{\frac{1}{n} \sum X^2}$$

The operation of squaring treats $-X$ and X alike, so the r.m.s. is essentially an average of the *magnitudes* (ignoring *signs*) of the X's and will lie in the midst of these magnitudes. If all the X's have the same magnitudes, their r.m.s. average will have that common value.

Example 3-10

In Example 3-5 the following seven numbers were averaged: 10, 6, -10, -2, 10, 15, 6. We compute as follows:

Magnitudes: 10, 6, 10, 2, 10, 15, 6

Squares: 100, 36, 100, 4, 100, 225, 36

Mean square: $\frac{1}{7}$ (100 + 36 + 100 + 4 + 100 + 225 + 36) = 85.86

r.m.s. = $\sqrt{\text{mean square}}$ = $\sqrt{85.86}$ = 9.266.

Thus, the magnitudes range from 2 to 15, and 9.266 is between these extremes. (The arithmetic mean of these magnitudes or absolute values is 59/7 = 8.43.) ▲

Root-mean-square averaging is encountered in many fields—notably in electricity (and other places where "waves" are encountered). An ac voltage alternates on either side of the value 0, and is 0 on the average. Its *effect*, however, is not zero, as anyone who has put his finger into a live light socket can attest. This effect is measured by an r.m.s. average. Similarly, audio power is expressed in terms of r.m.s., as seen in specifications for audio components.

The r.m.s. average of a set of numbers is different from their average magnitude, because a number contributes to the r.m.s. according to its square. The square of a large number is much larger (e.g., $100^2 = 10{,}000$), whereas the square of a number smaller than 1 is even smaller ($.1^2 = .01$). That is, r.m.s. averaging weights the numbers with the larger magnitudes more heavily than it does those with the smaller magnitudes.

3.5
DEVIATIONS AND THEIR AVERAGES

"Deviant" behavior is usually understood to be behavior that is different from normal or "average." It may lie in any direction from that "norm." If a person is not average or normal, it is often important to know *how much* different from average he or she is. Moreover, given a group of individuals, it may be useful to have a measure of typical variation about the average.

Example 3-11

Suppose that the average height of a population is 68 inches (173 centimeters). Knowing this would be a start in designing bathroom mirrors—one would know where to center them on the wall. But to know how high and low the mirror should extend so that most of the population can see itself, we would have to know how much variation or deviation about the average to expect. ▲

Example 3-12

The World Almanac gives "monthly normal temperatures" (in degrees Fahrenheit) for various cities, including these:

Month	San Francisco	Minneapolis
Jan.	49	12
Feb.	51	16
Mar.	53	27
Apr.	56	44
May	58	57
June	61	67
July	63	72
Aug.	63	70
Sept.	64	60
Oct.	61	49
Nov.	55	31
Dec.	50	18

The averages of these monthly temperatures are 57.0°F for San Francisco and 43.6°F for Minneapolis—not all that different. Anyone who has lived in both cities can attest that averages do not tell the whole story. It is the amount of variation about the averages that makes the climates so different with regard to temperature. ▲

Example 3-13   *Rain in the Dust Bowl*

Figures for the average rainfall in a given locality are often quoted in reference to agricultural production. The following is taken from William Lockeretz, "The Lessons of the Dust Bowl," *Amer. Scientist* **66**, 563 (1978):

> Great variations in precipitation, not just minor fluctuations, are the rule on the Plains. The concept of "average" is really only a

mathematical construction, not a very useful predictor of actual precipitation in a particular year. For example, from 1875 to 1936 the average annual precipitation at Dodge City, Kansas, was 20 inches, just about enough to produce crops. But in one out of five years it was above 25 inches, while in one out of six years it was less than 15 inches. This range represents the difference between bumper crops and virtual crop failure.

Clearly, variability about the mean can be as important a characteristic of a "random" quantity as the mean. ▲

The *deviation* of a number X in a set of numbers with mean $\bar{X}$ is simply the difference between the value X and the mean $\bar{X}$:

Deviation of X from $\bar{X}$:

$$X - \bar{X}.$$

Since $\bar{X}$, like all good averages, is in the *midst* of the X-values comprising the data, some of those X-values will lie to the right and some to the left of $\bar{X}$. That is, some deviations about $\bar{X}$ will be positive and some negative. The balance property of $\bar{X}$, described earlier, means that the total of the positive deviations has the same magnitude as the total of the negative deviations. So, when summed, together with algebraic signs, they will always exactly cancel!

The deviations of any n numbers $X_1, X_2, \ldots, X_n$ about their mean $\bar{X}$ add up to 0:

$$\sum (X - \bar{X}) = 0.$$

This assertion is easy to verify in general, along the lines of the following calculation for the case of three numbers:

$$(X_1 - \bar{X}) + (X_2 - \bar{X}) + (X_3 - \bar{X}) = X_1 + X_2 + X_3 - 3\bar{X}$$

$$= \sum X - 3 \sum \frac{X}{3}$$

$$= \sum X - \sum X = 0.$$

It then follows that the *mean* of the deviations about $\bar{X}$ is also 0, being their sum divided by n.

In describing the variability in a set of numbers by some kind of "typical" deviation from the mean, it is the size or magnitude of the deviations that is relevant. The magnitude or *absolute value* of the deviation of X from $\bar{X}$ is just the (unsigned) distance between the values X and $\bar{X}$ when plotted on the scale of X-values:

Absolute deviation from the mean:

$$|X - \bar{X}| = \text{distance between } X \text{ and } \bar{X}$$

(a nonnegative quantity).

One measure of typical deviation from the mean is the arithmetic average of the absolute deviations. This is often called the *mean deviation;* more accurately, it is the mean *absolute* deviation.

$$\text{Mean absolute deviation} = \frac{1}{n} \sum |X - \bar{X}|.$$

Although simple and intuitively appealing, this quantity is seldom used in modern statistical practice, so we do not bother to introduce a special notation for it. (Sometimes a "mean deviation" is defined as the average absolute deviation about the *median*, instead of about the mean.)

Example 3-14

The mean of the seven temperatures 10, 6, –10, –2, 10, 15, 6 (as seen in Example 3–5) is 5. The deviations about the mean value are, respectively,

$$5, \quad 1, \quad -15, \quad -7, \quad 5, \quad 10, \quad 1.$$

As promised, these deviations add up to 0, and their average is 0. The absolute deviations are

$$5, \quad 1, \quad 15, \quad 7, \quad 5, \quad 10, \quad 1,$$

with mean 6.29. This is the mean absolute deviation. ▲

The notion of "typical" deviation about the mean is most often defined as an r.m.s. average of the deviations, called the *standard deviation*, or *s.d.*, and often denoted by *S*. In dealing with more than one data set, the *S* may have a subscript; thus, the s.d. of $X_1, X_2, \ldots, X_n$ would be called S_X to distinguish it from the s.d. of, say, a set of *Y*'s.

Standard deviation (about the mean) of a set of numbers $X_1, X_2, \ldots, X_n$ with mean $\bar{X}$:

$$S = \sqrt{\frac{1}{n} \sum (X - \bar{X})^2}$$

= r.m.s. deviation about $\bar{X}$.

The term "standard deviation" is often abbreviated to *s.d.* or to *SD*.

Being a kind of average deviation, the standard deviation is a deviation of "ordinary" size, no larger than the largest deviation or smaller than the smallest deviation. And if all deviations are of the *same* magnitude, that common value is the standard deviation. (This can only happen if there are at most two distinct numbers in the data set, and these have equal freqencies.)

If all the numbers in a list are *equal*, their mean is that common value; and the deviations from it are all zero, as is then the standard deviation. Conversely, if a standard deviation is zero, the *only* way this can happen is for all deviations to be zero—for all *X*'s in the list to be equal.

The numbers $X_1, X_2, \ldots, X_n$ are all equal if and only if their standard deviation is zero. That is, S = 0 is equivalent to $X_1 = X_2 = \cdots = X_n$.

Example 3-15

To aid in calculating the s.d. of the seven temperatures of Examples 3-5 and 3-14, they are listed below in a column, with further columns for the deviations $X - \bar{X}$, and the squares of the deviations:

X	$X - \bar{X}$	$(X - \bar{X})^2$
-10	-15	225
-2	-7	49
6	1	1
6	1	1
10	5	25
10	5	25
15	10	100
Sums: 35	0	426
Averages: 5	0	426/7

The mean square deviation is 426/7. The root-mean-square or standard deviation is

$$S = \sqrt{\frac{426}{7}} = 7.80.$$

This should be checked to the extent of noting that it is a deviation of intermediate size [smaller than the largest (15) but larger than the smallest (1)]. It is also useful to mark off this deviation graphically on both sides of the mean (see Figure 3-3) and to visualize an s.d. in this way in *every* instance. ▲

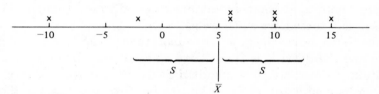

Figure 3-3.

There is a useful, alternative formula for standard deviation based on this expansion of a squared deviation:

$$(X - \bar{X})^2 = X^2 - 2X\bar{X} + (\bar{X})^2.$$

When both sides of this equation are summed, the result is

$$\sum(X - \bar{X})^2 = \sum X^2 - n(\bar{X})^2,$$

and upon dividing by n we obtain the average of the squared deviations called the *variance:*

$$S^2 = \frac{1}{n}\sum(X - \bar{X})^2 = \frac{1}{n}\sum X^2 - (\bar{X})^2.$$

This is perhaps more easily remembered using words:

Variance of a set of numbers

    = mean square deviation about the mean

    = mean of the squares – square of the mean.

The root-mean-square deviation, or *standard* deviation, is then given by this alternative formula:

Alternative formula for s.d.:

$$S = \sqrt{\frac{1}{n} \sum X^2 - (\bar{X})^2}.$$

(You have to pay attention to the *order* of the operations here: First, square each number; then sum those squares and divide the result by n; then subtract the square of the mean value and take the square root of the difference.)

Example 3-16

The calculation of s.d. in Example 3–15 can be accomplished using the alternative formula. The X-values and their squares are as follows:

	X	X^2
	-10	100
	-2	4
	6	36
	6	36
	10	100
	10	100
	15	225
Sums	35	601

The average square is the sum of squares (601) divided by the number of observations (7), and the variance is obtained by subtracting the square of the average (5^2) from the average square:

$$(s.d.)^2 = \frac{601}{7} - 5^2 = \frac{426}{7} = (7.80)^2.$$

This agrees with the result in Example 3-15. ▲

The alternative formulas for S^2 and S are convenient for hand computation; for even though numbers in a data set are fairly "round," and easy to work with, the mean (and hence the deviations $X - \bar{X}$) will usually be a decimal—somewhat more awkward.

The alternative formulas are particularly convenient for programming calculators. It is easy to program the accumulation of 1, X, and X^2 for each number as it is entered. Then after any entry, the calculator can be called on to divide $\sum X$ and $\sum X^2$ by $\sum 1 = n$, to obtain the mean and mean square of the data entered up to that point. However, despite their convenience, the alternative formulas are more sensitive to roundoff error. But with 10 to 14 places available in the computation, this is not ordinarily a problem. (See Problem 32.)

In comparing definitions of average deviation remember that a standard deviation weighs large deviations more heavily than small deviations. It is not clear that the s.d. is always better than the mean absolute deviation, which weighs all deviations equally. It depends, of course, on what is meant by "better," but there are sound statistical reasons to prefer the s.d. in many situations. It is by far the most commonly used measure of dispersion.

Example 3-17

High temperatures on four days in each of two localities were as follows:

Locality A: 50, 70, 70, 90

Locality B: 60, 60, 80, 80.

In *both* locations the mean is 70 and the mean absolute deviation is 10. But the s.d.'s are $S_A = 14.14$ and $S_B = 10$. In S_A the deviations of 20 and 0 have not averaged to 10 because of the squaring, which counts a single deviation of 20 as much as it counts four 10's. ▲

It should be mentioned that many follow the practice of dividing the sum of squared deviations by $n - 1$ instead of by n, to obtain this modified version of the s.d.:

$$\widetilde{S} = \sqrt{\frac{1}{n-1} \sum (X - \bar{X})^2}.$$

Many calculators are wired for this computation, and at least one gives both versions—a key for S^2 with the divisor n, and a key for $\widetilde{S}$ with the divisor $n - 1$. The reasons given for preferring $\widetilde{S}$ are debatable (in our view they are specious), but the debate will not be carried on here. Suffice it to say that both versions are in use, and you have to be aware which one is being used in any given situation. The one you prefer is easily obtained from the other:

$$S = \sqrt{\frac{n-1}{n}} \, \widetilde{S} \quad \text{and} \quad \widetilde{S} = \sqrt{\frac{n}{n-1}} \, S.$$

Moreover, when n is large, they are nearly equal.

3.6
COMPUTING S.D.'S FROM A FREQUENCY DISTRIBUTION

When a data set contains several observations of the same value x, the kind of collapsing that expresses a sum $x + x + \cdots + x$ as the product fx, where f is the frequency of x, can also be used in computing sums of squares or sums of squared deviations. Thus,

$$x^2 + x^2 + \cdots + x^2 = fx^2$$

and

$$(x - \bar{X})^2 + (x - \bar{X})^2 + \cdots + (x - \bar{X})^2 = f(x - \bar{X})^2,$$

where again f is the number of times the value x occurs in the data set. Collapsing sums according to the distinct values, then, yields the following formulas. These are not new definitions but merely revised instructions for the old definitions—instructions that take repeated values into account.

Standard deviation for data summarized in a frequency distribution:

$$S = \sqrt{\frac{1}{n} \sum f(x - \bar{X})^2}$$

or

$$S = \sqrt{\frac{1}{n} \sum fx^2 - (\bar{X})^2},$$

where f is the frequency of x and the sums extend over the *distinct* values x.

Example 3–18

The 15 counts of α partical emissions in 15 consecutive time intervals given in Example 3–6 are repeated here in a frequency-table summary. Included also in the table are columns for products fx as well as products fx^2.

Value, x	Frequency, f	Product, fx	Square, x^2	fx^2
0	2	0	0	0
1	3	3	1	3
2	4	8	4	16
3	3	9	9	27
4	2	8	16	32
7	1	7	49	49
Sums	15	35	—	127

The average is $\bar{X} = 35/15$, and the average of the squares is $127/15$. The average of the squares minus the square of the average is the variance, or square of the s.d.:

$$S^2 = \frac{1}{n} \sum fx^2 - (\bar{X})^2 = \frac{127}{15} - \left(\frac{35}{15}\right)^2 = (1.74)^2.$$

The s.d. is then $S = 1.74$.

Notice in the table above that the column of x's and the column of x^2's are not totaled. These totals are irrelevant! (There is no point to adding up the x's, that is, the distinct possible values; what is needed is the sum of the X's, the numbers in the data set.) The fx-column total (35) represents 15 observations: two 0's, three 1's, four 2's, and so on, whose sum is 35. Similarly, the fx^2-column total represents 15 squares: two 0^2's, three 1^2's, four 2^2's, and so on, whose sum is 127. It is these sums that are divided by 15, the number of observations, which appears as the total of the frequency column. ▲

Hand calculators used with statistical functions do not usually allow for entering grouped data, except by entering each x a number of separate times equal to its frequency—that is, by *not* exploiting the grouping, or repetition of values. A *programmable* calculator, of course, can be programmed to calculate the mean and s.d. from the values and frequencies of a frequency table.

3.7
CHANGING THE SCALE—CODING

Numerical quantities are often measured by different people on different scales, a fact that is pointed up by the campaign for conversion to the metric system. Even within a given system there is need to convert—from inches to feet, or from grams to kilograms, for instance.

The various familiar conversions are all of a type called *linear*, so named[†] because the geometric representations of the conversion equations are straight lines. Thus, the equation

$$C = \frac{5}{9}(F - 32),$$

which converts a temperature expressed in degrees Fahrenheit to a number of degrees Celsius, in a linear relationship. The constant term 32 shifts the origin of reference; the 0 on the Celsius scale corresponds to 32 on the Fahrenheit scale. The multiplier of $(F - 32)$ accounts for the fact that the unit of 1 degree is narrower on the F-scale than on the C-scale, by a factor 5/9. There are more Fahrenheit degrees (180) than Celsius degrees (100) between the freezing and boiling points of water. Moreover, an interval of 5 Celsius degrees is the same as 9 Fahrenheit degrees, anywhere on the scale. The multiplier 5/9 is called a "scale factor."

We shall need to know how the mean and s.d. are transformed when data are transformed by a linear relationship. For instance, when a set of temperatures Fahrenheit is converted to temperatures Celsius, this new set of numbers will have a new mean and new s.d. The mean, being just another value on the scale of values, simply undergoes the same transformation as the values comprising the data. But the s.d., a *deviation* from the mean, is unaffected by a change in reference origin. This is because the individual deviations do not

[†]The reader who is hazy on the topic of linear equations might find it helpful to refer to Section 11.1, which gives a brief review.

change. The s.d. *will* change with a change in the width of a unit—a scale-factor change.

If data $X_1, \ldots, X_n$ are converted to values $Y_1, \ldots, Y_n$ by the relation $Y = a + bX$, then the new mean and s.d., in terms of the old, are

$$\bar{Y} = a + b\bar{X}$$

$$S_Y = bS_X,$$

assuming b to be positive. (More generally, $S_Y = |b|S_X$.)

Example 3–19

The seven temperatures of Example 3–5 were Fahrenheit temperatures:

$$°F: \quad 10, 6, -10, -2, 10, 15, 6.$$

In degrees Celsius [with $C = 5(F - 32)/9$], they become

$$°C: \quad -12.22, -14.44, -23.33, -18.89, -12.22, -9.44, -14.44.$$

It was found earlier that $\bar{F} = 5$ and $S_F = 7.80$. Transforming $\bar{F}$ by the same relation that gives C from F, one obtains

$$\bar{C} = \frac{5}{9}(5 - 32) = -15,$$

whereas

$$S_C = \frac{5}{9}S_F = 4.33.$$

The average deviation (r.m.s. sense) is thus expressed as a smaller number of Celsius degrees than Fahrenheit degrees because the Fahrenheit unit is smaller. ▲

When data are equally spaced (as they ordinarily are in a frequency distribution), a linear transformation can be used to simplify the computation of mean and s.d. By picking an origin near what appears to be the middle value and using the spacing in the data to define a new unit, one can obtain transformed data consisting of small integers. This leads to the notion of *coding*, illustrated in the next example.

Example 3-20

The idea of coding is as easily explained with a few numbers as with a great mass of data. So consider just the numbers 75, 80, and 90. Clearly, the deviations about 80 are smaller than the numbers themselves. Also, only multiples of 5 appear, so deviations can be measured as so many 5's—using 5 as the new unit. Thus, if the old numbers are called X, new numbers Y given by

$$Y = \frac{X - 80}{5}$$

will be easier to work with. They are $-1, 0$, and 2. Their mean is

$$\bar{Y} = \frac{-1 + 0 + 2}{3} = \frac{1}{3}$$

and their s.d.,

$$S_Y = \sqrt{\frac{1^2 + 0^2 + 2^2}{3} - \left(\frac{1}{3}\right)^2} = 1.247$$

From these, the mean and s.d. for the original X's are easily obtained.

$$\bar{X} = 5\bar{Y} + 80 = 81.67$$

$$S_X = 5S_Y = 6.235. \quad \blacktriangle$$

Coding was more important before the advent of the conveniently available hand calculator. Even so, it is a handy device for spur-of-the-moment computations, and when your calculator charge runs down. Indeed, when using a calculator, coding may improve the efficiency and accuracy of the input.

A coding of values is useful when working with frequency distributions in which the class intervals are of equal size. The class marks are then equally spaced, and a convenient scale for them puts the 0 reference point at a class mark near the center of the data and uses the interval between class marks (i.e., the class-interval width) as a unit. The idea is best explained by example.

Example 3-21   *Cholesterol Levels*

The cholesterol levels of 100 males were measured and summarized in a frequency table, given here as Table 3-2. Also given in the table is a column of corresponding code values Y, which is the number of

Table 3-2   Cholesterol Levels

Class Interval	Class Mark, x	Frequency, f	$y = \dfrac{x - 237}{25}$	fy	fy^2
124.5-149.5	137	1	-4	-4	16
149.5-174.5	162	4	-3	-12	36
174.5-199.5	187	14	-2	-28	56
199.5-224.5	212	18	-1	-18	18
224.5-249.5	237	17	0	0	0
249.5-274.5	262	20	1	20	20
274.5-299.5	287	15	2	30	60
299.5-324.5	312	8	3	24	72
324.5-349.5	337	1	4	4	16
349.5-374.5	362	1	5	5	25
374.5-399.5	387	1	6	6	36
Sums	—	100	—	27	355

class intervals away from the one centered at 237, chosen arbitrarily as a class mark near the center of the data:

$$Y = \frac{X - 237}{25}.$$

Equivalently,

$$X = 237 + 25Y.$$

The mean and variance of the code values Y are obtained from the column totals as follows:

$$\bar{Y} = \frac{1}{100} \sum fy = .270$$

$$S_Y^2 = \frac{1}{100} \sum fy^2 - (\bar{Y})^2 = 3.55 - (.27)^2 = (1.865)^2.$$

The mean and s.d. of the original X-values are then easily recovered from those of the coded values:

$$\bar{X} = 237 + 25\bar{Y} = 237 + 6.75 = 243.75$$

$$S_X = 25 S_Y = 25(1.865) = 46.6.$$

Although these are correct for the frequency distribution in Table

3-2, the mean and s.d. of the original data (the "raw data"), not given here, happen to be $\bar{X} = 245.69$ and s.d. $= 46.0.$[†] ▲

3.8
LINEAR CORRELATION

We again take up the case, begun in Section 2.9, of data consisting of *pairs* of numerical observations (X, Y). Consider the two sets of data in Figures 3-4 and 3-5. In Figure 3-4, which exhibits snowfall data for Minneapolis-St. Paul (October and November versus December to May), some large X-values go with large Y-values, and some with small Y-values. [Here, "large" means to the right, for X, and toward the top, for Y; "small" means to the left, for X, and toward the bottom, for Y.] Thus, there seems to be little or no association between October and November snowfalls (X) and December to May snowfalls (Y).

This is in contrast with the situation depicted in Figure 3-5, in which the data points represent midquarter and final scores for students in a statistics class. Here, points with large X-coordinates tend also to have large Y-coordinates, and points with small X-coordinates tend to have small Y-coordinates. There does seem to be a relationship between midquarter and final scores.

It is useful to have an explicit measure of association, one that distinguishes between data like that of Figure 3-5, where there appears to be a strong relationship, and data like that of Figure 3-4, where there is little association. The most commonly used such measure will be developed next.

Since the location of the reference axes should be of no consequence in studying an association, we relocate the axes to intersect at the center of the data, that is, at the point $(\bar{X}, \bar{Y})$. For the data in Figures 3-4 and 3-5, this is done in Figures 3-6 and 3-7, respectively. The relocation amounts to measuring displacement horizontally from

---

[†]The inaccuracy in the s.d. introduced in the process of rounding off to class marks can be in either direction—yielding an s.d. that is either a little too large or a little too small, depending on the nature of the population being sampled. Some people like to subtract $h^2/12$ ("Sheppard's correction"), where h is the class-interval width, from the variance as calculated from a frequency distribution for grouped data. This correction works rather well for many data sets encountered in practice, as it does, indeed, in the present example.

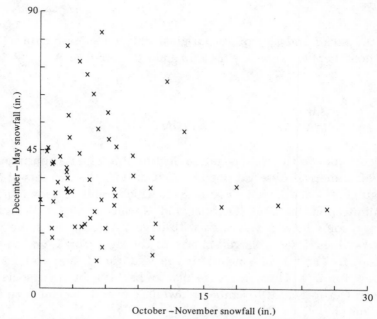

Figure 3-4. December to May snowfall versus October and November snowfall (since 1920, Twin Cities).

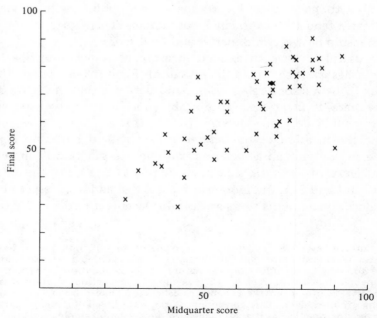

Figure 3-5. Composite grade versus score on second quiz (54 students in elementary statistics).

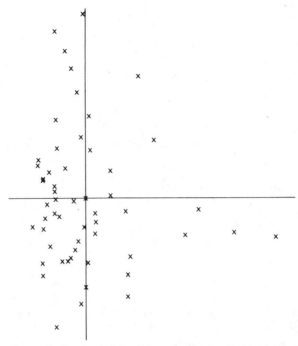

Figure 3-6. Snowfall data of Figure 3-4, plotted relative to axes through the mean point.

$\bar{X}$ and vertically from $\bar{Y}$, so that we deal with deviations $X - \bar{X}$ and $Y - \bar{Y}$ from the center point. Notice that in Figure 3-7, most points are in the first and third quadrants (numbered as shown), whereas in Figure 3-6, roughly the same number of points are in quadrants I and III as in II and IV.

The algebraic signs of the deviations in the various quadrants, and of their product in each case, are as follows:

Quadrant	$X - \bar{X}$	$Y - \bar{Y}$	$(X - \bar{X})(Y - \bar{Y})$
I	+	+	+
II	−	+	−
III	−	−	+
IV	+	−	−

In Figure 3-7, the products $(X - \bar{X})$ $(Y - \bar{Y})$ will be mostly positive, whereas in Figure 3-6, neither the positive nor the negative products greatly predominate.

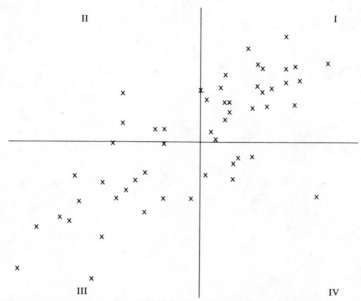

Figure 3-7. Composite versus midquarter scores, data of Figure
3-5 plotted relative to axes through the mean point.

The *covariance* of the X's and Y's in a set of bivariate data is defined to be the average product of the deviations about the means:

Covariance of $(X_1, Y_1), (X_2, Y_2), \ldots, (X_n, Y_n)$:

$$S_{X,Y} = \frac{1}{n} \sum (X - \bar{X})(Y - \bar{Y}).$$

In Figure 3-6 it is evident that the positive products will nearly cancel the negative ones, so the average product will be close to 0.
In Figure 3-7, however, there are only a few negative products, and
the average—the covariance—will be a large positive number.

Putting Y equal to X in the formula for covariance shows that the
covariance of a set of X's with itself—that is, the covariance of the
data points $(X_1, X_1), \ldots, (X_n, X_n)$—is the average squared deviation
of the X's about their mean, or S^2, the variance of the X's. Also,
just as there was an alternative formula for computing a variance, so
there is another way to compute a covariance:

Alternative formula for covariance:

$$S_{X,Y} = \frac{1}{n} \sum XY - (\bar{X})\,(\bar{Y})$$

> = average product minus the
> product of the averages.

(Like the formula for variance, this formula is easily derived by multiplying out the products of deviations before adding them up.)

As pointed out earlier, a deviation such as $X - \bar{X}$ is unchanged if we add any constant to each X, because that same constant is then added to the mean:

$$(X + k) - (\bar{X} + k) = X - \bar{X}.$$

(For example, the number of blocks from your house to school does not depend on where they start numbering the streets.) For this reason, the covariance of a set of data pairs (X, Y) will be the same as the covariance of the translated data set $(X + k, Y + m)$, obtained by adding k to each X and m to each Y:

$$S_{X,Y} = S_{X+k,\ Y+m}.$$

Although the covariance does not depend on the location of the reference axes, it is still not a good measure of association as it stands, because it *does* depend on the unit of measurement. Thus, if X denotes a length in inches, a covariance of X's and Y's is 12 times what it would be if the same length were measured in feet. To get around this difficulty, we measure the deviations $X - \bar{X}$ and $Y - \bar{Y}$ in "standard" units, as multiples of the units S_X and S_Y:

$$\text{Standardized } X\text{:} \quad \frac{X - \bar{X}}{S_X}$$

$$\text{Standardized } Y\text{:} \quad \frac{Y - \bar{Y}}{S_Y}\ .$$

Using these standardized values in the covariance computation yields what is called the *coefficient of correlation*, or more fully (for reasons that will emerge), the coefficient of *linear* correlation:

Correlation coefficient of X's and Y's in the data set $(X_1, Y_1), \ldots, (X_n, Y_n)$:

$$r_{X,Y} = \frac{1}{n} \sum \frac{X - \bar{X}}{S_X} \cdot \frac{Y - \bar{Y}}{S_Y}$$

$$= \frac{S_{X,Y}}{S_X S_Y} = \frac{\text{covariance}}{\text{product of s.d.'s}}.$$

Example 3-22

Heights and weights of five football players (on the 1978 Dallas Cowboys roster) are as follows:

Player Number	Height (in.)	Weight (lb)
12	74	202
33	71	190
44	70	215
72	81	270
63	77	252

Because the location of the reference axes is immaterial in calculating a correlation, we subtract 70 from each height and 200 from each weight, and compute the necessary averages:

	$h = H - 70$	$w = W - 200$	h^2	w^2	hw
	4	2	16	4	8
	1	-10	1	100	-10
	0	15	0	225	0
	11	70	121	4900	770
	7	52	49	2704	364
Sums	23	129	187	7933	1132
Means	4.6	25.8	37.4	1586.6	226.4

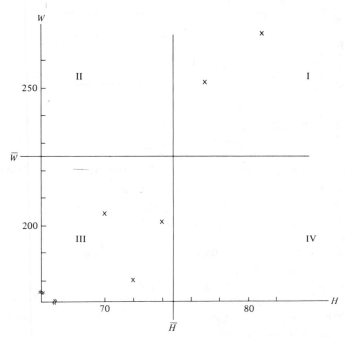

Figure 3-8. Plot of (height, weight) for five football players.

The correlation coefficient is then

$$r = \frac{S_{h,w}}{S_h S_w} = \frac{(\overline{hw}) - \bar{h} \cdot \bar{w}}{S_h S_w} = \frac{226.4 - 4.6 \times 25.8}{\sqrt{(37.4 - 4.6^2)(1586.6 - 25.8^2)}} = .88.$$

The five data points are shown in Figure 3-8, together with axes drawn through the mean point, (74.6, 225.8). Observe that the data points lie in the first and third quadrants of this mean-point coordinate system. Notice also that although the first three points (players numbered 12, 33, and 44) do not show much of an association between height and weight, the correlation for the five points is high because the other two players are both very tall and very heavy. ▲

The correlation coefficient is independent both of the location of the reference origin and of the units used. Thus, making any linear transformation on either X or Y will not alter the value of the correlation coefficient:

If $U = aX + b$, and $V = cY + d$, then (if $a > 0$, $c > 0$)

$$r_{U,V} = r_{X,Y}.$$

Example 3-23

Suppose that we have these three data points: $(10, 15)$, $(20, 20)$, and $(60, 40)$. The means are $\bar{X} = 30$ and $\bar{Y} = 25$. If, for each data point, we measure X as so many 10's away from 30:

$$U = \frac{X - 30}{10},$$

and Y as so many 5's away from 25:

$$V = \frac{Y - 25}{5},$$

the transformed coordinates (U, V) are $(-2, -2)$, $(-1, -1)$, and $(3, 3)$:

X	Y	U	V	U^2	V^2	UV
10	15	-2	-2	4	4	4
20	20	-1	-1	1	1	1
60	40	3	3	9	9	9
90	75	0	0	14	14	14

The covariance and s.d.'s for the (U, V)'s are computed as follows:

$$S_{U,V} = \frac{14}{3} - 0 \cdot 0 \qquad S_U^2 = S_V^2 = \frac{14}{3} - 0 \cdot 0$$

and then

$$r_{U,V} = \frac{14/3}{\sqrt{14/3} \ \sqrt{14/3}} = 1.$$

It therefore follows that, $r_{X,Y} = r_{U,V} = 1$. Clearly, the data points are on a *straight line*, whether we use the coordinates (U, V) or (X, Y), and this is the significance of the value 1 for the correlation. ▲

3.9
INTERPRETING A CORRELATION

If the points representing a set of bivariate data fall on a line, then $r = +1$ if the line slopes upward from left to right, and $r = -1$ if it slopes downward from left to right. (The correlation is undefined for points on a line that is either vertical or horizontal. The formula would give $0/0$ in either case.) Moreover, the value of r can never exceed 1 in magnitude, a fact that can be demonstrated using algebra. So it is a number between -1 and $+1$; and the closer it is to either extreme, the more nearly the data points fall on a straight line. It is in this sense that the correlation coefficient measures the degree of *linear* association.

A value of r near 0 can occur only if there are data points in all four quadrants of the axis system centered at the mean point, in such a way that the positive products nearly cancel the negative ones.

Example 3-24   *Bears and Wind Velocity*

Aerial surveys[†] on each of 20 days in a certain area of the Alaskan peninsula yielded counts of the number (Y) of black bears sighted. Also recorded was the average wind velocity (X). The 20 data pairs (X, Y) are as follows:

(2.1, 99)	(11.8, 76)	(10.5, 79)	(30.6, 47)
(16.7, 60)	(23.6, 43)	(18.6, 57)	(13.5, 73)
(21.1, 30)	(4.0, 89)	(20.3, 54)	(14.0, 72)
(15.9, 63)	(21.5, 49)	(11.9, 69)	(6.9, 84)
(4.9, 82)	(24.4, 36)	(6.9, 87)	(27.2, 23)

Calculating means, s.d.'s and the covariance gives these results:

$$\bar{X} = 14.82 \qquad \bar{Y} = 63.6$$

$$S_X = 7.18 \qquad S_Y = 20.47$$

$$S_{X,Y} = -141.5, \quad r_{X,Y} = -.96.$$

There is a high degree of linear correlation—the data points come close to falling on a straight line (see Figure 3-9). The correlation is

[†]Reported at a 1963 conference on wildlife and resources.

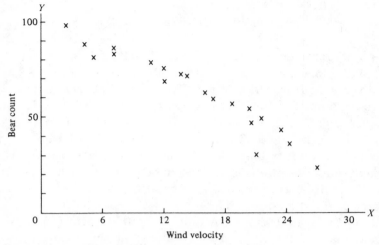

Figure 3-9.   Plot of bear count versus wind velocity.

negative because large bear counts tend to occur together with small wind velocities.

Although it is clear that outings of bears in large numbers do not cause the wind to die, it is tempting to conclude that bears like to stay inside on windy days. But all one can infer from the data is that there is an association.   ▲

Not all relationships are linear, and it can happen that two variables have a correlation that is very small in magnitude when, in fact, they are perfectly related by some nonlinear relationship. The low correlation would simply mean that the points do not come close to falling on a straight line. Figure 3-10 shows some points on a parabola for which $r = 0$.

Associating an amount of linear correlation that is not 1 or –1 with a particular scatter diagram is not easy unless some standard convention for scaling is used. The sets of data points shown in Figures 3-11 and 3-12, for example, come from populations with the same degree of linear correlation; they look different simply because the horizontal scale in one is stretched by a factor of 2. Without knowing the scales, one might guess that the correlation in Figure 3-12 is greater.

One can acquire a feeling for the meaning of r in a scatter diagram if such plots are always plotted with the same unit representing S_X on the horizontal axis as represents S_Y on the vertical axis. Figures 3-13 to 3-17 show scatter plots for data sets with various values of

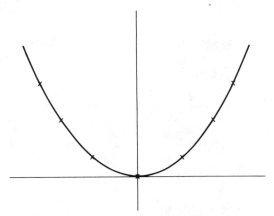

Figure 3-10. Some data points with $r = 0$, but which lie on a curve ($y = x^2$).

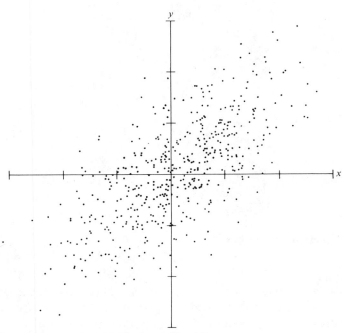

Figure 3-11. Data with correlation .7.

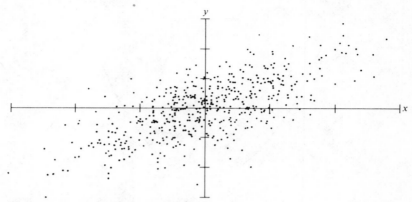

Figure 3-12.   Data with correlation .7.

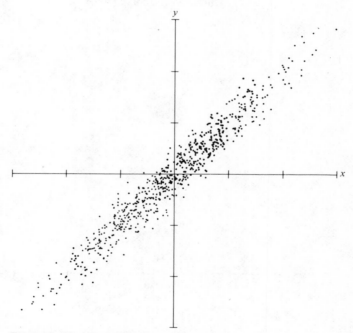

Figure 3-13.   Data with correlation .97.

r, with equal graphical scale units for the s.d.'s of X and Y. A set of data with an r near 1 will cluster rather tightly about the 45° line. (Pictures are drawn for the case of positive correlation. Corresponding pictures for negative correlations would appear the same except that the lines would slope upward to the left, the magnitude of r indicating the degree of clustering about the line with slope -1.)

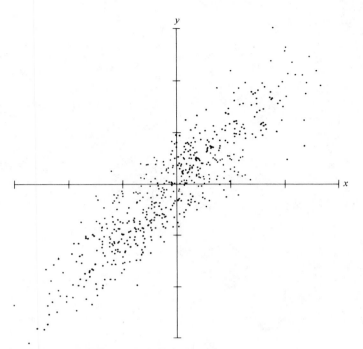

Figure 3-14. Data with correlation .9.

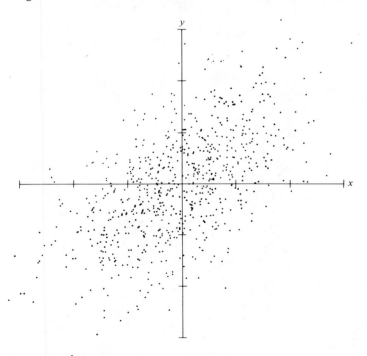

Figure 3-15. Data with correlation .6.

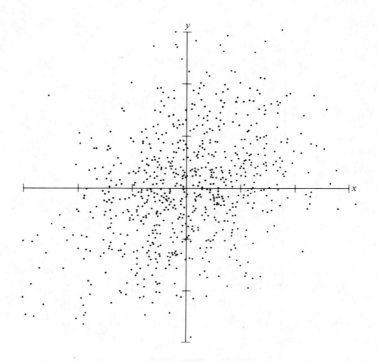

Figure 3-16. Data with correlation .3.

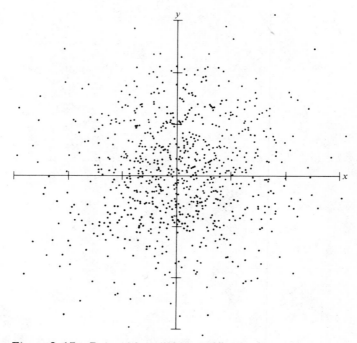

Figure 3-17. Data with correlation .05.

KEY VOCABULARY

Average
Measure of location
Mean
Trimmed mean
Outlier
R.m.s. average
Deviation
Mean deviation
Standard deviation
Variance
Linear transformation
Coding
Correlation
Covariance

QUESTIONS

1. Discuss the effect of extreme values in a data set on the median, and on the mean.
2. In finding the mean from a frequency distribution, how does the number of distinct values (or the number of class intervals) enter into the computation?
3. How do you find the sample size from a frequency distribution?
4. Why is trimming sometimes a good idea in calculating a mean?
5. In an r.m.s. average, why do we take the square root after finding the average of the squares?
6. Absolute deviations from the sample mean have to do with what aspect of a distribution?
7. Is the square of the average the same as the average of the squares?
8. What does a standard deviation of 0 tell you about the set of numbers?
9. What reasons might lead one to using "coding" in calculating the mean and standard deviation?
10. Is coding sometimes not so convenient? Explain.
11. What does a linear transformation of the scale of measurement do to the mean, and what does it do to the standard deviation of the data?

12. Why is the correlation coefficient easier to interpret as a measure of association than the covariance?

13. What is the covariance of a set of numbers with itself—that is, of the pairs (X_1, X_1), (X_2, X_2), . . . , (X_n, X_n)?

14. What does one linear transformation of the X's and another of the Y's do to the correlation coefficient of a set of bivariate data?

15. In what sense might the value –1 be thought of as a "perfect" correlation for a sample? For a population?

PROBLEMS

Sections 3.1–3.3

1. A running back in football carried the ball 12 times in a certain game, for runs of 8, 2, 4, –1, 5, 2, 17, 6, 3, 3, 0, and 7 yards. What is his average for the game?

2. Find the average number of cars per minute, given the following observations of the number of cars per minute, recorded (in 20 one-minute periods) between 12 and 12:30 P.M. at a certain point on a suburban country road:

> 29, 22, 10, 25, 26, 21, 30, 24, 16, 27,
> 23, 31, 17, 15, 21, 21, 30, 18, 33, 19.

3. Referring to temperature data given in Problem 24 of Chapter 2 (page 97):

    (a) Compute the average high and low temperatures in that month (i.e., the average of the daily highs and the average of the daily lows).

    (b) Would the average of the high and low temperatures for a given day be a good figure to give as the average temperature for the day?

    (c) Would you get the same "average" November temperatures by averaging the daily averages of high and low, as by averaging the monthly average high with the monthly average low?

4. Temperatures (in degrees Fahrenheit) were recorded hourly over a 24-hour period, as follows:

> 21 19 24 30 28 23
> 20 19 25 30 25 23

$$20 \quad 21 \quad 28 \quad 31 \quad 24 \quad 21$$
$$19 \quad 22 \quad 29 \quad 30 \quad 24 \quad 17$$

Calculate the average of these readings. (Is this a fair estimate of the "true" average for the day? How might you gather data to improve the estimate?)

*5. A newspaper headline (*Minneapolis Tribune*, July 22, 1978) stated: "Average price of home in area rises to $55,471." The article then gives the dollar volume of sales ($694,407,759), the number of sales recorded (12,615), and the average sale price in each of 30 metropolitan areas for the first 6 months of the year.

(a) The sum of the average sale prices given for the 30 areas is $1,621,080. Does dividing this by 30 yield the average sale price over the metropolitan area? (Why or why not?)

(b) How should the average sale price be computed from the information given—and how do you account for the headline quoted?

(c) Is the mean a useful measure of "typical" selling price, or would the median be better? Why do you suppose the mean was chosen (if it was)?

6. Table IX of Appendix B gives observations from a population with mean 0. Starting at randomly chosen points in the table, obtain two random samples of size $n = 10$ (proceeding down the column from your starting points), and compute the mean of each sample. (Are they equal? Is either equal to the population mean?)

*7. Ten raisin cookies are crumbled to count the number of raisins in each. The results are as follows:

Number of Raisins	Frequency
4	3
5	5
6	1
7	1

Compute the average number of raisins per cookie in this sample.

8. The following scores on a certain hole on a golf course have

been observed over a period of several years, for professional golfers:

Score	Relative Frequency
2 (eagle)	.02
3 (birdie)	.16
4 (par)	.68
5 (bogie)	.13
6 (double bogie)	.01

Determine the mean score.

*9. Data are summarized in the histogram in Figure 3–18.
   (a) How many observations are represented?
   (b) Make a frequency table.
   (c) Calculate the mean.

10. The histogram in Figure 3–19 gives the frequencies of tempera-ture readings (in degrees Fahrenheit) at noon in a certain city, for each day in August of a certain year. Calculate the mean.

11. Calculate the mean body temperature of the individuals listed in Data Set A (Appendix C) from the summary of the follow-ing frequency table. (Assuming the entries in this table to be correct, would a calculation of the mean from the raw data in Data Set A have come out differently?

Temperature	Frequency	Temperature	Frequency
96.8	1	99.0	16
97.2	1	99.2	18
97.8	2	99.4	14
98.0	3	99.6	14
98.2	10	99.8	3
98.4	3	100.0	8
98.6	33	100.2	2
98.8	12	100.4	2

*12. Find the midmean for the data in Problem 2.

13. Is there any point to trimming before finding a median—that is, finding the median of what is left in a sample after trimming (say) the top 10 percent and the bottom 10 percent of the ordered data? Explain.

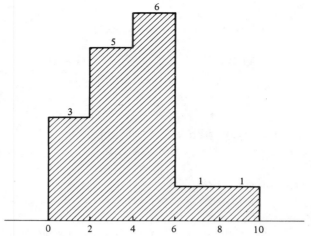

Figure 3-18. Histogram for Problem 9.

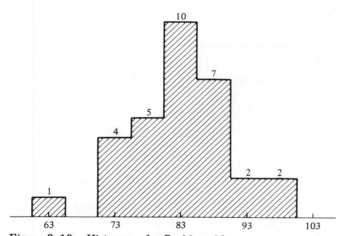

Figure 3-19. Histogram for Problem 10.

Sections 3.5–3.7

*14. Calculate the standard deviation of the data in Problem 1.

15. Calculate and compare the s.d.'s of the monthly temperature averages given in Example 3-12 for Minneapolis and San Francisco.

*16. Twenty observations were tallied and summarized in the following frequency table:

x	f
5	2
10	5
15	8
20	3
25	2

Find the mean and standard deviation.

*17. A sample of 25 observations $X_1, X_2, \ldots, X_{25}$ turns out to have sum $\sum X = 35$ and sum of squares $\sum X^2 = 200$. Calculate the standard deviation.

18. A set of ten numbers has mean 6 and s.d. 3. What is the sum of their squares?

*19. A chain has a store in each of the following cities along El Camino Real on the San Francisco Peninsula, listed along with distances (fudged to make them convenient) from San Francisco:

San Francisco	0
Daly City	5
San Mateo	15
Redwood City	30
Palo Alto	35
Sunnyvale	45
San Jose	55

(a) Determine the mean and median distances of these stores from San Francisco.

(b) In locating a warehouse, consideration is given to the distances from it to the several stores. (These distances are the absolute deviations from the warehouse location.) Calculate the mean absolute deviation about the mean, and also about the median point, to see which is preferred as a warehouse location. (Note that the average distance to the stores is smaller if the warehouse is in Redwood City, the median location. It happens that the median is always best for this purpose; and indeed, Redwood City would be best, in terms of average distance, even if the San Jose store were closed and moved to Carmel, 80 miles farther south.)

*20. The weight limit for luggage on an international flight is 20 kilograms (which is 44 pounds).

(a) If the total weight of the luggage of 100 passengers is 3520 pounds, what is the average luggage weight per passenger in kilograms?

(b) If the s.d. of the luggage weights of the 100 passengers is 3 kilograms, what is the s.d. in pounds?

21. A sample of 10 readings of temperature in degrees Fahrenheit has mean 68 and s.d. 9. Determine the mean and s.d. for the same data if they had been recorded in degrees Celsius.

*22. Calculate the s.d. of the data shown in Figure 3–18, using the frequency table obtained in Problem 9.

23. Calculate the s.d. of the nurses' temperatures summarized in the frequency distribution of Problem 11.

*24. Grade-point averages of students in each section of the class listed in Data Set B (Appendix C) are summarized in frequency tables as follows:

	Frequency	
GPA	Main Campus	Ag Campus
2.0	1	4
2.1	3	1
2.2	4	3
2.3	2	3
2.4	2	3
2.5	12	13
2.6	6	4
2.7	12	8
2.8	8	16
2.9	5	8
3.0	19	11
3.1	3	3
3.2	4	5
3.3	4	1
3.4	1	2
3.5	4	3
3.6	4	3
3.7	3	5
3.8	2	2
3.9	1	0

(a) Calculate the average GPA for each of the two groups.

(b) Calculate the standard deviations.

(c) In the case of the Main Campus section, regroup the data using class intervals 2.0 to 2.2, 2.3 to 2.5, and so on, and compare the mean and s.d. calculated from this distribution with those found in parts (a) and (b).

25. Find the mean and s.d. of the hole diameters given in Problem 34 of Chapter 2. (Use the frequency distribution obtained there unless you have a calculator with $\bar{X}$ and S keys.)

Section 3.8

*26. Given the data points (x, y) plotted in Figure 3–20:
  (a) Find which data point has the median x-value.
  (b) Which has the larger s.d., the set of x-values or the set of y-values?
  (c) Find the number of points with $x > y$.
  (d) Will the correlation coefficient turn out to be positive, negative, or zero?
  (e) Suppose that the whole picture is moved to the right, relative to the axes, by adding 2 to each x-value. How will the new value of r relate to the old?

27. Given these three data points for the pair of variables (X, Y): $(0, 3), (6,0), (3, 3)$:
  (a) Make a scatter diagram and predict whether r will turn out to be positive or negative.
  (b) Compute the means, s.d.'s, and the correlation r.

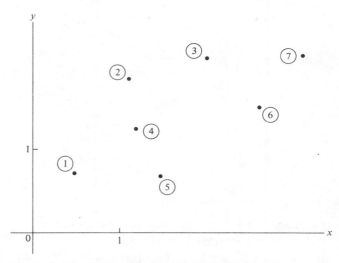

Figure 3-20.   Scatter diagram for Problem 26.

*28. The following are scores of 15 students in two midquarter examinations in an advanced statistics course:

Exam 1	Exam 2	Exam 1	Exam 2	Exam 1	Exam 2
97	100	64	85	50	46
50	67	57	47	65	67
50	46	45	50	78	78
81	89	40	49	60	58
94	85	47	53	60	41

Calculate the coefficient of correlation between scores on Examination 1 and scores on Examination 2.

29. Calculate the correlation coefficient for the data given in Problem 29 of Chapter 2, dealing with injuries on artificial turf.

*30. The GPA and weight of each of the first 10 students listed in Data Set B (Appendix C) are as follows:

GPA	3.0	2.7	2.4	2.5	2.7	2.1	2.7	2.7	2.7	3.6
Weight:	155	182	156	160	146	180	150	145	175	135

Calculate r. (HINT: Subtracting 2.7 from each GPA and 160 from each weight will make the calculations a little easier.)

REVIEW

31. The following are the weights (in pounds) of seven raccoons captured in a certain cornfield: 22.7, 24.8, 23.2, 21.4, 22.0, 18.9, and 23.1. Find the median and mean weights.

32. If you have a hand calculator with a key for S or S^2 (s.d. or variance) or can borrow one, try it out on these data sets:
    (a) 0, 2, 4.
    (b) 999990, 999992, 999994.
    The value of S should be the same in both cases. Does the calculator give the same value? If it does, try this set, which should also have the same S: 66666660, 66666662, 66666664.

33. In Problem 9 of Chapter 2 a frequency distribution of ages of 50 students was given, as follows:

Age	Frequency
17	4
18	16
19	14
20	10
21	4
22	1
23	1

Ages were given as of last birthdays, so that a "19-year-old," for example, is at least 19 but less than 20. Determine the mean and median age of the students.

34. Ten light bulbs were tested to see how long they would last. The times to burnout (in hours) were as follows:

    390, 705, 640, 882, 940, 480, 564, 690, 728, 1149.

    (a) Find the mean.
    (b) Find the median.
    (c) Find the mean after trimming the two largest and two smallest.
    (d) Find the range.
    (e) Find the mean absolute deviation about the mean.
    (f) Find the mean absolute deviation about the median.
    (g) Find the standard deviation.

35. The following is a record of counts of the number of α-particles emitted from a sample of carbon 14 in 25 five-second intervals:

    5, 10, 7, 9, 9, 8 10, 7, 3, 6, 7, 15, 8,

    9, 12, 10, 13, 5, 4, 6, 6, 12, 11, 11, 10.

    (a) Determine the sample range.
    (b) Make a frequency table, grouping values as follows: 3 to 5, 6 to 8, and so on.
    (c) Determine the median value.
    (d) Determine the mean and s.d. from part (b).
    (e) Draw a histogram corresponding to the table obtained in part (b).

36. Compute the average and standard deviation of systolic pressures for the individuals listed in Data Set A (Appendix C), summarized in the following frequency table:

Pressure	Frequency
90–96	3
98–104	21
106–112	40
114–120	32
122–128	17
130–136	21
138–144	5
146–152	1
154–160	1

37. Given below are test scores for three random samples of students—a score on Quiz 1 and a score on Quiz 2 for each student. Determine the correlation coefficient for each sample. (Two of them will become obvious from a quick plot. In calculating the other you may find it easier to choose a more convenient origin of reference.)

Sample 1		Sample 2		Sample 3	
Q_1	Q_2	Q_1	Q_2	Q_1	Q_2
80	71	60	65	40	50
83	71	65	75	40	40
83	73	70	85	50	50
85	73			50	40

38. Plot the pulse rates—apical versus radial—of the first 10 individuals listed in Data Set A, and repeated here:

Radial:	84	88	84	80	100	76	76	104	104	84
Apical:	86	90	98	79	96	71	75	105	102	86

The s.d.'s are nearly equal, so you could guess the value of r by comparison with the scatter diagrams of Figures 3–13 to 3–17. Try this with and without the pair (84, 98), which may be an "outlier." Calculate r.

39. The following is a stem-leaf summary of the butterfat data of Problem 36 of Chapter 2:

Depth	Stem	Leaves
2	3.7	4 4
4	3.8	9 8
6	3.9	6 7
9	4.0	0 3 5
13	4.1	0 5 6 1
8	4.2	3 8 4 9 9 0 7 4
9	4.3	2 3 3
6	4.4	9 8 2 2
	4.5	
2	4.6	7 7

(a) Find the mean of the middle two-thirds of the data.

(b) Find the standard deviation (about the mean).

40. Calculate the coefficient of correlation of the daily high and low temperatures given in Table 2.12 (page 97).

4

Probability Models

AS STATED IN CHAPTER 1, it is essential to use random selection in sampling a population, in order successfully to defend statements about the population that are based on the data in a sample. But then the defense can only be made in terms of *chance*—in terms of *probabilities* of getting misleading or nonrepresentative samples.

Anyone who has ever listened to weather reports has heard "probabilities" given for rain. Everyone who gambles must know about "odds." And everybody on the mailing list of *Reader's Digest* has received "six chances to win $100,000." In this chapter we examine the notions of probability, odds, and chances more closely, to make sure that we agree on their basic meanings, and to extend them to situations that may be new to you. (Chapter 5 will continue the discussion, taking up the case of continuous variables.)

4.1
RANDOM SELECTION MODELS

We have already talked about "equal chances" in describing a random selection of a chip from a bowl of chips, assuming that you have a practical appreciation of the meaning of "equal chances." Each chip in the bowl represents a "chance"—each has a chance of being the one selected, and as good a chance as any other chip in the bowl. This judgment is usually based on considerations of symmetry, for with thorough mixing and blind selection, all chips are treated equally, or symmetrically.

To say that chances are equal is a mathematical statement that defines an ideal situation, a *model* for the experiment. When we say that a chip is *selected at random*, we are assuming that all chips in the bowl have the same chance of being chosen. And as in dealing with a mathematical model for any real-world situation, we never know for certain that this model faithfully represents an actual selection process. But we can test it—by doing the experiment being modeled—to see if it fits well enough to be a useful description. In science, models are continually being replaced by models that explain recent data better.

Unequal chances are often described in terms of *odds*. Suppose, for instance, that you own 3 of the 10 chips in a bowl. Then when one is drawn at random, 3 of the (equal) chances are yours, but 7 are not. The odds on drawing one of your chips are said to be 3 to 7, or 3:7. In everyday language, the odds are said to be *against* you when you have fewer chances of winning than of losing, and *for* you if you have more chances of winning than of losing. The odds are said to be *even* when your chances of winning equal your chances of losing.

Of course, it is the odds *ratio* that counts—odds of 2:4 are the same as odds of 1:2, in that you have the same chance of winning in both cases. This ratio is often reexpressed in terms of a *probability* of your winning, by taking the ratio of the number of your chances or chips to the total number of chances or chips in the bowl. Thus, an odds ratio of 3:7 becomes a probability of

$$\frac{3}{3+7} \quad \text{or} \quad \frac{3}{10},$$

sometimes expressed as 30 percent. "Odds of 50–50," a phrase often used to describe the situation of even odds, means that there is a 50 percent chance for each of the two possible outcomes (you win, or you lose).

Example 4-1 *The Reader's Digest Lottery*

When you receive, in a mass mailing, an envelope marked, dramatically, "Six chances to win!", do you feel happier than if you had received only *one* chance to win? If everyone in the lottery has six chances—six chips in the bowl—then everyone has the *same* chance to win. You can calculate odds on your winning or the probability that you win if you know the *denominator*—the other part of the odds ratio or the probability ratio. This denominator would be the total number of chances that are mailed back for the drawing. (If you encouraged all your friends not to mail theirs back, you would increase your odds slightly.) ▲

If a chip is selected at random from a bowl containing N chips, then

the *probability* of each is $1/N$

the *odds* on a given chip are $1:(N-1)$.

If k of the N chips are identified as yours, and you win if one of them is drawn, then

the *probability* of your winning is $p = k/N$

the *odds* on your winning are k to $N - k$, or, equivalently, p to $1 - p$.

The odds are *against* you if $p < 1/2$, *for* you if $p > 1/2$, and *even* if $p = 1/2$.

Sets of chips may be distinguished in other ways than by saying they are "yours." Suppose that, of 10 chips in a bowl, 3 are red, 5 are black, and 2 are white. The condition that the outcome is (say) a red chip, is an example of an *event*. The event can be thought of as either the condition of redness, or as the set of chips that are red. Clearly, the probability of the event that red "wins" or is selected is 3/10.

If a chip is selected at random from a bowl of N chips, and if the event E defines a set of k of the N chips, then the probability of E is k/N, written

$$P(E) = \frac{k}{N}.$$

Example 4-2

The 52 cards in a standard deck of playing cards can be thought of as chips in a bowl. Dealing a card after thorough shuffling selects one of these "chips" at random. Some of the events one might associate with this random selection are these:

H: The card is a heart.

10:   The card is a ten.

R: The card is red.

RA: The card is a red Ace.

L: The card has a number less than 10.

The corresponding probabilities are found by counting the number of cards in each event—the number satisfying the given condition—and dividing each by 52:

$$P(H) = 13/52$$
$$P(10) = 4/52$$
$$P(R) = 26/52$$
$$P(RA) = 2/52$$
$$P(L) = 32/52. \quad \blacktriangle$$

In working with probabilities, there are several fundamental properties that are useful to know. The first is that a probability—a proportion of the whole—is a number between† 0 and 1:

$$0 \leqslant P(E) \leqslant 1.$$

So if in calculating a probability you come up with 2.4, say, you *know* that something is wrong!

A second property, as important as it is obvious, is that of additivity:

$$P(E \text{ or } F) = P(E) + P(F) \qquad \text{if } E \text{ and } F \text{ cannot occur together.}$$

This is true because when two conditions define distinct, nonoverlapping sets of chips, the number in *one or the other* is the *sum* of the number in one and the number in the other.

---

†Recall that $A \leqslant B$, which we read "A is less than or equal to B," means that $B - A$ is positive or zero. Inequalities in a string are *simultaneous* conditions. Thus, $A < B < C$ means that A is less than B, which, in turn, is less than C.

A special case of the additivity property is often useful, namely, the case in which F consists of all the chips not included in E. This event F, or "not E," is called the *complement* of E; it is complementary to E in the sense that together they account for all the chips:

$$P(E) + P(\text{not } E) = 1$$

or

$$P(\text{not } E) = 1 - P(E).$$

Example 4–3

In the random selection of a card from a deck of cards (as in Example 4–2), the events R (red) and B (black) are complementary:

$$\frac{26}{52} = P(R) = P(\text{not } B) = 1 - P(B) = 1 - \frac{26}{52}.$$

And, as might be of interest to one playing the game of hearts:

$$P(\text{heart or queen of spades}) = P(\text{heart}) + P(\text{queen of spades})$$

$$= \frac{13}{52} + \frac{1}{52} = \frac{14}{52}. \quad \blacktriangle$$

Selecting chips from a bowl is an experiment that involves symmetries. Some other experiments involving symmetries are the toss of a coin, the toss of a die, and the spin of a wheel of fortune or a roulette wheel. In each of these, the symmetries lead people to agree that the various possible outcomes are equally likely, and this equal likelihood of outcomes is taken as defining the *ideal* or *model*, for each experiment. If the experiment is ideal, there is an *equivalent* experiment involving the selection of a chip from a bowl supplied with marked chips. A substitute for the toss of a die, for instance, would be the selection of a chip at random from a bowl containing equal numbers of chips marked 1, 2, 3, 4, 5, and 6. This kind of equivalence was mentioned in Chapter 1.

The probability of an event in *any* experiment with equally likely outcomes is defined as it would be for the substitute, chip-from-bowl experiment:

Probability of an event E in an experiment with equally likely outcomes:

$$P(E) = \frac{\text{number of outcomes in } E}{\text{number of possible outcomes}}.$$

There are some experiments of chance, however, in which there are no symmetries to guide the formulation of a model—an ideal distribution of chances. An example is the toss of a thumbtack, discussed in Section 1.5. Although it is not clear in this case what the odds are, or whether they favor point up (U) or point down (D), we imagine that there *is* an appropriate odds ratio—that there is a probability for U and a probability for D. It happens that we do not know these probabilities and cannot deduce them by examining the tack or the way it is tossed. But whatever these "true" odds may be, there is an essentially equivalent[†] chip-from-bowl experiment. For instance, if $P(U)$ = 3/5, the selection of a chip at random from a bowl with three chips marked U and two marked D could be substituted for a toss of the thumbtack.

4.2
LONG-RUN PROPORTIONS

The experiments used as illustrations in the preceding discussion—the toss of a coin or die, the random selection of a chip from a bowl, and even the toss of a thumbtack—are *repeatable*. One can, over and over again, carry out the instruction, "Toss this coin in the air, giving it a good spin." To be sure, one toss is not *exactly* like another in every detail, because we cannot control the initial conditions of the toss. Differences in these conditions contribute to the "randomness," and all one can do is to give the coin, each time, a "random" spin and a "random" impulse. (Indeed, randomness might be thought of as a combination of the elements of an experiment that are not taken into account.)

[†]A mathematical purist might argue that no bowl with finitely many chips could mimic the thumbtack if the probability of U is an irrational number, such as $2/\pi$. But one could come arbitrarily close with a sufficiently large number of chips.

Similarly, one can repeatedly follow these instructions: "Mix the chips in this bowl thoroughly, and then reach in blindly and draw one out." For this to be a repetition of the same experiment, in repeated drawings, there would have to be the additional instruction, "and replace the chip you draw after recording the result."

Now, it is found that people continue to accept the model of equally likely outcomes for the toss of a coin, and bet accordingly, even after considerable experience with repeated tosses. This is because in a long sequence of repeated tosses, although there may be no predictable pattern to the sequence of heads and tails, they find the *proportion* of heads and tails to be nearly equal. This makes the tossing seem "fair." In a way, it confirms their intuitive judgment, based on symmetry of the coin, that heads and tails have an "equal chance" of occurring at a given toss. The same kind of phenomenon is experienced in drawing chips at random from a bowl, and in other experiments involving symmetries.

Example 4-4

A coin that seemed symmetrical was tossed 25 times to see whether the "fair coin" model could adequately describe the experiment. The following sequence of heads (H) and tails (T) was observed:

H   H   T   H   H   T   T   H   H   H   H   H   H

H   T   H   T   H   H   T   T   T   T   H   T.

The *proportion* of heads, calculated after each trial for the trials up to that point, is

1/1, after 1st toss

2/2, after 2nd toss

2/3, after 3rd toss

3/4, after 4th toss

4/5, after 5th toss

4/6, after 6th toss,

and so on. These proportions are plotted, as a function of the toss number, in Figure 4-1, in which the observed sequence of H's and T's is shown along the axis of toss number. (Lines are drawn between the points of the plot for ease in following the progress of the trials with the eye.)

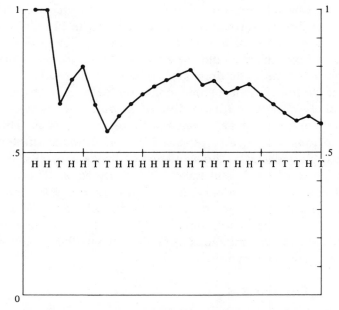

HHTHHTTHHHHHHHHTHTHHTTTTHT

Figure 4-1. Proportion of heads to that point, after each
toss, corresponding to sequence of 25 out-
comes as shown.

The proportion varies unpredictably and erratically; but even in
these 25 trials, the proportion seems to be heading for the ideal proportion, 1/2. However, 25 trials is not exactly a "long" run; perhaps
we should look at longer sequences. ▲

Example 4-5 *Odd Versus Even*

A chip was drawn, repeatedly and with replacement between drawings, from a bowl† containing 10 chips numbered 0, 1, 2, . . . , 9.
The proportion of odd integers encountered up to any given draw
was recorded and plotted (as in Example 4-4) as a function of the
trial number, out to 2000 trials. The plot is shown in Figure 4-2.
Two features of the plot deserve attention:

1. The proportion of odd integers after 2000 trials is about .49,

†To be perfectly honest, what we *really* did was to program a computer to
produce a sequence of integers whose behavior would be indistinguishable from
a sequence of actual random draws from a real bowl of numbered chips.

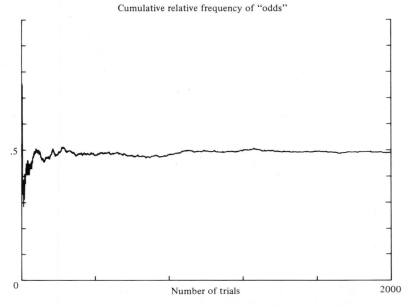

Figure 4-2. Relative frequency of odd integers (Example 4-5).

close to the probability of an odd integer in a single draw in the model with equally likely integers.

2. The initial behavior is unpredictable and erratic, but as sampling continues, the proportion of odd integers begins to *stabilize*— approach a limiting value.

Figure 4-3 shows the record of the proportion of odd integers in 10,000 trials. The proportion of odds after these 10,000 trials seems to be closer to 1/2 than it was after 2000 trials, and the stability is more evident—as much as can be detected in the midst of the fuzziness of the felt-tipped recording pen. Figure 4-4 shows an enlarged section of a record of 10,000 trials (enlarged in both horizontal and vertical directions), indicating the sort of thing that is going on within the fuzziness of Figure 4-3. Figure 4-5 shows an enlarged section of a record of 100,000 trials, on the same scale as in Figure 4-4. The proportion of odds is closer to 1/2 than it was after 10,000 trials; but there are still fluctuations, although their magnitudes are small, owing to the large denominator (near 100,000).

In the long run there is a tendency toward a limiting value, but

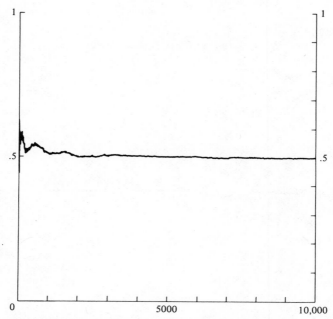

Figure 4-3. Proportion of "successes" recorded during 10,000 trials.

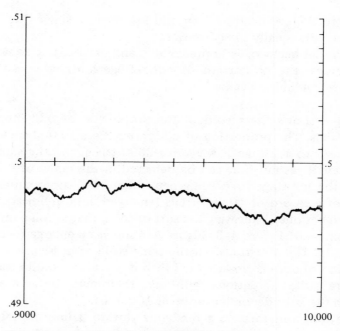

Figure 4-4. Trials 9000 to 10,000 (Example 4-5).

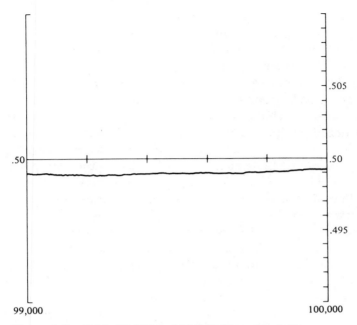

Figure 4-5. Trials 99,000 to 100,000 (Example 4-5).

there is no guarantee that the limiting value is precisely 1/2. That is, there is no guarantee that the experiment is perfectly symmetrical with regard to odd and even integers. ▲

Even when an experiment has no symmetries to suggest that certain sets of outcomes are equally likely, the phenomenon of long-run stability is observed. Thus, when a thumbtack is tossed repeatedly, the proportion of tosses in which the tack lands with point up stabilizes, after many tosses, apparently approaching a limiting value.

Law of averages (informal statement):
    In a large number of independent trials of a repeatable experiment, the relative frequency of occurrence of an event E will tend to stabilize at a limiting value.

This "law," also called the *law of large numbers*, is a "law" in the sense of describing an observable phenomenon. If the experiment involved has symmetries that permit formulation of a model with equally likely outcomes, the observed limiting proportion (at least to the extent that it *can* be observed in only a finite number of trials) will be close to the probability of the event E in the model.

[The law of averages is found as a mathematical "theorem" in axiomatic developments of probability theory. There, one begins by assuming certain simple requirements for probabilities and an appropriate definition of independence of trials. Mathematical reasoning then leads inexorably to the conclusion that relative frequencies must converge to the assigned probabilities, in the long run. The fact that in practice, relative frequencies are observed to stabilize about some limiting value means that the mathematics seems to be representing what is going on.]

If the long-run relative frequency of E should turn out to be different from its probability in an assumed model, then either (1) the probabilities in the model are not quite right (perhaps the die is not perfectly symmetric, for example), or (2) the sequence of trials fails in some way to be a sequence of independent trials of the same experiment.

It is the law of averages that furnishes a way of checking out an assumed model. For example, if one claims that a certain probability $P(E)$ is, say, .40, this claim will tend to be believed if a long sequence of repeated trials of the experiment results in E close to 40 percent of the time. If the limiting value appears to be closer to 30 percent, however, this would be reason to revise one's judgment as to the likelihood of E—and perhaps to look at the experiment more critically to see how his earlier analysis led him astray.

The law of averages is often misrepresented and misapplied. People often think that if they have been having a run of bad luck, then Nature will tend to make up for that bad luck in the next few trials. In particular, they seem to believe that a string of heads in tosses of a coin will increase the probability of tails on the next toss! However, if the experiment is really the same each time, and if the trials are independent, then the odds remain the same, no matter what the recent past history happens to be.

Example 4-6 *Lions Versus Vikings*

After a long string of victories over the Detroit Lions, the Minnesota Vikings coach was questioned by a newspaper reporter: "Ten times in a row—what are the odds on eleven?" The reporter's account continued:

"If you flipped a coin and it came up heads ten straight times," said Viking coach Bud Grant, "what would be the chance of it coming up heads the 11th time?" He answered his own question. "Fifty-fifty."

The coach was right about the coin—it has no memory. And if the coin is fair, the probability of heads at *any* given toss is 1/2. This is as true on the 11th toss as on the first, no matter how many heads in a row are observed before the 11th toss. However, a series of Vikings–Lions games is not like a series of tosses of a coin, and is not adequately modeled as a sequence of independent trials of an experiment. The odds are not necessarily even, and they do not remain the same. Players have memories and learn from experience, and teams change their personnel over the years.

There is another way to look at the coach's question about 10 heads in a row in 10 tosses of a coin. The coin may *not* be fair! Indeed, when any experiment turns out one particular way 10 times in a row, this can be interpreted (as will be seen in Chapter 8) as strong evidence that the probability of that outcome is actually greater than 1/2. And in that case the odds for the 11th trial would not be 50-50, even if the trials are independent. ▲

Example 4-7 *Odd Versus Even (continued)*

We refer back to Figure 4-5, showing the relative frequency of odd integers in a sequence of trials, from trial 99,000 to trial 100,000. After 99,000 trials, "even" is ahead of "odd" by 200, for there were 49,400 odds and 49,600 evens. In the next 1000 trials, there is scarcely any change in this lead. (If there were *no* change, that is, if even were still ahead of odd by 200 at 100,000 trials, the relative frequency of odd would be closer to 1/2 than after 99,000 trials because of the larger denominator.) As seen in Figure 4-4, the lead can widen over any group of trials, because at any point the process of producing odds and evens starts over again—with no memory of past trials. ▲

4.3
SUBJECTIVE PROBABILITY

We have been looking at some experiments of chance that can be repeated. In some of these, symmetries suggested a model; but in all of them, the possibility of repetition furnished a means of checking the usefulness or validity of the model. However, some experiments with uncertain outcomes cannot or will not ever be repeated. Consider these questions:

- Is there life on Mars?
- Will Jerry Brown ever attain the presidency?
- Will the Vikings beat the Lions twice next season?
- Will this particular patient recover from open-heart surgery?
- Will it rain tomorrow?
- Will my house burn down during the coming year?

People often give probabilities in situations like these. They may even bet on the outcome. They do so according to their *degree of belief* in a given outcome. Their probabilities, then, are a matter of individual judgment—they are personal or *subjective*. They are certainly not characteristics of the experiment alone but depend as well on the person making the assessment.

Example 4-8   *Recession, or Not?*

In the summer of 1978, in testimony before the Senate Finance Committee, Raymond Saulnier (economic adviser in the Eisenhower administration) stated that there was "no more than a 50–50 chance we can avoid a recession." Walter Heller (economic adviser to Presidents Kennedy and Johnson) said to the same committee that the chances of recession were "only 1 in 3."

Although these evaluations of the chances differ, they are both legitimåte as personal judgments—as subjective probabilities. It is possible, of course, that they may have different definitions of "recession" and different notions of time span in which the recession might occur. But even if they were to agree on these points, their subjective probabilities could differ. And, of course, they can change with time. In October 1978, Heller and an economist with the Brookings Institute agreed "that developments in the last several months have made a recession in 1979 an even bet." (Curiously, the newspaper account continued, "but they disagree over whether it will occur." They seem to have discovered even odds that are not even!) ▲

When a person gives even odds that boxer *A* will win a certain bout, he is saying that *in his judgment* the bout is like the toss of a fair coin—heads *A* wins, and tails he loses. The person's subjective probability of *A*'s winning is 1/2. On the other hand, another person's probability that *A* wins may be 2/3. Differences in personal probabilities often account for an eagerness to bet on the part of both parties.

It is when people's subjective probabilities tend to agree that we find almost universally accepted models, such as those with equally likely outcomes encountered in tossing coins or dice, or selecting chips at random. In such cases the agreed-on probabilities are often thought of as *objective*—as characteristics of the experiment that are not subject to personal opinions. Even so, a person may have a subjective probability that a tossed coin falls heads, for instance, which does *not* agree with the value 1/2 that describes an ideal toss. This differing belief may be quite rational, based on something he perceives in the particular mode of tossing that suggests a bias to him.

Example 4-9   *A Biased Coin*

Experience suggests that if you stand a quarter on end with its face visible and in an upright position and then spin it with a flick of the finger, the side that turns up when it slows down and falls over is more apt to be heads than tails. It is apparently not like the toss of the same coin in the air. (We tried it 20 times with one coin and got 19 heads. Another sequence of spins with a different coin resulted in 60 heads in 100 tries.) Some people refuse to believe this, even in the face of data that support it. Their subjective probability remains at 1/2 for heads, and "biased" coins, for them, do not exist. (If you are such a person, certain examples in this book will not make much sense to you.) ▲

Example 4-10   *Medical Prognoses*

Suppose your physician informs you that only 7 out of 10 people with your particular condition have survived, or will recover from, a type of surgery with which you are faced. This may be a fairly accurate picture of the whole population of individuals in that predicament, and it may be relevant to some purposes, medical or actuarial. But the question is: Does this population picture apply to *you* in particular? You only "go around once," and there are circumstances special to you, known to you and/or your physician, that may make your degree of belief in survival different from the 7 to 3 odds for the population. Indeed, your probability that you recover may differ from that of your physician! ▲

Subjective probabilities, being personal, are not easy to ferret out. They are reflected in how one might bet, but betting involves money and adds the complication of how valuable different amounts of money are to a person. One approach for assessing a person's subjective probability for an event is this: First, find a repeatable experiment such as die-tossing or the selection of chips, with probabilities that are generally agreed on. Then use these known odds as a basis for determining the personal odds on the event in question by comparison. For example, if a forthcoming Vikings-Lions game is the experiment in question, we might ask: "Would you rather receive $10 if a coin falls heads or receive $10 if the Vikings beat the Lions next Sunday?" Indifference would place the person's probability of the Vikings' winning at 1/2. If it is not a matter of indifference, the probability is not 1/2; but it can be pinned down in similar fashion by comparing the game with chip-from-bowl experiments having various proportions of chips labeled "win" and "lose." Finding a proportion of chips such that the person is indifferent to whether he gets the $10 if the Vikings win or if a chip marked "win" is drawn determines the subjective probability of a win.

However they may be arrived at, subjective probabilities should be like relative frequencies—nonnegative fractions that, when all possible outcomes are taken into account, add up to 1, or 100 percent, and that have the addition property for mutually exclusive events. But there is no guarantee that a given individual's subjective probabilities will have these properties.

4.4
BETTING ODDS

In a betting situation, the language used is usually that of odds instead of probabilities. The "odds" in a bet may refer to objective odds—the ideal odds that people agree on, as in dice games, roulette, and cards. Or it may refer to odds determined by amounts of money bet, reflecting a general subjective evaluation, as in a horse race. Or it may be the ratio of stakes people agree on when they enter a two-person bet, which may or may not agree with anyone's subjective odds.

Example 4–11

When someone is willing to give 5 to 1 odds on a certain event, it means he would be willing to pay you $5 if the event does not happen provided that you agree to pay him $1 if the event happens. The total amount being bet is $6. He is suggesting that, for him, the probability of the event is at least 5/6. If you are a betting person, you would accept his offer if your probability of the event in question is at most 5/6.

If you assess the odds to be 5 to 1, then the bet in which your opponent puts up $5 to your $1 is a *fair* bet according to your assessment. If you think the odds are more like 4 to 1, it would be reasonable to take the 5 to 1 bet; and if your opponent thinks the odds are 6 to 1, it would be reasonable for him to offer the bet. You would both consider the bet to be favorable. ▲

Example 4–12 *Betting on the World Series*

It was generally felt, among baseball aficionados, that the Los Angeles Dodgers should be favored over the New York Yankees in the 1978 World Series. The generally accepted "fair" odds were 6 to 5, corresponding to a probability of Los Angeles' winning of 6/(6 + 5) = 6/11. If you had called your neighborhood bookie to make a bet on New York, you may have been offered only even odds; and if you wanted to bet on Los Angeles, you may have had to give 7 to 5 odds. This would allow the bookie to make money no matter which team wins. For example, suppose that he accepts a total of $6000 bet on New York and $7000 on Los Angeles. If Los Angeles wins, he collects $6000 and pays out 5/7 of $7000, or $5000, netting $1000. If New York wins, he collects $7000 and pays out $6000, again netting $1000. His motive for setting the odds to be as fair as possible is that he wants to balance the number of people willing to bet (or, better, to balance the amount of money bet) on both sides. Although it is probably true that the odds he sets are what he perceives as being fair, it would be more accurate to say that they are what he thinks the "average bettor" perceives as being fair! ▲

Example 4–13 *Horse Race Odds*

In pari-mutuel betting at race tracks, odds are determined by the amounts of money bet on the individual horses. A simplified version, something like the European system, would work like this. Suppose

Table 4-1

Horse	Amount Bet	"Probability"	Odds	Posted Odds	Equivalent Probability	$2 Ticket Pays
A	$ 880	.181	819–181	4–1	1/5	$ 9.95
B	2150	.442	558–442	Even	1/2	4.07
C	572	.117	883–117	7–1	1/8	15.31
D	1264	.260	740–260	9–4	4/13	6.93
	$4866	1.000			1.133	
	– 486.6	(for the track)				
	$4379.6	(available to split)				

that there are four horses, and total amounts bet are as shown in Table 4-1. Then the odds against A's winning are 3986 to 880, and the corresponding subjective probability—reflecting the bettors' collective judgment—is 880/4866 = .181. Suppose that the track takes 10 percent. The posted odds for A would be something close to those corresponding to a probability of

$$\frac{880}{4866 - 486.6} = .201,$$

or 4–1, say. (These equivalent probabilities would add up to more than 1 because of subtracting the track's take from the denominator.) A $2 ticket, 1/440 of the amount bet on A, would pay 1/440 of the amount available to be paid out:

$$\frac{\$4379.6}{440} = \$9.95.$$

A more detailed account of actual pari-mutuel systems used in North America and in Europe may be found in the *Encyclopaedia Britannica.* ▲

Example 4-14 *Air Travel Insurance*

Suppose that you pay 50 cents for a $10,000 policy against loss of life on a trip by air, say from Minneapolis to New York—on the order of 1000 miles. In effect, you are betting with odds of .50 to 9,999.50 that you will *not* survive. The insuring company is betting, with odds 19,999 to 1, that you *will* survive.

Suppose (as was the case in one year) that there is .3 fatality per 100,000,000 passenger-miles, or per 100,000 passenger trips of length 1000 miles. This suggests odds of .3 to 100,000 that you will not survive. You are clearly paying too much, 50 cents instead of 3 cents. (It is conceded, though, that some of the premium is for paperwork.) People generally do not mind the unfair bet because they scarcely miss 50 cents. A person does not make enough trips to feel the monetary loss. The insurance company, which collects thousands (millions?) in 50-cent premiums, would appear to have a good thing going. ▲

People often take unfair bets. They do this for various reasons, including these:

- Their subjective assessment of odds may make the bet seem fair, or even favorable.
- There is, for some, a certain pleasure (possibly addictive) in betting.
- The money to be won may be proportionately much more valuable to a person than the money invested.

4.5
RANDOM VARIABLES

Any variable associated with an individual drawn at random from a population is called a *random variable*. As discussed in Chapter 1, some variables are categorical and others are numerical. In either case, any particular category or particular value of the variable is an *event*—an event whose probability is the proportion of individuals in that category or having that value.

Example 4-15

A city has four TV channels—one affiliated with NBC, one with ABC, one with CBS, and one with PBS. Each member of the population of potential TV viewers in that city can be classified, at a given moment, as watching one of these channels or as not watching at all. The corresponding population proportions of each category are probabilities—p_A, the probability that an individual selected at random is watching ABC; p_N, the probability the individual is watching NBC; and so on:

Channel Watched	Probability
ABC	p_A
NBC	p_N
CBS	p_C
PBS	p_P
None	p_0
Sum	1

These probabilities show how the population of viewers is "distributed" among the five categories of the random variable V = "channel watched."

It would be typical that the probabilities or population proportions are unknown, and that a random sample of individuals would be drawn as a means of finding out something about these probabilities and, therefore, about the characteristic of interest. ▲

Example 4–16 *Multiple Births*

The number of babies given birth at one time by, say, a white American mother, can be thought of, or "modeled," as a random variable. The possible values are integers: 1, 2, . . . , and each of these has a probability, or population proportion, which again happens to be unknown:

Number of Babies	Probability
1	p_1
2	p_2
3	p_3
⋮	⋮
Sum	1

The actual population of all such mothers is not available, but there is an equivalent chip-from-bowl model—a bowl with chips numbered 1, 2, 3, . . . , in numbers proportional to the probabilities $p_1, p_2, \ldots$. A random selection from this bowl, or equivalently, the table of values and probabilities given above, provides a model for the random variable Y = number of babies in a birth.

(This model, although applicable to the population as a whole, does not take into account any relevant special information that one might have. For instance, if the couple's parents have experienced multiple births, or if the mother had been taking fertility pills, this additional information would narrow the population, and so alter the odds from what they might be more generally.) ▲

Model for a discrete random variable X:

$$\text{Values: } x_1, x_2, \ldots$$

$$\text{Probabilities: } p_1, p_2, \ldots,$$

where $p_i = P(X = x_i)$, and

1. $0 \leqslant p_i \leqslant 1$,   for each i.
2. $\sum p = 1$,   summed over all values (or categories).

A model of this type is referred to as a *discrete probability distribution*, because it shows how chances (totaling 1, or 100 percent) are distributed among the various possible values of X. (The qualifier "discrete" refers to the fact that the list of possible values is just that, a "list" or set of discrete values—as opposed, for example, to all possible numbers between 0 and 1, which are too numerous† to be listed, even in an infinite list.)

4.6
NUMERICAL RANDOM VARIABLES— MEAN AND S.D.

As in the case of a set of numerical data, so here in the case of a population of numbers, present in various proportions or probabilities, the notions of mean, median, standard deviation, and range (to name the most commonly used descriptive measures) are applicable. How-

---

†These are said to be *uncountably infinite*. The numbers in an infinite list are said to be *countably infinite*.

ever, in dealing with populations that may be infinite, or not available for counting, the computation of measures involving division by n (the sample size) requires some explanation.

Recall that, for a set of numbers summarized in a frequency distribution—values x_1, x_2, . . . , with corresponding frequencies f_1, f_2, . . . , the *mean* is given by the formula

$$\frac{1}{n} \sum fx = \sum x\left(\frac{f}{n}\right).$$

As suggested by the right-hand side, the division by n can be done *before* summing, just as well as after, with values x being multiplied by *relative* frequencies. So, really, the sample size does not matter as long as we have relative frequencies. But this is exactly what we do have when a population distribution is given in terms of probabilities. Thus,

$$\text{population mean} = \sum xp.$$

The mean of a population of X-values will be denoted by m.v.(X), read "mean value of X," and sometimes by the Greek version of "m," namely, μ (mu). [The mean is also called the *average value* of X, the *mathematical expectation* of X, or the *expected value* of X, the last phrase giving rise to the common notation $E(X)$.] The mean value of X, or the mean of its distribution, is a number that "locates" the distribution. It is usually unknown in a practical situation, and you can compute it from the given formula only if you are willing to assume a particular model. The mean value may or may not be one of the possible values of X. It is a typical or middle value of the distribution—the *sort* of value you might expect, although sometimes X is bigger and sometimes smaller.

The standard deviation of a population of X-values is computed by the s.d. formula of Chapter 3, except that now the multiplying frequencies are replaced by relative frequencies or probabilities, and again there is no division by n. And, of course, the mean about which the deviations are taken is the *population* mean, μ. A population s.d. is a number that describes the spread of the distribution of probability among the possible values, that is, the variability in the population variable X. It is often denoted by the Greek letter σ (lowercase sigma).

Thus, the mean and standard deviation of a population of values have the same interpretation as the sample mean $\bar{X}$ and sample s.d. S

for a list of numbers. The computations are essentially the same.

Mean value of a random variable X (population mean):

$$\mu = \text{m.v.}(X) = \sum xp.$$

Standard deviation of a random variable X (population standard deviation):

$$\sigma = \text{s.d.}(X) = \sqrt{\text{m.v.}[(X - \mu)^2]}$$
$$= \sqrt{\sum (x - \mu)^2 p}.$$

The alternative method of computing an average squared deviation of a list of numbers about the mean of the list, as the average square minus the square of the average, again provides an alternative formula for s.d.:

Alternative formula for standard deviation:

$$\text{s.d.}(X) = \sqrt{\text{m.v.}(X^2) - [\text{m.v.}(X)]^2}$$
$$= \sqrt{\sum x^2 p - \mu^2}.$$

Example 4–17

When two coins are tossed, the result is one of these four outcomes:

    HH: heads on both coins
    HT: heads on coin 1, tails on coin 2
    TH: tails on coin 1, heads on coin 2
    TT: tails on both coins.

These four outcomes are equally likely in the model that seems to represent best what happens when two "fair" coins are tossed independently of one another. Suppose that X denotes the number of heads that show in a toss of the two coins. This variable has the value

2 for HH, the value 1 for HT or TH, and the value 0 for TT. The population distribution for X is as follows:

x	p
0	1/4
1	2/4
2	1/4

The same model would apply to the selection of a chip at random from a bowl with chips numbered 0, 1, and 2 in the ratio $1:2:1$.

The mean and s.d. depend only on the proportions or probabilities, not on the total number of chips in the chip-from-bowl model. Thus,

$$\mu = \text{m.v.}(X) = \sum xp$$
$$= 0 \cdot \frac{1}{4} + 1 \cdot \frac{2}{4} + 2 \cdot \frac{1}{4} = 1,$$

and the average square is

$$\sum x^2 p = 0^2 \cdot \frac{1}{4} + 1^2 \cdot \frac{2}{4} + 2^2 \cdot \frac{1}{4} = \frac{3}{2}.$$

From these we compute the s.d.:

$$\text{s.d.}(X) = \sqrt{\frac{3}{2} - 1^2} = .707.$$

Figure 4–6 shows the probability distribution graphically, with rods of heights in the ratio $1:2:1$. It is clear, by symmetry, that 1

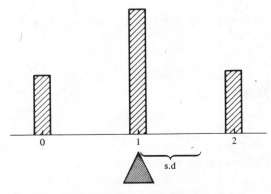

Figure 4–6. Distribution for Example 4–17.

is the balance point for the distribution—the mean value. And the value .707, as an "average" (r.m.s. type) deviation, appears to be reasonable; it is less than the largest deviation, which is 1, and larger than the smallest deviation, which is 0. ▲

Example 4-18

Suppose that half of the families in a certain city have one car, one-fourth have two cars, one-eighth have three cars, and the rest, another eighth, have no cars. If a family is picked at random from the population of families, the number of cars it owns is a random variable, call it X. The probability distribution of X is as follows:

Number of Cars, x	Probability, p	xp	x^2p
0	1/8	0	0
1	1/2	1/2	1/2
2	1/4	2/4	4/4
3	1/8	3/8	9/8
Sums	1	11/8	21/8

The mean number of cars per family in this population is then

$$\text{m.v.}(X) = \sum xp = 0 \cdot \frac{1}{8} + 1 \cdot \frac{1}{2} + 2 \cdot \frac{1}{4} + 3 \cdot \frac{1}{8} = 1.375.$$

The standard deviation about this mean (shown in Figure 4–7) is

$$\text{s.d.}(X) = \sqrt{\frac{21}{8} - \left(\frac{11}{8}\right)^2} = \sqrt{\frac{79}{64}} = .857. \quad ▲$$

4.7
INDEPENDENCE

In Chapter 1 we introduced independence as an intuitive concept. The basic idea is that, for independence, knowing values of some variables in a set of random variables should not affect the odds for the others.

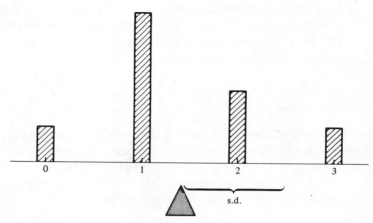

Figure 4-7. Distribution for Example 4-18.

Example 4-19 *One-Eyed Face Cards*

A deck of cards includes 12 face cards, some faces in profile so that
only one eye shows. Consider the population of just these 12 face
cards, and the variables X = suit, Y = number of eyes showing. The
12 cards are distributed, with respect to these two variables, as
shown in the following two-way frequency table:

		Suit, X				
		♠	♥	♦	♣	
Number of Eyes, Y	1	1	1	1	0	3
	2	2	2	2	3	9
		3	3	3	3	12

When a card is drawn at random from the face cards, X and Y are
random variables, with probabilities for each obtained by dividing
the marginal totals by 12. For instance,

$$P(Y = 1) \; = \; \frac{3}{12} \, .$$

However, if it is somehow known that the card drawn is a club, the
relevant population is only the set of clubs; and one should not bet
anything that the card is one-eyed, because there are no one-eyed

clubs. We express this as a *conditional* probability, the probability of $Y = 1$ computed when it is known that X = club, and write

$$P(Y = 1 | X = \text{club}) = 0.$$

(The vertical bar is read "given that.") Similarly, among the spades, one card has one eye, so the conditional probability of $Y = 1$ given that the card drawn is a spade is

$$P(Y = 1 | X = \text{spade}) = \frac{1}{3}.$$

Clearly, the information about X alters the probabilities for Y from what they are in the marginal distribution.

Similarly, since

$$P(X = \text{club} | Y = 1) = 0 \neq \frac{3}{9} = P(X = \text{club} | Y = 2),$$

we see that information about Y can alter the odds for X.

Suppose, next, that we modify the experiment and consider the population consisting of just the six "major" suit face cards (spades and hearts). The distribution of X and Y for this population of six cards is given by the following frequency table:

		Suit, X		
		♥	♠	
Number of Eyes, Y	1	1	1	2
	2	2	2	4
		3	3	6

If, here, it is known that the card drawn is a heart, the odds on its having one eye are $1 : 2$, the same as the odds $(2 : 4)$ when the suit is *not* known. The proportionality of rows (or equivalently, of columns) shows that any information about one variable does *not* affect the odds for the other. In this case, X and Y are said to be *independent*.

One can reason equally well in terms of probabilities or proportions by dividing the population frequencies by the population size. For the case of the six major suit face cards, the probabilities are as follows:

		Suit, X		
		♥	♠	
	1	1/6	1/6	2/6
Number of Eyes, Y	2	2/6	2/6	4/6
		3/6	3/6	1

The row sums in the right margin are probabilities for Y, and the column sums in the lower margin are probabilities for X. The proportionality of rows (or of columns), required for independence, is now equivalent to the following multiplicative property:

$$P(\text{one-eyed heart}) = \frac{1}{6} = \frac{2}{6} \times \frac{3}{6} = P(\text{one-eye}) \cdot P(\text{heart})$$

$$P(\text{two-eyed spade}) = \frac{2}{6} = \frac{4}{6} \times \frac{3}{6} = P(\text{two-eyes}) \cdot P(\text{spade}),$$

and so on, which holds for every cell of the two-way table. ▲

Random variables X and Y are *independent* when and only when

$$P(X = x \text{ and } Y = y) = P(X = x) \cdot P(Y = y)$$

for every value x of X and every value y of Y.

The multiplication formula in this definition of independence is most often used to *calculate* a "joint" probability $P(X = x \text{ and } Y = y)$ from assumed marginal probabilities for X alone and Y alone when it appears that these variables are independent in the experiment at hand. (The assumption of independence yields the simplest kind of joint model, but like any assumption, it would be open to challenge on the basis of sample evidence. This will be looked at in Chapter 10.)

There is a natural extension to more than two variables. For independence, the joint probability of events relating to the individual variables must be the product of the marginal probabilities of those events.

If the event E_1 relates to a random variable X_1, the event E_2 relates to X_2, and so on, then the n random variables $X_1, X_2, \ldots, X_n$ are *independent* if and only if for all such events $E_1, \ldots, E_n$,

$$P(E_1 \text{ and } E_2 \cdots \text{ and } E_n) = P(E_1)P(E_2) \cdots P(E_n).$$

Example 4-20   *Independent Tosses of a Die*

A fair die is tossed three times in independent tosses. That is, they are tossed in such a way that the result of one toss has no bearing on the outcomes of another toss or other tosses. To incorporate this independence into a model for the sequence of tosses, we *define* probabilities by use of the multiplication formula. For instance, the sequence of outcomes 2, 4, 5 is assigned the probability

$$P(2, 4, 5) = P(2 \text{ on toss } 1) \times P(4 \text{ on toss } 2) \times P(5 \text{ on toss } 3)$$

$$= \frac{1}{6} \times \frac{1}{6} \times \frac{1}{6} = \frac{1}{216}.$$

Similarly, each of the 216 possible sequences has probability $1/216$—they are *equally likely*. ▲

The definition of independent variables given above does not involve their numerical nature, and would apply equally well to categorical variables. In particular, a category for variable A is an *event;* so variables A (with categories $A_1, \ldots, A_k$) and B (with categories $B_1, \ldots, B_m$) are independent provided that

$$P(A = A_i \text{ and } B = B_j) = P(A = A_i)P(B = B_j)$$

for *every* pair of categories A_i and B_j. And again, this multiplication rule is usually used for the purpose of constructing the model for a pair of variables that are to be independent. The two-way table of probabilities for the two variables will then automatically have proportional rows and proportional columns.

Example 4-21 *The Number of Heads*

When four coins are tossed (or when one coin is tossed four times), they are usually thought to fall in such a way that if you know how one or more of them land, you do not change how you would bet on the others. That is, they are assumed to fall *independently*. The model is then constructed by assigning to any outcome such as HTTT (heads on coin or toss number 1, tails on numbers 2, 3, and 4) a probability that is the *product* of probabilities for the coins (or tosses) taken individually:

$$P(\text{HTTT}) = P(\text{H on 1}) \times P(\text{T on 2}) \times P(\text{T on 3}) \times P(\text{T on 4})$$

$$= \frac{1}{2} \times \frac{1}{2} \times \frac{1}{2} \times \frac{1}{2} = \frac{1}{16}.$$

Now, suppose that we want to find the odds on getting exactly one head among the four coins. This can happen in any of four ways: HTTT, THTT, TTHT, or TTTH. The probability for each of these sequences is 1/16 (computed as we computed the probability of HTTT). So

$$P(\text{one head in 4 trials}) = P(\text{HTTT}) + P(\text{THTT}) + P(\text{TTHT}) + P(\text{TTTH})$$

$$= \frac{1}{16} + \frac{1}{16} + \frac{1}{16} + \frac{1}{16} = \frac{1}{4}. \quad \blacktriangle$$

Example 4-22 *A Use of Probability in Criminal Court*

An elderly woman was assaulted and robbed in an alley in San Pedro, California. A witness saw a blonde woman with a ponytail run out of the alley and get into a yellow car driven by a bearded black male with a mustache. A couple answering that description was arrested and brought to trial.

The prosecutor called in a mathematician, who explained the theory of probability. In his summary the prosecutor used what he termed "conservative estimates" of the chances of each coincidental characteristic, as follows:

$$P(\text{blonde}) = 1/3$$

$$P(\text{pony tail}) = 1/10$$

$$P(\text{mustache}) = 1/4$$

$$P(\text{beard}) = 1/10$$

$$P(\text{yellow car}) = 1/10$$

$$P(\text{black male with white woman}) = 1/1000.$$

Multiplying these, he computed the probability that any other couple shared these characteristics to be 1 in 12 million. The verdict was "guilty."

The California Supreme Court reversed the conviction. One of the several reasons given was that the factors multiplied to obtain the "1 in 12 million" chance did not relate to *independent* phenomena. For example, suppose that everyone with a beard had a mustache. Then we would find

$$P(\text{beard and mustache}) = P(\text{beard}) = 1/10,$$

rather than the 1/40 that would be obtained by multiplication. Of course, some men with beards do not have mustaches, but surely the odds on a mustache increase with the information that a man has a beard. And this means that using the multiplication rule for independent events would give too small a probability for the event "beard and mustache." Similarly, the probability of both beard and mustache would be changed by the fact that the man was a black, and so on.

[A much more convincing reason for reversal, not particularly relevant here, was that the probability given, had it been correctly computed, would be the probability of the *existence* of such a couple. But one was *known* to exist. What should have been calculated was the probability that there might be at least one *other* couple of the same type in the area. This probability computation would depend on various assumptions about the area, but was computed by the Court to be 41 percent. This would surely be the basis for a "reasonable doubt" of guilt.]

[The prosecutor's calculation of odds is reminiscent of the question people often ask a mathematician, when something happens that strikes them as unusual: "Gee, what's the probability *that* would happen?" The point is that any event can seem to be preposterously unlikely by describing it in sufficient detail. Thus, the poker hand consisting of A, K, Q, J, 10 of spades has probability (in a random deal):

.507   when described as having no pairs

.00355   when described as a "straight"—a sequence not all
            of one suit

.00198   when described as a "flush," being all of one suit

.00001385   when described as being a straight flush

.00000154   when described as being a royal flush

.000000385 when described as being a royal flush in spades.] ▲

Example 4-23 *Hitting Streaks*

At one point in the 1978 National League baseball season, Pete Rose
had hit in 37 consecutive games, tying Tommy Holmes in the record
book. He needed 20 more to break Joe DiMaggio's major league
record of 56 straight games. Las Vegas' odds that he would be able
to do it were about 1 to 99. Perhaps they computed them in this way:

If each time at bat were a trial of an experiment with $P(\text{hit}) = 1/3$
and $P(\text{no hit}) = 2/3$, then the probability of at least one hit in the
(approximately) four times at bat in a game is

$$P(\text{at least 1 hit}) = 1 - P(\text{no hits})$$

$$= 1 - [P(\text{no hit in a single try})]^4$$

$$= 1 - \left(\frac{2}{3}\right)^4 = .8025.$$

The multiplication

$$\frac{2}{3} \cdot \frac{2}{3} \cdot \frac{2}{3} \cdot \frac{2}{3}$$

assumes his four at-bats to be independent. Then, making the further
assumption of independence of games, we find that

$$P(\text{at least 1 hit in each of 20 games}) = (.8025)^{20} = .0123.$$

It is doubtful that a string of times at bat produces results that are
exactly like the independent trials of an experiment, but the calculation is interesting—and may even have been done in Las Vegas to get
the odds. The point is that you cannot *calculate* a probability without postulating a model; and the assumption of independence, although not exactly appropriate, may be approximately correct.

On August 1, 1978, Pete Rose struck out his final time up, without
a hit, ending a 44-game hitting streak. ▲

KEY VOCABULARY

Random selection
Probability model
Odds
Complement of an event
Probability
Law of averages
Subjective probability
Fair bet
Random variable
Discrete probability distribution
Population mean
Population standard deviation
Conditional probability
Independent random variables

QUESTIONS

1. What is a probability "model?"
2. What is the model for a random selection of an individual from a finite population?
3. What is essential for interpreting the number of "chances" you hold in a lottery?
4. What do we mean by a "chip-from-bowl" model for experiments such as those involving dice or counts of traffic, where there is no actual selection from an actual population of objects?
5. Can any number be a probability?
6. What is meant by the "complement" of an event?
7. What is the fundamental assumption in calculating the probability of an event as the ratio of the number of favorable outcomes to the total number of outcomes?
8. A baseball player has been hitless in his last 20 times at bat, and a radio announcer says that the player is "due for a hit." This suggests that the probability of a hit is greater than it would be if he had not recently gone hitless. Is this a correct interpretation of the law of averages?
9. How do probabilities of events show up in a long series of independent trials of an experiment?

10. Will a long-run relative frequency necessarily agree with one's subjective probability of an event?
11. How is your subjective probability of an event related to the way you would bet on the event?
12. What makes a variable a "random" variable?
13. How do the mean and s.d. of a population differ from those of a sample?
14. What is the intuitive meaning of the notion of *independence* of two random variables?

PROBLEMS

Section 4.1–4.4

*1. A card is picked at random from a deck of playing cards. Ace counts high. What is the probability that:
   (a) The card is a 9 or smaller?
   (b) It is a 9 or smaller, if you know it is a heart?
   (c) The card is a face card (King, Queen, or Jack)?

2. Two chips are selected simultaneously and at random from a bowl containing five chips, marked A, B, C, D, and E. There are 10 possible selections:

   AB, AC, AD, AE, BC, BD, BE, CD, CE, DE.

   If these are equally likely, what is the probability that
   (a) The selection includes A?
   (b) The selection does not include B?
   (c) The selection includes neither A nor B?

*3. A chip is selected at random from five chips, marked A, B, C, D, and E, and then a second chip is selected at random from the four chips not chosen first. The possible sequences of chips drawn are these:

   AB, AC, AD, AE, BC, BD, BE, CD, CE, DE,

   BA, CA, DA, EA, CB, DB, EB, DC, EC, ED.

   Given that these 20 sequences are equally likely, determine the probability that
   (a) The sequence includes A.

(b) The sequence does not include A.

(c) The sequence includes neither A nor B.

(d) The first chip is A.

(e) The second chip is A.

(f) The second chip follows the first in the alphabet (e.g., CD).

4. A bingo game has 75 chips numbered from 1 to 75. Chips 1 to 15 go under B, 16 to 30 go under I, 31 to 45 under N, 46 to 60 under G, and 61 to 75 under 0.

    (a) When a chip is drawn at random, what is the probability that it goes in the B category (1 to 15)?

    (b) Suppose that the first chip does fall in the B category; what is the probability that the next chip drawn (from those remaining) is not a B?

    (c) What is the probability that neither the first nor the second draw results in a B or an I? [See Problem 3(c).]

*5. You are offered, in a certain bet, \$5 against your \$2.

    (a) If the bet is fair, what is the probability that you lose?

    (b) If you think the probability of your winning is 1/5, should you take the bet?

*6. In Reno, Nevada, Harrah's "Race and Sports Book" listed these odds for the 1978 World Series, just prior to the playoffs:

Team	Odds
Dodgers	7–5
Yankees	8–5
Royals	2–1
Phillies	5–1

Thus, if you bet \$5 on the Dodgers, you would win \$7 if the Dodgers went on to win the series, implying a 5/12 "probability" for the Dodgers.

    (a) Calculate the other "probabilities."

    (b) Explain why these "probabilities" add up to more than 1.

7. In September 1978, Las Vegas oddsmakers said the odds were 2 to 5 on Muhammed Ali's regaining the WBA heavyweight title. An Associated Press report stated: "A \$5 bet would win \$2 if Ali wins. The line on Spinks . . . was 2 to 1, meaning a \$1 bet would win \$2 if Spinks retained the crown." There seem to

be two probabilities of Ali's winning implied in these odds. What are they? (The oddsmaker's "true" subjective probability is between these.) Why are they different?

Sections 4.5 and 4.6

*8. In McMurray, Pennsylvania, a couple set out to bet their life savings in the Pennsylvania Instant Bingo lottery for lifetime security. Their savings amounted to approximately $20,000. The state expected to sell 35,000,000 tickets for $1 each and to return about $15,400,000 in prizes.
   (a) Calculate how much the couple would get back, on the average.
   (b) A lottery spokesman said, "We don't think it's a good investment. Even if a person bought all the tickets, he'd be a loser." (We'd say: "*Especially* if a person bought all the tickets, he'd be a loser.") If a person bought only 20 percent of the tickets, is he a sure loser?

9. A roulette wheel has 38 equally likely compartments, numbered 0, 00, 1, 2, . . . , 36. Suppose, in a certain game, that when an odd number comes up you lose $10; if an even number (other than 0 or 00) comes up you lose $20; if "0" comes up, you lose $50; and if "00" comes up you win $500. What are your "expected" winnings—that is, how much do you win on the average? (Losses are negative amounts won.)

*10. The distribution for the total number of points thrown using two fair dice in a crap game is as follows:

$$P(2) = P(12) = 1/36$$
$$P(3) = P(11) = 2/36$$
$$P(4) = P(10) = 3/36$$
$$P(5) = P(9) = 4/36$$
$$P(6) = P(8) = 5/36$$
$$P(7) = 6/36$$

Determine the mean and standard deviation of the total number of points thrown.

11. The following table gives relative frequencies of scores on a

certain hole on a golf course, observed over a period of several years, for professional golfers and for club members:

Score	Relative Frequency Pro	Club
2	.02	.00
3	.16	.03
4	.68	.22
5	.13	.27
6	.01	.27
7	.00	.13
8	.00	.05
9	.00	.02
10	.00	.01

Interpreting the relative frequencies as probabilities, compute the means and s.d.'s of pro and of member scores.

*12. In the Goren point-count system for bidding in contract bridge, cards are assigned points as follows:

Ace	4
King	3
Queen	2
Jack	1
Other	0

(a) Give the probabilities that a card drawn at random will have 0, 1, 2, 3, and 4 points, respectively.

(b) Compute the mean value of the number of points assigned to a card selected at random.

(c) Find the average number of points in a hand of 13 cards dealt at random. (HINT: Determine the total number of points in the deck and decide how they would be distributed among the four players, on the average.)

13. A contestant on the TV game show "Let's Make a Deal" had won about $7000 in prizes and was then offered the option of trading for a choice of one of three doors. Behind one door was a prize worth $1000, behind another was one worth $9000, and behind the third, one worth $20,000—but the contestant did not know which prize was behind which door.

(a) What is the mean value of this option to choose a door?

(b) Should the contestant take the trade? Explain.

Section 4.7

*14. There are two pairs of black, two pairs of brown, and one pair of blue socks in a drawer. If one sock is picked at random from the drawer and then another, the table of probabilities for the various outcomes is as follows:

		First Sock		
		Black	Brown	Blue
Second	Black	6/45	8/45	4/45
Sock	Brown	8/45	6/45	4/45
	Blue	4/45	4/45	1/45

(a) Determine the distribution of probability among the colors for the second sock drawn. Is it different from the distribution for the first sock?

(b) Are the color of the first sock and the color of the second sock independent variables?

(c) From the table, determine the probability that the second sock is black given that the first one is black. Is this consistent with the way the experiment is described?

(d) Suppose that the first sock were placed and mixed in with the rest before the second selection; determine the probability table giving the distribution among the outcome pairs.

*15. In a draft lottery (such as that described in Example 1–7) suppose that men born on 250 of the 366 days are to be called for the draft. Suppose also that a man's birthday is February 6 and his brother's is April 18. Assuming random selection, what is the probability that neither will be drafted?

16. A two-digit number is constructed by picking the ten's digit at random (from 0, 1, ..., 9) and then the unit's digit at random in an independent selection.

(a) What is the probability that the two-digit number is 47?

(b) What is the distribution of probability among the integers 00, 01, 02, ..., 99? (This kind of calculation justifies using a sequence of "random digits" to generate k-digit random integers.)

*17. A fair coin is tossed repeatedly in independent tosses.
   (a) Find the probability that the sequence starts out T, T, T, H.
   (b) Find the probability that there are k tails before the first heads appears.
   (c) Find the probability of *at most* three tails before the first heads appears.

18. In a carnival game sometimes called "Razzle Dazzle" you throw eight dice several times. If you throw a total of *eight* on any toss, you win immediately. What is the probability that you throw a total of 8 in a single toss of the eight dice, assuming independence?

*19. The numbers on a roulette wheel (see Problem 9) are assigned colors. The zero and double zero are green, and the others are half red, half black. A betting table is laid out so that you can put money on a single number, on a color, or simultaneously on 2, 3, 4, 5, 6, or 12 numbers. (The only group of five numbers possible consists of 0, 00, 1, 2, and 3.) The payoffs are as follows:

A $1 Bet on:	Pays (if you win):
Any single number	$35
Red (or black)	1
Odd (or even)	1
2-split	17
3-split	11
4-split	8
5-split	6
6-split	5
12-split	2

Compute the mean payoff for each bet. (You get back your $1 along with the amount shown as the payoff.)

Review

20. A fair die is tossed repeatedly, and after each toss one computes (1) the relative frequency of 1's up to that point, and (2) the average number of points per toss thrown up to that point.

(a) What is the limiting value of the relative frequency in (1) as the number of trials increases without limit?

(b) What is the limiting value of the average in (2) as the number of trials increases without limit?

21. In Example 1–16 a method of sampling to avoid embarrassment was described. Each person interviewed would be asked to toss a coin, and if it landed heads, to answer Question 1, otherwise Question 2:

Question 1. Have you had an extramarital affair?

Question 2. Is the next-to-last digit of your phone number an even number?

Let p denote the probability that a person picked at random has had an affair, and assume that the probability of an even digit in Question 2 is 1/2. If the coin toss is independent of the phone number and independent of having affairs, calculate the probability of a "yes" response in terms of p.

22. Given the distribution of the total number of points showing in the toss of two dice defined in Problem 10:

(a) Find the probability of "crapping out" by throwing a 2, 3, or 12.

(b) Find the probability of winning immediately by throwing a 7 or an 11.

(c) If you throw something other than 2, 3, 7, 11, or 12, you must "make your point" by throwing that number again before throwing a 7. Which "point" would you like to have?

23. A model for the emission of an α-particle from a radioactive substance specifies the probability of x counts in a 1-second interval as proportional to

$$\frac{m^x}{x(x-1)(x-2)\cdots 3 \cdot 2 \cdot 1}$$

for $x = 0, 1, 2, 3, \ldots$, where m is a constant that depends on the substance. For the case $m = 1$, the probabilities are as follows:

x:	0	1	2	3	4	5	6
Prob.:	.368	.368	.184	.061	.015	.003	.001

(For $x = 7, 8, \ldots$, the probabilities add up to .000; i.e., to zero to three decimal places.)

(a) Make a histogram (with bars wide enough to touch), and shade the area representing the probability of *at most* one count in a 1-second interval.

(b) Find the mean number of counts in a 1-second interval.

(c) Find the s.d. of the number of counts in a 1-second interval. [In parts (b) and (c) you can only get approximate values, based on the probabilities given for x = 0, 1, 2, 3, 4, 5, 6.]

24. With reference to the same bout as in Problem 7, the Reno Turf Club "set the line at 9–5 for Spinks and 5–11 for Ali." These are different from the Las Vegas odds; you might then expect to be able, by betting on Ali in Reno and on Spinks in Las Vegas, to come out ahead no matter who wins. (This is called a "Dutch Book.") Is this possible with the odds as given? (HINT: Calculate your winnings under both outcomes for a variety of possible betting combinations.)

25. A newspaper ad for a "cash bingo" promotion in a grocery chain listed these odds for prizes of various amounts:

Prize	Odds for One Store Visit
$2000	306,000 to 1
1000	153,000 to 1
200	32,553 to 1
100	16,277 to 1
50	10,066 to 1
25	5,033 to 1
10	3,888 to 1
5	1,943 to 1
2	106 to 1

(a) Calculate the average value of a visit to the store for a person who takes part in this "game."

(b) The ad also gives 2504 to 1 as "odds for 13 store visits," for the $200 prize. What might this mean, and how might it have been computed?

26. The data shown below are from an actual football prediction game similar to other games that circulate illegally in places of business, bars, golf courses, and so on. The player picks a number of football games from a set of 14 particular college games or from 14 particular professional games, and predicts winners against preset "point spreads." The spreads are intended to make predicting a given game a 50–50 proposition.

The list gives the return for each $1 bet. For instance, if the player picks three games and his predictions for those games turn out to be correct, he gets back $5 for his dollar bet. If he chooses to predict all 14 games, he gets back $175 if his predictions are all correct. $30 if 13 out of 14 are correct, and $5 if 12 out of 14 are correct. The average or mean return per dollar bet, when predicting 14 games, turns out to be only $.064—and this is being generous because ties actually lose! (The computation is based on the assumption that the probability of a correct prediction is 1/2.) Calculate the average return for each of the other possible number of games predicted. (Use Table I of Appendix B.)

3 for	3	pays	5 for 1
4 for	4	pays	11 for 1
5 for	5	pays	17 for 1
6 for	6	pays	22 for 1
7 for	7	pays	30 for 1
8 for	8	pays	50 for 1
9 for	9	pays	75 for 1
{ 10 for 10		pays	100 for 1
{ 9 for 10		pays	20 for 1
(14 for 14		pays	175 for 1
{ 13 for 14		pays	30 for 1
(12 for 14		pays	5 for 1

27. A certain prospect pays 11 to 1 if it happens. You bet $1 that it will.
    (a) What is your expected or average gain, if the probability of the prospect is p? (Express this in terms of p.)
    (b) For what probability p would your expected gain be 0?
28. If, referring to Example 4-12 (page 171), you bet $35 on one team, find your expected or mean gain assuming that the probability that Los Angeles wins is 6/11,
    (a) From betting the $35 on New York (at even odds).
    (b) From betting the $35 on Los Angeles (at 5-7).

5
Continuous Random Variables

THERE ARE SITUATIONS that are not described well by models introduced in Chapter 4. Suppose that one is interested in a heifer's weight, a turtle's age, a car's mileage rating, or some other variable which is the result of a *measurement*. If these measurements are at all accurate, the number of possible outcomes is enormous. For example, if the heifer is weighed to the nearest kilogram, then there are hundreds of outcomes that must be assigned positive probability by any reasonable model.

Regarding the measurements as infinitely accurate means that we must reckon with an infinite number of outcomes. Somewhat paradoxically, this approach leads to a class of models that are very manageable and are commonly assumed in statistics. In such a model an individual outcome must have probability 0, but it is *intervals* that are generally the events of interest. This is quite natural; for example, if a hog is said to weigh 547 kilograms, what is usually meant is that the hog's weight is in the *interval* of values from 546.5 to 547.5 kilograms.

A full treatment of *continuous* models is very mathematical, employing the calculus. Our discussion will be less mathematical and more intuitive. We begin with an experiment involving symmetry, a continuous version of "selecting at random."

Example 5-1   *The Spinning Pointer*

A pointer is spun in a horizontal plane around a pivot, and its final position after coming to rest is given by a number on a continuous scale, say from 0 to 1 (see Figure 5-1). If the pointer is spun vigorously, so that it revolves many times before stopping, the number at which it stops is best thought of as a random variable. Call it X.

Imagine, now, that the tip of the pointer is an ideal point—not blunted at all, and that the scale can be read with infinite precision. Under these ideal circumstances, how much would *you* bet that $X = .3$—that the pointer stops at 0.3, on the nose? Remember, there are infinitely many stopping positions, only one of which is 0.3! If all stopping positions were "equally likely," as symmetry might suggest, the odds would be infinity to 1 against observing $X = .3$. In betting with someone who can measure with sufficient accuracy, you are bound to lose! We say that the probability that $X = .3$ is 0:

$$P(X = .3) = 0.$$

Is it then enough, in modeling the pointer spin, to say only that there are infinitely many possible values, each with probability 0? This is of little use in dealing with an event such as $.1 < X < .3$, which says that the pointer stops somewhere between .1 and .3. Yet it is quite reasonable to ask for probabilities of such events; they are not 0, and cannot be found by adding infinitely many 0's (corresponding to the infinitely many possible values on an interval). ▲

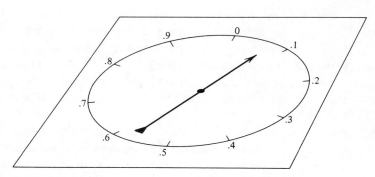

Figure 5-1.

5.1
SMOOTH HISTOGRAMS

It was seen in Chapter 2 that to record data from a continuous variable it was necessary to invent a scheme of class intervals; such a scheme is equivalent to a roundoff process. In looking for the ideal description of a continuous variable by taking more and more data, it is clear that the particular choice of roundoff, which is arbitrary, should play no role. Indeed, taking infinitely many observations with a given roundoff scheme would reveal ideal proportions (or probabilities) for *that* roundoff scheme; and these would constitute a discrete model. To get at the continuous nature of the variable, it is necessary to use another kind of limiting process, taking a successively finer mesh of roundoff points—equivalent to using more and more, narrower and narrower class intervals.

Example 5-2   *Back to the Pointer*

Suppose that we begin by using *ten* class intervals to record results from successive spins of the pointer of Example 5-1. We spun a simulated pointer many times to illustrate what happens in the long run.[†] The histogram for a sequence of 500 spins and another for a sequence of 10,000 spins are shown in Figure 5-2. (The vertical sides shared by adjacent bars have been omitted in that figure, since they contribute little to the picture.) Observe that the top of the histogram with the larger number of spins is flatter. Indeed, as the symmetry of the circle suggests, the limiting relative frequency for each class interval is the same, namely, 1/10; this is the probability of each of the 10 equally likely class intervals.

Figure 5-3(a) shows a similar histogram for 4000 trials recorded using 100 class intervals. Again, the flat-top histogram is suggested. The histogram in Figure 5-3(b), summarizing the results of 100,000 trials, is even smoother and flatter. It seems apparent that as we take a larger and larger number of observations and simultaneously use more and more class intervals (of width tending to zero), the histogram approaches the flat-top graph of Figure 5-4. It is this smooth

---

[†]The "pointer" we used was actually a random-number generator on a programmable calculator, one that produces eight-digit decimals—numbers between 0 and 1. The calculator was programmed to put these numbers into the proper class intervals and then to construct the corresponding histogram.

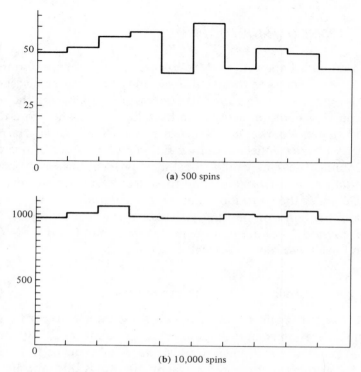

Figure 5-2. Pointer spins, in 10 class intervals: (a) 500 spins;
(b) 10,000 spins.

histogram that is taken as the *model* for the ideal spinning pointer.
From the fact that the height of the graph is constant, we infer that
there are no preferred values or regions of values; and this is what
symmetry would suggest. It is in this sense that the values might be
thought of as being "equally likely."

At each stage of the limiting process, the sample proportion of
observations between .1 and .3, say, is the *relative area* under the
top of the histogram and between vertical lines at .1 and .3 (and
above the horizontal axis). Similarly, the corresponding relative area
under the limiting, smooth histogram gives the *probability* (or
population proportion of values) between .1 and .3. This area, indicated in Figure 5-5, is 2/10 of the total area; so for the result X of
a single spin,

$$P(.1 < X < .3) = \frac{2}{10}. \quad \blacktriangle$$

The spinning pointer is special because of its symmetry, which

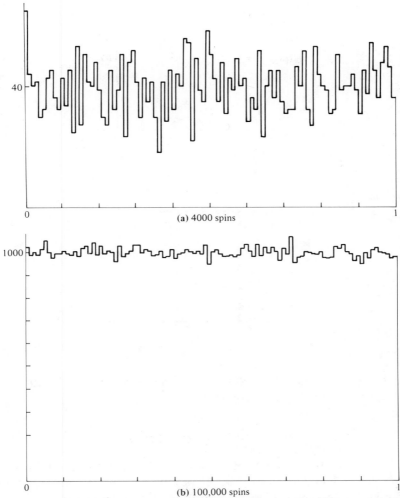

40

0   (a) 4000 spins   1

1000

0   (b) 100,000 spins   1

Figure 5-3. Pointer spins, in 100 class intervals: (a) 4000 spins; (b) 100,000 spins.

makes it easy to calculate the probabilities that might be of interest. Moreover, the number of class intervals used seems to be playing no role—the limiting histogram is as smooth and flat with 10 class intervals as with 100. However, in developing a smooth histogram as a model for a more general kind of continuous variable, it is necessary to use more and more class intervals as we take more and more observations in order to reveal the fine structure as the sampling fluctuations are smoothed out. That is, the very word "continuous" suggests that there *may* be something different going on at $x + .001$

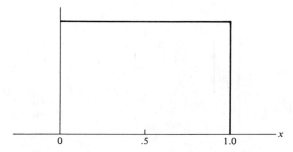

Figure 5-4. Model for the spinning pointer.

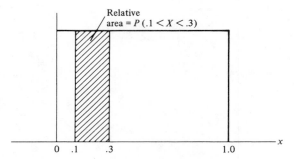

Figure 5-5. Probability of an interval.

than at x, say; and if we always used class intervals of width 0.1, we would never see such differences.

Example 5-3 *A Triangular Distribution*

Let Y denote the *average* of two successive spins of the pointer of Examples 5-1 and 5-2. In thinking about what to expect, ask yourself how an average of .9 can be obtained. Surely, if the first spin is less than .8, the average of it and the next spin cannot be .9! On the other hand, to get an average of .5, the first spin can be anything from 0 to 1. Thus, an average near .5 is about five times as likely as an average near .9.

A sample of n values of Y is obtained by spinning the pointer $2n$ times, twice for each Y-value in the sample. Using a simulated pointer, we did this to obtain a sample of size 1000 (which required 2000 spins), a sample of size 5000, and a sample of size 100,000. Histograms for these samples are shown in Figure 5-6. Observe that with a small number of class intervals, even a large number of trials does not smooth the histogram, because rounding off into class

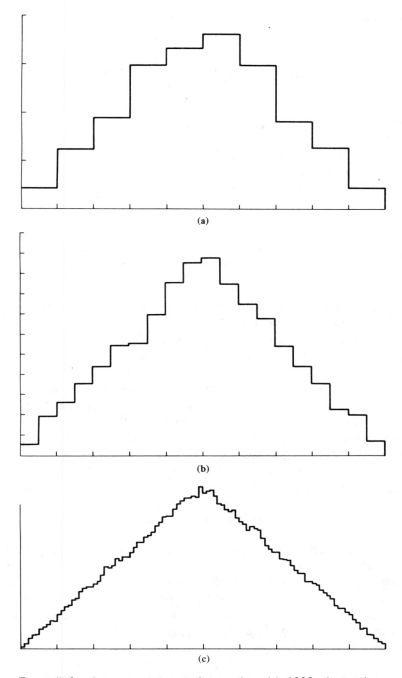

Figure 5-6. Averages at two pointer spins: (a) 1000 observations in 10 class intervals; (b) 5000 observations in 20 class intervals; (c) 100,000 observations in 100 class intervals.

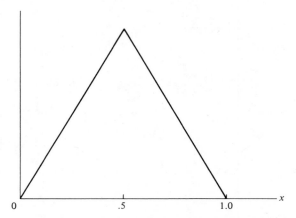

Figure 5-7. Model for the average of two-pointer
spins.

intervals makes even the ideal histogram jagged. (That is, the probabilities of the class intervals are not equal.) However, as the number
of class intervals increases along with the number of trials, the
jaggedness tends to be smoothed out. If, as we assume, there is a
continuous model for the experiment, this model would be given
by a *smooth* histogram—the triangular one shown in Figure 5-7.

The hump in the smooth histogram reflects the situation described
at the outset, namely, that middle values (those around .5) will
occur much more often than values close to 0 or close to 1. Even so,
you may not have guessed that the model would have an exactly
triangular shape. That it does can be shown mathematically, by
developing the reasoning given at the beginning of this example. ▲

Example 5-4 *Students' Grades*

The final numerical grades of 570 students in one of the author's
classes in statistical theory over a period of many years are presented
as a histogram, in Figure 5-8. These were recorded in his grade book
as percentages to the nearest integer; this is not exactly a continuous
scale, but a continuous model seems simpler—and possibly more
appropriate, since the roundoff was just a practical necessity. Imagin-
ing an infinite population of all possible students taking the course—
one that is not changing over the years—we would model it using a
smooth histogram. We can only guess what the model really is; and
we have made such a guess in drawing the smooth curve in Figure
5-9, superposed on the histogram of the 570 grades. ▲

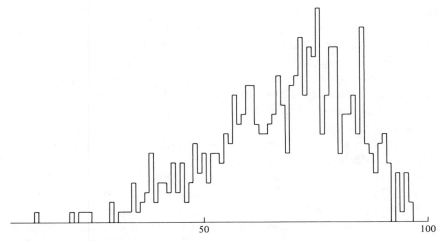

Figure 5-8.   Histogram of 570 final grades.

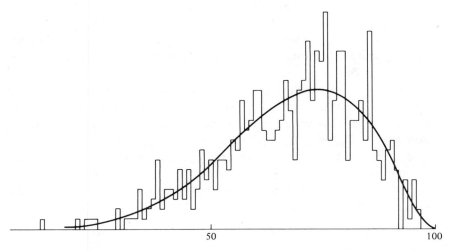

Figure 5-9.   Histogram of 570 final grades with smooth histogram superposed.

We shall be using smooth histograms as models for continuous random variables, but you can also imagine a more basic chip-from-bowl model. There would have to be infinitely many chips in the bowl, somehow numbered with possible values of the random variable in such a way that values in any interval are present in numbers proportional to its probability of occurrence. (That is where the imagination comes in!) An actual bowl with finitely many chips could serve up to a point—up to the point of specified degrees of roundoff and accuracy in the probabilities. The 1500 entries in Table IX of Appendix B, for instance, marked on 1500 chips, would provide an approximate chip-from-bowl model for the particular population sampled there.

For a continuous random variable, the smooth histogram plays the role of a probability table, allowing one to calculate probabilities of events that might be of interest. As suggested in Example 5-2, this is done in the same way that frequencies are found from a histogram; that is, a relative frequency is to a histogram as a probability is to a smooth histogram.

It was pointed out in Chapter 2 that when a histogram is properly drawn, the *area* of a bar is in proportion to the frequency it represents. And the frequency of observations in a region consisting of several class intervals is proportional to the total area of the corresponding bars. The *relative* frequency is then that area divided by the total area in *all* the bars—a proportion of the whole area under the histogram. In the typical histogram of Figure 5-10, the frequency of values between *a* and *b* is proportional to the shaded area—the area under the histogram between vertical lines at *a* and *b*. The relative frequency is the ratio of the shaded area to the area of the whole histogram. (This does not require equal class intervals when the bar heights are properly adjusted so that area represents frequency.)

Probability is represented in the smooth histogram model for a continuous variable in exactly the same way. In the typical smooth histogram of Figure 5-11, the probability that a value of X between *a* and *b* is observed is the shaded area—the area between vertical lines at *a* and *b* under the histogram's top outline—divided by the area under the whole histogram. It is that *fraction* of the whole area which lies between *a* and *b*.

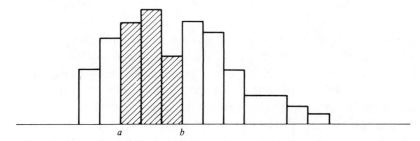

Figure 5-10. Relative frequency as relative area.

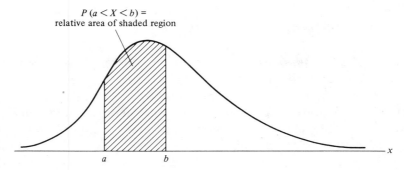

Figure 5-11.   Probability as relative area.

Example 5-5

Again, let X denote the average of two successive pointer spins (as in Example 5-3) with a triangular smooth histogram as its model. The probability that X will have a value between, say, .25 and .75 is the proportion of the total area that lies between those values of the histogram. Figure 5-12 shows the smooth histogram, with the region between .25 and .75 shaded in. It is a matter of elementary geometry to calculate that

$$P(.25 < X < .75) = \frac{\text{shaded area}}{\text{whole area}} = \frac{3}{4}.$$

(See Figure 5-12. The dotted lines divide the region into eight component triangles with equal areas, six of which lie between .25 and .75. So the odds are 6 to 2, or 3 to 1.) ▲

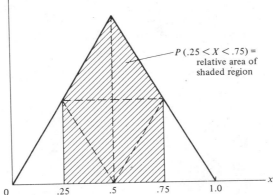

Figure 5-12.   Geometric calculation of a probability (Example 5-5).

5.2
PROBABILITIES FROM TABLES

Smooth histograms whose outlines are not as simple as those of a triangle do not lend themselves to calculations as easy as the one just carried out in Example 5-5. In general, an irregular area can be approximated by methods of numerical analysis, which involve decomposing the area into many small regions of some simple type, such as rectangles or triangles. Fortunately, *tables* of probabilities (relative areas) are available for the distributions that will concern us.

Such tables are generally of one of two types, although both types serve to relate probabilities to values of the random variable. The difference is a matter of table format—whether the table is arranged for convenience in finding a probability (corresponding to a given value of the random variable) or in finding a value (corresponding to a given probability).

Tables of *cumulative probabilities* give the probabilities of open-ended intervals—usually open-ended on the left—corresponding to conveniently spaced values of the random variable. Such a table is explained in the next example.

Example 5-6   *The Standard Normal Distribution*

Table II of Appendix B gives probabilities for a random variable Z whose distribution is called *standard normal*. A brief extract is given here, for convenience, as Table 5-1. The standard normal distribution is given by the smooth bell-shaped histogram shown in Figure 5-13. Table II of Appendix B (or Table 5-1) gives the relative area under the curve to the left of a value z, and this relative area is

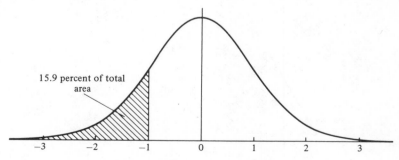

Figure 5-13.   A standard normal curve.

Table 5-1 Standard Normal Distribution

z	$P(Z \leqslant z)$
-3.5	.0002
-3.0	.0013
-2.5	.0062
-2.0	.023
-1.5	.067
-1.0	.159
-.5	.308
0	.500
.5	.692
1.0	.841
1.5	.933
2.0	.977
2.5	.9938
3.0	.9987
3.5	.9998

the probability that the random variable Z takes on a value to the left of z. Thus, the probability that Z will be less than -1 is

$$P(Z \leqslant -1) = \frac{\text{area to left of } -1}{\text{total area}} = .159,$$

so the shaded area in Figure 5-13 is 15.9 percent of the total. Similarly, for example,

$$P(Z \leqslant 2) = .977$$
$$P(Z > 1) = 1 - P(Z \leqslant 1) = .159$$

and

$$P(-1 < Z < 1) = P(Z \leqslant 1) - P(Z \leqslant -1)$$
$$= .841 - .159 = .682.$$

These probabilities give the relative area to the left of 2, to the right of 1, and between -1 and 1, respectively. (It is not necessary to be concerned with the distinction between $<$ and $\leqslant$ when dealing with continuous variables, because the probability of an individual value is 0.)

Table II of Appendix B appears more complicated than Table 5-1 because provision is made for entering the table at Z-values given to two decimal places—and interpolating to three. The left margin of the table gives the integer part and the first decimal place, and the number in the second decimal place is read as a column heading in the top margin. For instance, to enter at $z = 1.63$, go to the row with 1.6 at the left and then across in that row to the entry under 3:

$$P(Z \leqslant 1.63) = .9484.$$

The rows labeled –3. and 3. are exceptions; the column headings for these give the first decimal place. (For example, the entry for 3.2 is .9993.) ▲

The other type of table gives *values* of the random variable being described corresponding to conveniently chosen probabilities. One enters such a table at a probability p and reads out the value with that probability to its left. When the probabilities are chosen as multiples of .01, the corresponding values are called *percentiles*. For instance, the 45th percentile is the value such that the probability is .45 of observing a value no larger (and .55 of observing a value no smaller). The 50th percentile is the *median* value of the distribution.

Of course, a table with conveniently spaced values of a variable X can be used to obtain percentiles of X, and a table giving percentiles can be used to read out probabilities. However, it is usually less awkward to use the table designed for the purpose at hand.

Example 5-7

Table IIa of Appendix B gives percentiles for the standard normal distribution—the distribution whose cumulative probabilities are given in Table II. For instance, entering Table IIa at the probability .60 we find the Z-value .2533:

$$P(Z \leqslant .2533) = .60.$$

This is found (with a little more work) in Table II by entering at $z = .25$ and $z = .26$:

$$P(Z \leqslant .25) = .5987$$
$$P(Z \leqslant .26) = .6026,$$

and interpolating one-third of the way from .5987 to .6026 to obtain

$$P(Z \leqslant .2533) \doteq .5987 + \frac{1}{3}(.6026 - .5987) = .6000. \quad ▲$$

Example 5-8

Table V of Appendix B gives percentiles for each of several distributions of similar type, distributions that will be found useful later in analyzing certain sampling procedures. Each line refers to one of these distributions. Letting X denote the variable whose percentiles are given in the fifth line of Table V, we repeat here the percentiles of X as x-values:

Table 5-2   Cumulative Probabilities

$P(X \leqslant x)$	x
.010	0.554
.025	0.831
.05	1.15
.10	1.61
.25	2.67
.50	4.35
.75	6.63
.90	9.24
.95	11.1
.975	12.8
.99	15.1

Entering at .90 in the left-hand column we see that the probability that X takes a value less than 9.24 is 90 percent. The value 9.24 is the 90th percentile of the distribution.

From this table of percentiles we can read the *median*—or the value such that half the probability is to its left (and half to its right); thus, since

$$P(X \leqslant 4.35) = .50,$$

the median of the distribution is 4.35. The first quartile, similarly, is the entry opposite .25, or 2.67. The interquartile range (distance between quartiles) is 6.63 - 2.67 = 3.96. And so on. The smooth histogram for X is shown in Figure 5-14, with the quartiles marked. (The quartiles divide the total area into four sections of equal area.)

A table of percentiles can be used to calculate probabilities, although using the table in this backward mode is not so convenient. For instance, the probability $P(X < 5)$ is seen to be between .50 (for $X < 4.35$) and .75 (for $X < 6.63$), but the table is really too sparse to interpolate very accurately. ▲

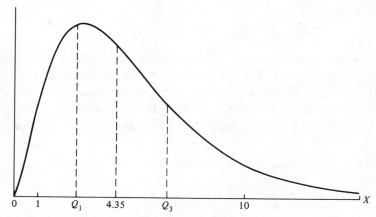

Figure 5-14. Histogram for Example 5-8.

5.3
THE MEAN AND S.D.

The notions of mean and average deviation about the mean, given earlier for frequency distributions and their histograms, extend to probability distributions and their smooth histograms. The actual calculation of these parameters requires either numerical approximations, based on approximating the smooth histogram using a system of class intervals, or methods of calculus—the mathematical tool for dealing with continuous models. It will be enough for us to understand the concepts intuitively as describing a smooth histogram in the same way they describe histograms in the case of a sample, or that of a discrete probability model.

The *mean value*, or *expected value*, of a random variable X will be symbolized (as in the discrete case) by m.v.(X). When thinking of this as the mean value of a *population*, we shall denote it also by the Greek letter μ, as in the discrete case.

The mean is the middle in the sense of center of gravity—the balance point. In particular, then, if a distribution is symmetrical about some value, that center of symmetry is the mean value.

The mean of a symmetrical, continuous distribution of values is the value at the center of symmetry.

The *standard deviation* of a continuous random variable X will be denoted, as in the discrete case, by s.d.(X), or sometimes, when thinking of X as a population variable, by σ.

Example 5-9 *The Spinning Pointer*

Let X denote the value, on the interval from 0 to 1, at which a spinning pointer stops (with the scale from 0 to 1 around the circumference, as we have been assuming right along for our spinning pointers). Inspection of the smooth histogram model for X, with its flat top, shows that the distribution is symmetric about $x = .5$ (see Figure 5-15). So this is the mean value of X:

$$\text{m.v.}(X) = .5.$$

The absolute deviations about the mean value of .5 range from 0 (when $x = .5$) to .5 (when $x = 0$ or 1). The r.m.s. average deviation about the mean would have to be a number between 0 and .5. It can be shown (by calculus) that the "typical" deviation called the standard deviation is

$$\text{s.d.}(X) = \frac{1}{\sqrt{12}} = .289$$

(see Figure 5-15). ▲

Example 5-10 *The Triangular Distribution (again)*

Let Y denote the *average* of two successive pointer spins, as in Example 5-3. The smooth histogram for Y was seen there to be

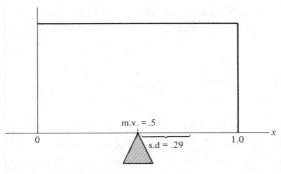

Figure 5-15. Balance point and s.d. for the pointer model.

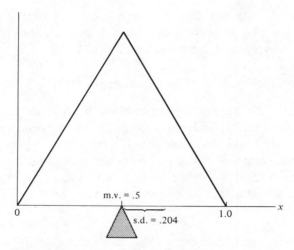

Figure 5-16. Balance point and s.d. for a triangular model.

triangular, symmetrical about the value .5. This is then the mean value:

$$m.v.(Y) = .5.$$

It is shown in Figure 5-16 as the balance point, and the standard deviation marked off there is the value

$$s.d.(Y) = \frac{1}{\sqrt{24}} = .204$$

(again found by methods of calculus). This s.d. or typical deviation is less than in the previous example (where X = single pointer spin), indicating that probability is more tightly distributed about the mean. That is, Y-values near the mean are more common than those far from it. ▲

Example 5-11 *Mean and S.D. of the Standard Normal Distribution*

Probabilities for the distribution introduced in Example 5-6 are defined by its smooth histogram, given in Figure 5-13. A formula for it will be given in Section 5.4, but using that formula to obtain the various important characteristics of the distribution requires (usually) methods of calculus. An exception is that the mean value can be deduced from the symmetry of the histogram—it is symmetric about the value 0, which is therefore the mean:

$$\text{m.v.}(Z) = 0.$$

The standard deviation turns out to be

$$\text{s.d.}(Z) = 1,$$

so the unit along the Z-axis is one s.d. As is seen in Figure 5–13, this is the distance (along the scale of values) from the center of symmetry out to the "point of inflection" on either side—the point where the curve changes direction of bending.

Probabilities for Z, found by numerical methods, are obtainable from Table II of Appendix B, an excerpt from which was given in Table 5–1 and discussed in Example 5–6. From the table we can find the probability between the points ±1, or one s.d. on either side of the mean:

$$P(-1 < Z < 1) = P(Z < 1) - P(Z < -1)$$
$$= .841 - .159 = .682.$$

That is, a standard normal variable is within one s.d. of its mean about 68 percent of the time.

Similarly, the chances of observing a value of Z within *two* s.d.'s of its mean are about 19 in 20:

$$P(-2 < Z < 2) = P(Z < 2) - P(Z < -2)$$
$$= .977 - .023 = .954.$$

And the chances of getting a value farther from the mean than *three* s.d.'s are small indeed:

$$P(\,|Z| > 3) = 1 - P(-3 < Z < 3) = .0026.$$

Scanning Table IX of Appendix B, which gives thousands of observations on Z, will help you to learn what to expect in samples of Z-values. In particular, you will notice that (1) positive and negative Z-values are about equally frequent; (2) values smaller than 1 in magnitude predominate, about 2 to 1; and (3) values bigger than 3 in magnitude are rare. ▲

5.4
NORMAL DISTRIBUTIONS

The *standard* normal distribution (Examples 5–6 and 5–11) is called "standard" because its mean and s.d. have what are called "standard" values—0 and 1, respectively. The histogram for this distribution is

defined mathematically by a simple formula, given here for information; it will not be explicitly used or referred to again:

Standard normal distribution:

Formula[†] for histogram: $e^{-z^2/2}$

Mean: m.v.$(Z) = 0$

Standard deviation: s.d.$(Z) = 1$

Probabilities: given in Table II.

During the course of decades of collecting and summarizing data,
it has been found that many variables, occurring in diverse fields,
have histograms with shapes that are very close to the shape of the
standard normal histogram (Figure 5–13). The term *normal distribution* will apply to any symmetrical smooth histogram that can be
made to coincide with the standard normal histogram by a shift of
the origin to the center of symmetry and choosing one s.d. as the
unit of measurement. Thus, using μ to denote the mean value of a
normally distributed variable X and σ to denote its standard deviation, we have this correspondence of X-values with standard normal,
or Z-values:

$$\mu \longleftarrow 0$$
$$\mu \pm \sigma \longleftarrow \pm 1.$$

The *linear* relationship (a translation combined with a stretching)
that achieves this correspondence is

$$X = \mu + \sigma Z,$$

or (upon solving for Z):

$$Z = \frac{X - \mu}{\sigma}.$$

So the Z-scale locates a value of X according to its deviation from μ
as so many σ's. Figure 5–17 shows a normal histogram for X with a
Z-scale drawn alongside the X-scale.

[†]The symbol e is the base of the natural logarithm system, an irrational number: $e \doteq 2.71828$. The area under this histogram curve is $\sqrt{2\pi}$. Many people
like to scale the vertical axis so that the total area under the curve is 1 (representing the total probability). The formula for the curve with total area 1 would
have the extra factor $1/\sqrt{2\pi}$.

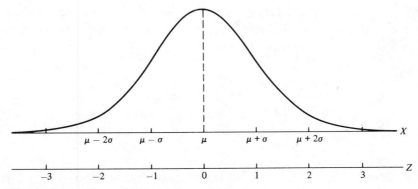

Figure 5-17. General normal curve, with added Z-scale.

The value of the standard Z corresponding to a value of a normally distributed variable X is referred to as a Z-score, or as a *standardized* score:

Z-score for the value x of a normally distributed random variable X:

$$z = \frac{x - \text{m.v.}(X)}{\text{s.d.}(X)}.$$

As seen by referring to Figure 5-17, the Z-score of $\mu + 2\sigma$ is 2, the Z-score of $\mu - 3\sigma$ is –3, and so on. A Z-score of +1.7 means that the X-value is 1.7 σ's to the right of μ.

To find the probability, say, that X is to the left of a particular value x, we first calculate the corresponding Z-score. Entering Table II of Appendix B with this Z-score we find the probability (or relative area) to the left of the Z-score on the Z-scale, or equivalently, to the left of the value x on the X-scale.[†]

It will be convenient to have a shorthand way of saying "look in Table II to find the cumulative probability to the left of the value z in a standard normal distribution." This probability can be called

[†]By now the student may be confused about the use of lower- and uppercase letters. The convention we should like to follow is that capital letters (X, Z, etc.) denote random variables and lowercase letters (x, z, etc.) denote specific, particular values that the random variable may take on. Thus, the event written $X = x$ means the event that the random variable (perhaps a measured weight) has the particular value x. We write "Z-score" with a capital Z, referring to the standard normal variable. The Z-score corresponding to a particular value x of the random variable X would be a particular Z-value, z.

$P(Z \leqslant z)$, if we have defined Z as a standard normal variable; but it will be convenient to have the simpler and commonly used notation $\Phi(z)$ for this probability. Thus, for instance:

$\Phi(2.4) = P(Z \leqslant 2.4)$

= entry in Table II opposite 2.4

= probability that any normal variable is not greater than 2.4 s.d.'s to the right of its mean

= .9918.

(As noted before, a normal variable is continuous, and the probability that $Z \leqslant 2.4$ is the same as the probability that $Z < 2.4$.)

Example 5-12 *Light-Bulb Life*

The length of time X an ordinary light bulb will last before burning out is a random variable—no two bulbs last exactly the same time, and you cannot tell ahead of time just how long one will last. Bulbs of a given type have a life-length pattern that seem to be pretty well described by a normal curve. Assuming that the normal model is in fact correct, and that the mean life is m.v.(X) = 700 hours, with an s.d. of 50 hours, one can calculate probabilities, such as

P(bulb burns out before 625 hours) = $P(X < 625)$.

To find this probability, the relative area under the histogram to the left of 625, it is only necessary to compute the z-score corresponding to 625:

$$z = \frac{625 - \text{m.v.}(X)}{\text{s.d.}(X)} = \frac{625 - 700}{50} = -1.5,$$

and look this up in Table II:

$$\Phi(-1.5) = .067 \doteq 7 \text{ percent.}$$

The Z-score -1.5 says that the value 625 is one and one-half s.d.'s to the left of the mean value of 700 (see Figure 5-18).

Similarly, 68 percent of the bulbs would last between 650 and 750 hours—between one s.d. on either side of 700 (see Example 5-11).

The third quartile, say, would be the value x_3 of X such that

$$P(X < x_3) = .75.$$

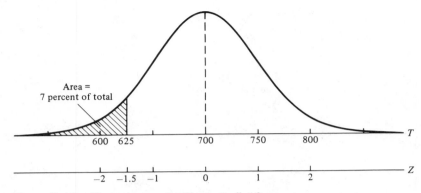

Figure 5-18. Normal curve for Example 5-12.

This would be the X-value corresponding to the 75th percentile of a standard normal X:

$$z_3 = \frac{x_3 - \text{m.v.}(X)}{\text{s.d.}(X)} = \frac{x_3 - 700}{50} = .675,$$

found in Table IIa opposite .75. And then

$$x_3 = 700 + 50 \times .675 = 733.7.$$

By symmetry, the first quartile is $x_1 = 700 - 50 \times .675 = 666.3$. Thus, 50 percent of the bulbs would last between about 666 and 734 hours. ▲

Example 5-13

Test scores in widely administered aptitude exams are often distributed according to a normal curve. Suppose that two students get their scores back, as follows:

> Student 1: Score = 523, 86th percentile.
>
> Student 2: Score = 475, 45th percentile.

Question: What are the mean and s.d. of all test scores?

The 45th and 86th percentiles of the standard normal distribution are found using Table II by entering the body of the table to find probabilities close to .4500 and .8600 and reading the Z-value at the margin:

$$z_{.45} = -.126$$

$$z_{.86} = 1.08.$$

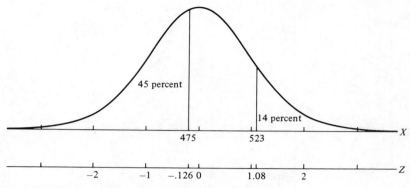

Figure 5-19. Normal curve for Example 5-13.

These Z-scores would be related to the students' reported scores by the standardizing formula:

$$1.08 = \frac{523 - \text{m.v.}(Z)}{\text{s.d.}(Z)}$$

$$-.126 = \frac{475 - \text{m.v.}(Z).}{\text{s.d.}(Z)}$$

This set of simultaneous equations can be solved to obtain the mean and standard deviation (which you can check by substituting them into the equations):

$$\text{m.v.}(Z) = 480$$

$$\text{s.d.}(Z) = 39.8.$$

The students' scores and corresponding Z-scores are shown in Figure 5-19. ▲

5.5
BIVARIATE MODELS

Models for a pair of continuous random variables (X, Y) are analogous to univariate models, with the complication of an added dimension. A cloud of data points in the plane can be summarized in a frequency distribution by rounding off each variable in a system of class intervals, creating thereby a grid or two-way array of compartments. A three-dimensional histogram, with solid bars over the rectangular compartments proportional to their frequencies, provides a visual representation of the distribution—albeit somewhat awk-

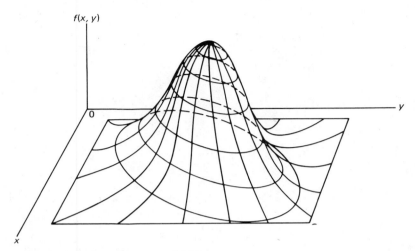

Figure 5-20.   Bivariate normal histogram.

ward to picture on a flat page. (See Section 2.9, in particular, Figure 2-15.)

Again, a continuous *model*—an ideal distribution—is given by a smooth histogram, a surface above the region of points (X, Y). This is usually a rolling surface, with one or more hills. The model can be thought of as the result of a limiting process—increasing the number of data points ($n \to \infty$) while simultaneously refining the class interval structure, so that the class interval widths tend to 0—just as the smooth histogram for a single variable was developed in Section 5.1.

Smooth histogram surfaces for a bivariate model are just as hard (or harder) to draw as the discrete histogram, and the visual representations are not often used. A smooth histogram of the type called *bivariate normal* is shown in Figure 5-20. Its bell shape is a two-dimensional version of the univariate, bell-shaped normal curve. There are symmetries, but its contour lines are not circular (as they would be for real bells) but elliptical. The hilltop is centered above the mean point [m.v.(X), m.v.(Y)], and the correlation coefficient of X and Y in the particular model pictured is about .7.

A cloud of data points from such a distribution is shown in Figure 5-21, with a set of axes drawn through the mean point. The concentration of data points is highest about that point, which corresponds to the peak of the smooth histogram—the region of greatest probability.

Computations of probabilities and population averages from a

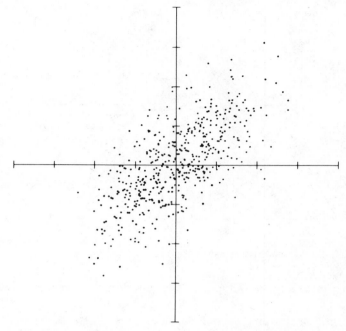

Figure 5-21.   Data from a bivariate normal distribution.

bivariate smooth histogram is nontrivial and will not be attempted. We have introduced such models just to point out that, while the two-dimensional case is more complicated than the one-dimensional case, the models and statistical problems are conceptually the same.

KEY VOCABULARY

Continuous random variable
Smooth histogram
Cumulative probability
Percentile
Median
Mean
Standard deviation
Normal distribution
Standard normal distribution
Symmetrical distribution
Z-score
Standardizing
Bivariate normal model

QUESTIONS

1. How does one account for assigning probability 0 to a particular value of a continuous variable, when that value *can* occur?
2. What geometric quantity in a smooth histogram gives the probability assigned to an interval of values?
3. How can a table of cumulative probabilities be used to get the probability of an interval?
4. How do the notions of percentile and median of a population relate to the corresponding notions for a frequency distribution of a data set?
5. How much area, in a smooth histogram, lies between two given percentiles of a distribution?
6. The standard deviation of a random variable is a kind of average or typical deviation. A deviation of what, and from what?
7. What is "standard" about a standard normal distribution?
8. What is "normal" about a normal distribution?
9. The normal distribution is never *exactly* the correct model for any phenomenon of our "real" world. Why not?
10. We could have got by with just *one* of the two pages of the normal table (Table II). How would we have managed this?
11. What does it mean to "standardize" a random variable?

PROBLEMS

Sections 5.1 and 5.2

*1. The smooth histogram shown in Figure 5-22 describes the population of scores in the first midterm exam in an elementary statistics course. By counting squares, determine approximately:

(a) The proportion of scores between 60 and 70.

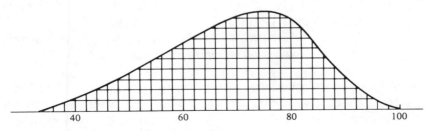

Figure 5-22. Histogram for Problem 1.

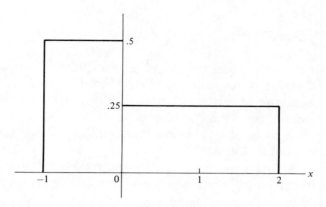

Figure 5-23. Histogram for Problem 2.

(b) The proportion of scores greater than 80.
(c) The median score.
2. The random variable X is continuous, with distribution given by the histogram in Figure 5-23. Determine the following:
(a) $P(X \leqslant 1)$.   (e) $P(|X| < 1/2)$.
(b) $P(X \leqslant -1/2)$.   (f) $P(|X - 1| > 1/2)$.
(c) $P(0 < X < 1)$.   (g) $P(X = 1)$.
(d) $P(X < 0 \text{ or } X > 1)$.
(This variable X can be realized as follows: Pick a random number Y between 0 and 1, perhaps using a pointer spin or a table of random digits. Then toss a coin, and if it falls heads, multiply the number Y by 2; if it falls tails, multiply Y by -1. The resulting number is the desired X.)
*3. A random variable Z has a distribution defined by the histogram of Figure 5-24. (This distribution can be generated by taking the distance between two independent random numbers X and Y on the interval from 0 to 1—two pointer spins.)
(a) Find $P(Z > 1/2)$. (This is the probability that X and Y differ by more than $1/2$.)
(b) Find $P(Z < .1)$. (This is the probability that X and Y differ by less than .1.)
(c) Find the median of Z. (HINT: Call it m, and set the area of the triangle to the right of m equal to $1/2$ the total area.)
4. A TV pinball game was played many times with scores summarized in the frequency distribution in Table 5-3. Draw a histogram for this distribution, and superpose a smooth histo-

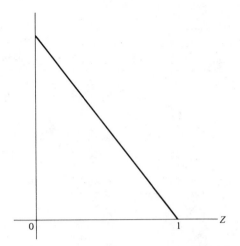

Figure 5-24.   Histogram for Z in Problem 3.

Table 5-3

Scores (100's)	Frequency	Scores (100's)	Frequency
18–20	7	57–59	34
21–23	10	60–62	22
24–26	36	63–65	18
27–29	60	66–68	14
30–32	65	69–71	9
33–35	92	72–74	7
36–38	92	75–77	2
39–41	123	78–80	4
42–44	109	81–83	0
45–47	104	84–86	1
48–50	82	87–89	0
51–53	53	90–92	1
54–56	63		

gram that you think would represent the ideal or population
distribution.

*5. Table III lists the percentiles for several continuous distribu-
tions, one in each row. Each distribution is symmetric about
the value 0. The third row gives these percentiles.

p	.60	.70	.80	.85	.90	.95	.975	.99	.995
(100p)th Percentile	.277	.584	.978	1.25	1.64	2.35	3.18	4.54	5.84

Table 5-4

Score	Percentile Rank	Score	Percentile Rank
740	99	460	28
720	97	440	24
700	94	420	21
680	91	400	17
660	86	380	14
640	81	360	11
620	74	340	9
600	70	320	7
580	62	300	5
560	56	280	4
540	51	260	3
520	44	240	2
500	39	220	1
480	34		

(a) Determine the probability between the values 1.64 and 4.54.

(b) Determine the probability between the values -1.64 and 1.64 (Use the symmetry of the distribution; the histogram for negative values is the mirror image of the histogram for positive values.)

(c) Find the probability between -1.64 and 4.54.

(d) Estimate the 75th percentile.

(e) Estimate the probability to the left of the value 1.

*6. Graduate Record Examination scores of 155,623 applicants in the United States during the 1977 to 1978 academic year are summarized by percentile rank, for the section on "analytical ability," in Table 5-4. (The percentile rank of a score is the percentage of scores less than it.)

Determine the proportion of scores:

(a) Greater than 640.

(b) Less than 440.

(c) Between 500 and 600.

(d) Between 400 and 700.

(e) More than 570 (estimate this).

7. Scores of 112,241 foreign students who took the TOEFL (Test of English as a Foreign Language) examination in the 1976 to 1977 academic year are summarized by percentile ranks in Table 5-5.

Table 5-5

Score	Percentile Rank	Score	Percentile Rank
660	99	480	38
640	98	460	29
620	94	440	21
600	90	420	14
580	84	400	9
560	76	380	5
540	67	360	2
520	58	340	1
500	48		

Find the proportion of the scores:
(a) Greater than 580.
(b) Between 500 and 600.
(c) Between 400 and 570.

Sections 5.3 and 5.4

*8. Given the histogram of Problem 1 (Figure 5-22):
   (a) Is the mean closest to 60, 70, 80, or 90?
   (b) Is the s.d. closest to 4, 12, 20, or 28?

*9. By looking at the histogram for Problem 2 (Figure 5-23), estimate the location of the mean. The mean can be calculated as follows. Imagine all the values between 0 and 2 rounded off to the value 1, and assign this the probability 1/2; and round off all the values between –1 and 0 to –1/2, assigning this the probability 1/2. Then compute the mean of this two-point distribution of probability. (The mean of such a histogram can be computed by locating the mean of the several sections and assuming that the probability for each section is concentrated at the mean of that section.)

10. Referring to Figure 5-24, giving the histogram for Z in Problem 3, decide whether:
   (a) The mean value of Z is closest to 1/3, 1/2, or 2/3.
   (b) The s.d. of Z is closest to .02, .24, .48, or .96.

*11. The histogram for the average of two independent spins of the pointer of Examples 5-1 and 5-2 was given in Figure 5-16. The sum U of the two spins is just twice their average, so the same model can be used if the horizontal scale is multiplied by 2.

(a) What is the probability that two spins yield numbers adding up to less than .5?

(b) What is the mean value of U?

(c) Referring to Example 5-10, determine the s.d. of U.

*12. The Graduate Record Examination scores in 1977 to 1978
were approximately normally distributed with mean 521 and
s.d. 123. Assuming this distribution, find (approximately, at
least):

(a) The 90th percentile score.

(b) The proportion of scores between 500 and 600.

(c) The proportion of scores within 1 s.d. of the mean.

(d) The proportion of scores farther than two s.d.'s from the
mean.

13. The amount of coffee in a 1-pound can varies from can to can.
If the mean weight is 1.05 pounds and weights are normally
distributed about this mean with s.d. = .04, determine the
probability that

(a) A can contains less than 1.00 pound.

(b) A can contains more than 1.10 pound.

(c) The coffee in a can weighs within one s.d. of the mean
weight.

*14. The mean and s.d. of the GRE scores described in Problem 6
are $\bar{X}$ = 521 and S = 123. Taking these as population mean
and s.d., respectively, calculate Z-scores for several of the
listed scores and (using Table II) the corresponding normal
percentile ranks. Observe the degree of fit.

15. The mean and s.d. of the TOEFL scores described in Problem 7
are 504 and 74, respectively. Using these as population parameters, determine Z-scores for several of the listed scores, and
(using Table II) the corresponding normal percentile ranks.
Observe the close fit.

Review

16. The area under the graph of

$$y = \frac{1}{1 + x^2},$$

shown in Figure 5-25, is π. If this graph is taken as the histogram for a continuous variable X, approximate the probability

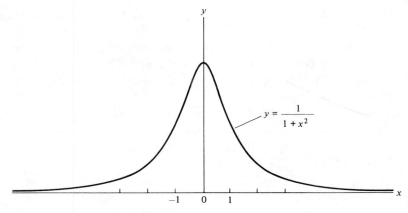

Figure 5–25. Histogram for Problem 16.

that X takes on a value between .5 and 1.5 by calculating the area of a trapezoid—the trapezoid whose "bases" are the ordinates of the curve at .5 and at 1.5. (A trapezoid is a quadrilateral with two opposite sides parallel. Its area is the average of the parallel sides times the distance between them.)

17. The blood platelet count in a population of humans is approximately normally distributed, with mean 235 and standard deviation 44 (in thousands per cubic millimeter).
    (a) Find the 30th percentile of the distribution.
    (b) Find the proportion of the population with counts outside the range 180 to 290.
    (c) What interval centered at the mean count would include 70 percent of the population?

18. Use Table II to verify:
    (a) The 20th percentile of Z shown in Table IIa.
    (b) The entry opposite .15 in Table IIb.

19. Make a stem-leaf diagram for the weights of the men in the Main Campus section listed in Data Set B (Appendix C).
    (a) Does the distribution appear to be normal?
    (b) The mean and s.d. of the weights are 159.8 and 19.94, respectively. In a normal distribution with these values as population mean and s.d., what proportion of the weights would lie between 128 and 148 pounds? What proportion of the 71 weights listed in the stem-leaf diagram are between these values?

(c) Compare the third quartiles of the sample and of a normal distribution with mean 159.8 and 19.94.

(d) Make a stem-leaf diagram of the weights of the women in the Ag Campus section of Data Set B. Do these weights appear to be any more normally distributed than those of the men in part (a)?

20. The standard normal distribution can be approximated (somewhat crudely) by dividing the range of possible values into class intervals and representing each by a central value, in Table 5-6.

Table 5-6

Class Interval	Class Mark	Probability
< -3.5	?	.0002
-3.5 to -3.0	-3.25	.0011
-3.0 to -2.5	-2.75	.0049
-2.5 to -2.0	-2.25	.0168
-2.0 to -1.5	-1.75	.044
-1.5 to -1.0	-1.25	.092
-1.0 to -.5	-.75	.149
-.5 to 0	-.25	.192
0 to .5	.25	.192
.5 to 1.0	.75	.149
1.0 to 1.5	1.25	.092
1.5 to 2.0	1.75	.044
2.0 to 2.5	2.25	.0168
2.5 to 3.0	2.75	.0049
3.0 to 3.5	3.25	.0011
> 3.5	?	.0002

(The probabilities given were found using Table II.)

(a) Verify a few of the listed probabilities.

(b) Approximate the s.d. of the standard normal distribution by finding the s.d. of this list of values (class marks) and corresponding relative frequencies (probabilities). [Because the mean is 0, what is wanted is the square root of the average square. You can cut the task in half by exploiting the symmetry: $(-1.25)^2 = (1.25)^2$, for example. Use 4 and -4 as the class marks for the open-ended intervals; this will not introduce much error.]

21. Given that the 67th percentile of test scores is 540 and the

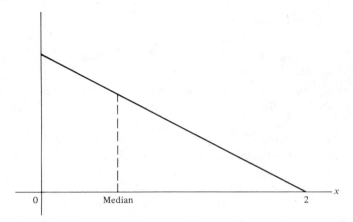

Figure 5-26.

14th percentile is 420 (as in Problem 7), find the mean and s.d. of test scores assuming that they are normally distributed.

22. Find the 75th percentile of a random variable that is normally distributed with mean 50 and standard deviation 10.

23. A continuous variable X has the smooth histogram shown in Figure 5-26. This is an example of a distribution that is *skewed* to the left—probability is concentrated toward the left side of the interval of possible values, rather than symmetrically distributed.

    (a) Find the *median* value of X. (HINT: The line through the median will divide the area in half provided that the sides of the smaller triangle in the figure are each $1/\sqrt{2}$ times the corresponding sides of the larger triangle.)

    (b) Try your intuition: On which side of the median does the average or mean value lie? (The relationship here is typical of distributions that are skewed to the left.) In the case of a triangle, the balance point is two-thirds of the way from a vertex to the center of the opposite side, so the mean value here is two-thirds of the way from $X = 2$ to $X = 0$.

24. A newspaper article based on Bureau of Standards data states that (1) 90 percent of all women are between 59.5 and 67.5 inches tall, (2) the average height for women is 64 inches, (3) about 36 percent of all women are taller than average, (4) about 10 percent of the female population is over 67 inches tall, and (5) at least 5 percent "grow into the 5 feet nine inches to over 6 feet category."

(a) Given statement (1), and if the heights of women were normally distributed, what would be the mean and standard deviation, assuming that the interval 59.5 to 67.5 includes the *middle* 90 percent?

(b) In view of (3), can the distribution of heights be symmetric? If not, which way is it skewed? (See Problem 23 above.)

(c) Show that if the interval 59.5 to 67.5 inches were the middle 90 percent, there would be no one between 67 and 67.5 inches.

(d) What percentage of women are between 64 and 67 inches?

25. A county traffic engineer told a city council that over a two-day period in December, 84 percent of vehicular speeds on a certain county road in the city were in the 30 to 40 mile per hour range. He said that 85 percent of all vehicles were traveling at, or less than, 38 mph. Further, he said, "the 85th percentile is the standard measure for determining speed limit."

(a) Show by calculation that if speeds are normally distributed with mean 34.37 and s.d. 3.50, the percentages given would be correct.

(b) Under the assumption of part (a), what fraction of vehicles were exceeding the posted limit of 35 mph?

6
Estimating
Population
Proportions

IN SURVEY WORK the purpose of sampling is often to determine the *proportion* of individuals of a particular type in the population. The proportion of males, the proportion of Republicans, and the proportion of smokers are examples. In other situations we are often interested in assessing the *probability* of a particular outcome—for example, the probability that a die shows six or not, the probability that an inocculation takes or not, or the probability that an unborn child will be male.

Probabilities can be thought of as proportions, and proportions as probabilities. Thus, the probability that a manufactured article is good is the proportion of chips marked "good" in the corresponding chip-from-bowl model. And the proportion of individuals in a population who are watching a particular TV channel is the probability of picking such a person in a random selection from the population.

Whatever the event of interest in such problems, we find it convenient to think of its occurrence as "success" and its nonoccurrence as "failure." The probability of success (or the population proportion of successes) will be denoted by p. The probability of failure (or the proportion of failures) is then $1 - p$, which will be called q. The model is then as follows:

Outcome	Probability
Success	p
Failure	q
Sum	1

This chapter deals with the use of sample results in estimating the probability or population proportion p. If a sample is "representative" of the population, the *sample proportion* of successes should be close to the population proportion to be estimated. We denote this sample proportion by $\hat{p}$.

Notation:

p = population proportion of successes

$\hat{p}$ = sample proportion of successes.

Sampling at random has a good chance of producing a sample that is close to representative. Moreover, when sampling is done at random, it is possible to assess the chances of a specified degree of closeness. The distributions and error formulas to be presented here will be based on the assumption of random sampling—usually with replacement, if the population is finite, although the case of sampling without replacement is included if the population is large compared to the sample size. Other kinds of probability sampling are often used in survey work, out of necessity, but the formulas we take up do not apply to those methods—nor to samples that are not obtained by probability methods.

Using random sampling, we are assured by the law of large numbers that the sample proportion $\hat{p}$ will be close to the population proportion p, if the sample is large. *How* close depends on the sample size, although with any given sample size, some samples will do better than others.

6.1
THE SAMPLING DISTRIBUTION OF $\hat{p}$

A sample proportion $\hat{p}$ *varies* from sample to sample. It is a random variable, its value depending on the particular sample one happens to draw. As we do with any random quantity, we describe its variation in terms of the "population" of its possible values. Thus, we have a new model to consider—the population of $\hat{p}$-values, or sample proportions, corresponding to all the various possible samples (of given size) that might be drawn. The distribution of probability among these $\hat{p}$-values is called the *sampling distribution* of $\hat{p}$.

Example 6-1

Tossing a coin five times is equivalent to drawing a chip at random five times, with replacement and mixing, from a population of chips that are marked half with "heads" and half with "tails." The *population* proportion—the probability of heads—is 1/2. The sample proportion of heads can be 0, 1/5, 2/5, 3/5, 4/5, or 1. Some of these values of $\hat{p}$ are more probable than others, and the probabilities are to be found in Table I of Appendix B. They are as follows:

Sample Proportion	Probability
0	.0312
1/5	.1562
2/5	.3125
3/5	.3125
4/5	.1562
1	.0312

This table gives the sampling distribution of $\hat{p}$. It could also be represented by a bowl of chips—one chip marked $\hat{p} = 0$, five marked $\hat{p} = 1/5$, ten marked $\hat{p} = 2/5$, ten marked $\hat{p} = 3/5$, five marked $\hat{p} = 4/5$, and one marked 5/5. (The possible values of $\hat{p}$ would then be present in the bowl in proportion to their probabilities.) One can have this bowl in mind in thinking of the "population" of $\hat{p}$-values.

Problem 3 at the end of the chapter asks you to do some actual sampling—to obtain many samples of size $n = 5$, in order that you can observe firsthand the variation of $\hat{p}$ from the sample to sample. ▲

The sampling distribution of a sample proportion $\hat{p}$ is a distribution on its possible *values*—0, $1/n$, $2/n$, . . . , n/n. These $\hat{p}$-values have a *mean* value, computed using the corresponding probabilities in the usual formula for a population mean. It can be shown mathematically that this mean value is always p, the population proportion, when the sampling is *random*.

Mean value of a sample proportion:

$$m.v.(\hat{p}) = p,$$

where p is the population proportion, and provided that the sampling is random (with or without replacement).

Example 6-2

A bowl contains 25 chips, of which 10 are red and the rest white. The population proportion of red chips is 10/25, or 2/5. Five chips are to be selected from this bowl or population, one at a time, at random. The proportion $\hat{p}$ of red chips among the 5 chips drawn is a random variable whose distribution depends on whether or not there is replacement:

	Probability	
Sample Proportion, $\hat{p}$	Sampling With Replacement	Sampling Without Replacement
0	.078	.057
1/5	.259	.257
2/5	.346	.385
3/5	.230	.237
4/5	.077	.059
1	.010	.005

(The probabilities in the middle column come from Table I, using $n = 5$ and $p = .4$; those in the last column were calculated by counting combinations.) The mean value of $\hat{p}$ is 2/5, the population mean, for either sampling with or sampling without replacement:

$$0 \times .078 + \frac{1}{5} \times .259 + \frac{2}{5} \times .346 + \frac{3}{5} \times .230 +$$

$$\frac{4}{5} \times .077 + 1 \times .010 = .400$$

$$0 \times .057 + \frac{1}{5} \times .257 + \frac{2}{5} \times .385 + \frac{3}{5} \times .237 +$$

$$\frac{4}{5} \times .059 + 1 \times .005 = .400.$$

Thus, we have verified the formula m.v.($\hat{p}$) = p in this case. ▲

The *standard deviation* of the sampling distribution of $\hat{p}$ is also expressible in terms of the population parameter p:

When sampling is done randomly, the standard deviation of a sample proportion is

$$\text{s.d.}(\hat{p}) = \begin{cases} \sqrt{\dfrac{pq}{n}}, & \text{if sampling is with replacement} \\[2ex] \sqrt{\dfrac{pq}{n} \cdot \dfrac{N-n}{N-1}}, & \text{if there is no replacement,} \end{cases}$$

where $q = 1 - p$, and N is the population size.

The factor

$$\sqrt{\frac{N-n}{N-1}},$$

a "correction factor" for finite populations, is nearly 1 if the population is much larger than the sample. In such a case the formula $\sqrt{pq/n}$ can be used whether or not there is replacement.

Example 6-3

We return to the population of Example 6-2, a bowl with 10 red and 15 white chips. The standard deviation of the proportion of red chips among the chips in a random sample of size 5 is

$$\text{s.d.}(\hat{p}) = \begin{cases} \sqrt{\dfrac{.4 \times .6}{5}} = .219, & \text{if sampling is with replacement} \\[2ex] \sqrt{\dfrac{.4 \times .6}{5} \times \dfrac{25-5}{25-1}} = .200, & \text{if there is no replacement.} \end{cases}$$

These can be checked by using the formula of Chapter 4 for a population s.d.—the mean square minus the square of the mean. Thus, in the case of replacement,

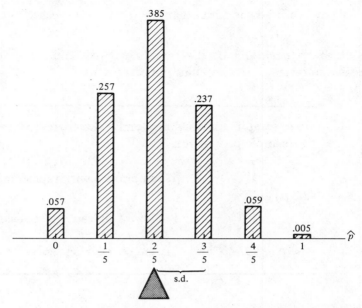

Figure 6-1. Distribution of $\hat{p}$, simple random sample from population with $p = 2/5$; $N = 25$, $n = 5$.

$$\text{mean square} = 0^2 \times .078 + \left(\frac{1}{5}\right)^2 \times .259 + \left(\frac{2}{5}\right)^2 \times .346$$

$$+ \left(\frac{3}{5}\right)^2 \times .230 + \left(\frac{4}{5}\right)^2 \times .077 + 1^2 \times .010 = .208,$$

so that s.d.$(\hat{p}) = \sqrt{.208 - (.4)^2} = .219$.

Figure 6-1 gives a histogram for the distribution of $\hat{p}$ under sampling without replacement. The balance point is the value 2/5 (or p), which appears reasonable, as does the typical deviation of .200. ▲

Example 6-4

Suppose that a sample of size 1600 is drawn at random from the population of voters in a large city, and that 40 percent of those voters actually favor candidate S. The sample proportion $\hat{p}$ favoring S will depend on who happens to be chosen in the sample. It could be 40 percent, also, but more often than not it will be a little higher or a little lower. On the average—over the population of $\hat{p}$-values—it will be 40 percent:

$$\text{m.v.}(\hat{p}) = p = .40,$$

and its r.m.s. deviation from this average is

$$\text{s.d.}(\hat{p}) = \sqrt{\frac{pq}{n}} = \sqrt{\frac{.4 \times .6}{1600}} = .01225.$$

So, in samples of 1600, the sample proportion can differ from 40 percent by one or two percentage points—or more.

Although voter surveys are ordinarily done without replacement, we used the s.d. formula that applies, strictly speaking, to the case of random sampling with replacement. If there are 250,000 voters, for example, the "correction factor" is

$$\sqrt{\frac{250,000 - 1600}{250,000 - 1}} = .997,$$

or nearly 1. Using the more precise formula for s.d. would have given the value .01221. ▲

6.2
STANDARD ERROR

When a population proportion p is estimated to be $\hat{p}$, the difference between these values is called the *error* of estimation:

$$\text{error} = \hat{p} - p.$$

This error will be sometimes negative, sometimes positive, and occasionally 0. When sampling is random, the *average* error will be 0, because then the average of $\hat{p}$ is p. What is really of interest, however, is the *absolute* error, or the *magnitude* of the error:

$$\text{absolute error} = |\hat{p} - p|.$$

Thus, if the value of p is actually 1/2, a sample proportion of .483 would be in error by the amount .483 – .500, or –.017. The absolute error is obtained by just dropping the minus sign; it is .017.

In general, the error of estimation has two components: an error attributable to sampling variability—the luck of the draw—called *sampling error*, and an error attributable to poor sampling methods, termed *bias*. The bias of a nonrandom sample will usually introduce a large error in the estimation of p by $\hat{p}$, even if the sample is huge. There is no way of assessing such errors without information from sources other than the sample being used.

The error attributable to random sampling *can* be assessed. It is this sampling error on which we now focus our attention; and we assume that sampling is random, so that the formulas that are applicable to random sampling can be used.

The absolute error $|\hat{p} - p|$ will vary from sample to sample—it will be sometimes large, sometimes small. We take its r.m.s. average as a measure of the reliability of $\hat{p}$ as an estimate of p. This r.m.s. average is the standard deviation of the error, and is equal to the standard deviation of $\hat{p}$ itself (because $\hat{p}$ differs from the error by the constant p):

$$\text{r.m.s. average error} = \text{s.d.(error)}$$

$$= \text{s.d.}(\hat{p}) = \sqrt{\frac{pq}{n}}.$$

Unfortunately, in statistical practice p is not known; for if it were, there would be no estimation problem. So the s.d. of $\hat{p}$ cannot be evaluated—but it *can* be approximated, simply by replacing the unknown p with a good guess, $\hat{p}$. Doing so yields the *standard error*[†] of $\hat{p}$:

Standard error of a sample proportion, in estimating a population proportion p:

$$\text{s.e.}(\hat{p}) = \sqrt{\frac{\hat{p}(1 - \hat{p})}{n}},$$

assuming a random sample of size n.

(There would be a correction factor, as given earlier for the s.d., if the sampling is without replacement from a finite population.) The quantity $1 - \hat{p}$ estimates $1 - p = q$, so we shall call it $\hat{q}$ and write the s.e. as $\sqrt{\hat{p}\hat{q}/n}$.

Example 6-5 *Dropping a Tack*

A thumbtack was dropped on a hard wooden surface 100 times and landed with the point up 56 times. The probability p that it lands with point up at a single toss is estimated to be $p = 56/100$, with standard error

[†]Many people refer to the s.d. of p as the standard error, and to what we have termed standard error as *estimated standard error*.

$$\text{s.e.}(\hat{p}) = \sqrt{\frac{\hat{p}\hat{q}}{n}} = \sqrt{\frac{.56 \times .44}{100}} = .0496.$$

The estimate $\hat{p}$ = .56, or 56 percent, can easily be off by as much as 5 percentage points—either way (too large or too small)—or more, since the s.e. is only a kind of average or typical absolute error. ▲

Example 6-6 *TV Ratings*

Suppose that a random sample of 993 viewers shows that 199 are watching a particular TV program. Taking the sample proportion p = 199/993, or 20 percent, as an estimate of the population proportion watching that program is apt to involve an error. The standard error of p, an approximation to its s.d., is

$$\text{s.e.} = \sqrt{\frac{.20(1 - .20)}{993}} = .0127,$$

or 1.27 "percentage points." You should then not be surprised if the estimate of 20 percent were off by one or two percentage points. (But you would never have the occasion to register surprise, since the true error will never be known.)

The formula for standard error, which is valid for random samples, is still pretty good for a simple random sample of size 1000 or so from the millions in the population of viewers. In the case of the Nielsen ratings described in Chapter 1 (Examples 1-1 and 1-2), the viewers polled were *not* drawn as a simple random sample. The Nielsen organization claims that the formula $\sqrt{\hat{p}\hat{q}/n}$ applies as an approximation, but this is questionable. ▲

The accuracy of the sample proportion in estimating p, as measured by s.d.$(\hat{p})$, clearly depends on the sample size. The s.d. of $\hat{p}$ is inversely proportional to the square root of the sample size. Thus, to cut a standard error in *half*, the sample size would have to be *increased* by a factor of 4; to cut it by a factor of 3, one would increase the sample size ninefold; and so on.

Example 6-7

How do samples of size 100 and 2500 compare in providing an estimate of a population proportion?

Intuitively, one would expect that the larger sample would give the more reliable estimate of p; and, indeed, the sample proportion $\hat{p}$ for the larger sample has the smaller s.d. The denominator in the formula for s.d.$(\hat{p})$ is $\sqrt{100}$ = 10 when n = 100, and is $\sqrt{2500}$ = 50 when n = 2500. Thus, the s.d. of $\hat{p}$ when n = 2500 is only 1/5 of what it is when n = 100 (where 1/5 = 10/50 = $\sqrt{100/2500}$). ▲

The relation between sample size and standard error can help in the practical matter of setting the sample size. "How big a sample do I need?" is a question often asked by an investigator; and the answer depends on what kind of accuracy he wants—and can afford. If desired accuracy is specified as an amount of standard error, such as s.e.$(\hat{p})$ = .01 (or one percentage point), one can equate this to the expression that gives the standard error:

$$\sqrt{\frac{\hat{p}\hat{q}}{n}} = .01.$$

However, neither p *nor* the estimate $\hat{p}$ is known when the experiment is being planned. Still, something can be done. A sample size can be found so that, with it, the specified s.e. will not be exceeded. This is done as follows.

The quantity $\sqrt{\hat{p}\hat{q}} = \sqrt{\hat{p}(1 - \hat{p})}$ is at most 1/2 (its value when $\hat{p}$ = 1/2) and varies slowly when $\hat{p}$ is moderate, as seen in its graph in Figure 6-2. So for any given n, the largest possible standard error is $.5/\sqrt{n}$. Moreover, if one happens to be confident, a priori, that $p < .2$ (say), then the maximum of $\hat{p}(1 - \hat{p})$ is less than $.2 \times .8$, and the largest s.e. is less than $.4/\sqrt{n}$.

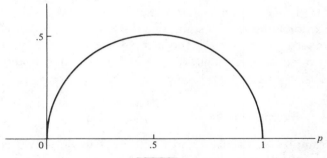

Figure 6-2.   Graph of $\sqrt{p(1 - p)}$.

Example 6–8   *How Large a Sample?*

Suppose that a standard error of at most .01 is required, in estimating p by $\hat{p}$ (from a random sample). Equating this to the largest possible value of the s.e. for given n:

$$.01 = \frac{.5}{\sqrt{n}},$$

we find $\sqrt{n} = 50$, or $n = 2500$. Using a random sample of size 2500 will guarantee that the standard deviation of the estimate $\hat{p}$ does not exceed .01; but, of course, it may turn out to be less. If 2500 observations is more than one can afford, then accuracy would have to be sacrificed. To ensure an s.e. of no more than .02, for example, one would need only one-fourth as many observations, or 625. But an even smaller sample might suffice if something were known about the range of possible values of p. Thus, if it happened to be known that $p > .9$, then the maximum s.e. is about $.3/\sqrt{n}$; and to achieve a standard error no larger than .02 one would need only 225 observations. ▲

6.3
THE LARGE-SAMPLE DISTRIBUTION OF $\hat{p}$

As we saw earlier, the calculation of probabilities for a sample proportion $\hat{p}$ can get rather tedious when the sample size is large. It should come as good news, then, that the large-sample case can be handled rather easily with the aid of a table of normal probabilities such as Table II of Appendix B. The way to use this table is prescribed by the following remarkable result:

Central Limit Theorem (special case):

> The sampling distribution of $\hat{p}$ in a large random sample is approximately *normal*, with mean p and standard deviation $\sqrt{pq/n}$.

This is one of the most important results in statistics, yet it is also one of the oldest. It was discovered and proved by DeMoivre (1667–1754) and extended by Laplace (1749–1827).

Whereas the central limit theorem gives a limiting distribution for $\hat{p}$ as the sample size becomes infinite, it does not address the question of just how good the normal approximation is in the case of a sample of given size. Although the question is important, its answer depends on the value of p (the population proportion) and on the sample size. Experience is helpful, but perhaps a rule of thumb is in order. For a population proportion between .1 and .9, a sample of size 50 will be adequate for the kind of accuracy that is usually desired (probabilities to a couple of decimal places). If p is close to .5, a sample of 10 or even less may be adequate. But if p is very small or very close to 1, as many as several hundred observations may be needed for a good approximation. A simple rule here is that when p is near 0 or 1, then n should be large enough so that npq exceeds 5.

Example 6–9

Suppose that p = .001 and n = 1000, so that $npq \doteq 1$. In many situations 1000 observations would be considered a large sample, but here the distribution of $\hat{p}$ is highly skewed. The probabilities for the possible values of $\hat{p}$ are, in part, as follows:

Sample Proportion	Probability
0/1000	.37
1/1000	.37
2/1000	.18
3/1000	.06

The average number of successes in 1000 trials is np = 1, and the distribution is decidedly nonsymmetric about 1/1000. It is far from normal. ▲

The central limit theorem asserts that the distribution of $\hat{p}$, which is discrete, can be approximated by a normal distribution, which is for continuous variables and has a *smooth* histogram. This may seem disconcerting until one realizes that the possible proportions in a sample of size n, which are $0/n$, $1/n$, $2/n$, . . . , n/n, become more and more dense on the interval from 0 to 1 as n increases. That is, $\hat{p}$

becomes more and more like a continuous variable. As it turns out, an excellent approximation is possible, if one is careful, even when the sample size is so small that the discrete histogram for $\hat{p}$ is poorly approximated by the smooth normal curve. We illustrate the approximation in the case of a sample small enough that Table I permits us to compare the exact and approximate probabilities.

Example 6-10

Consider a population with $p = 1/2$. In a random sample of size $n = 8$ there can be any number of successes from 0 to 8. The proportion $\hat{p}$ is a random variable whose values and probabilities (copied from Table I) are given in Table 6-1.

Table 6-1. Probabilities for $\hat{p}$

$8\hat{p}$	Probability
0	.004
1	.031
2	.109
3	.219
4	.273
5	.219
6	.109
7	.031
8	.004
Sum	1

This distribution is shown graphically in Figure 6-3. Anticipating the approximation by a continuous distribution, we redraw the graph with bars widened to form a histogram, in Figure 6-4. Next, in Figure 6-5, a normal curve with the same area is superimposed, one with the same mean and s.d. as $\hat{p}$:

$$\text{m.v.}(\hat{p}) = .5, \qquad \text{s.d.}(\hat{p}) = \sqrt{\frac{pq}{n}} = \sqrt{\frac{.5 \times .5}{8}} = .177.$$

(We are only interested in relative areas, so we may as well assume that the vertical scale is adjusted to make the total area 1 in each case. Then probabilities are not just proportional to areas, but equal to areas.) The area of a bar is seen to be approximately the area under that part of the normal curve over the same base interval. Thus,[†]

[†]We use the symbol $\doteq$ for approximate equality.

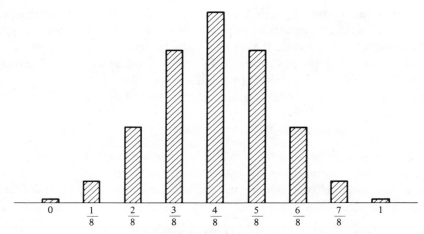

Figure 6-3. Distribution of the proportion of heads in eight tosses of a coin.

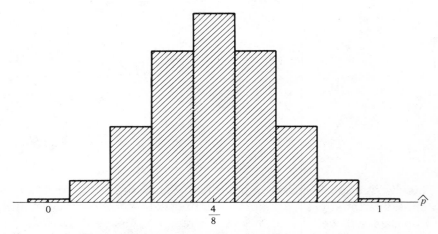

Figure 6-4. Histogram for the proportion of heads in eight tosses of a coin.

$P(\hat{p} = 3/8)$ = area of bar centered at 3/8

$\doteq$ area under normal curve between 2.5/8 and 3.5/8

$$= \Phi\left(\frac{.4375 - .5}{.177}\right) - \Phi\left(\frac{.3125 - .5}{.177}\right) = .214,$$

where the values of $\Phi(z)$ are obtained from Table II of Appendix B. This is not far from the actual value (.219) given in Table 6-1.

A more common kind of question in statistical applications is of this sort: What is the probability of three *or fewer* heads in eight trials? That is,

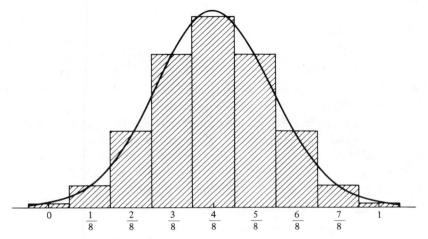

Figure 6-5. Normal curve approximating the histogram for the proportion of heads in eight tosses.

$$P(\hat{p} \leqslant 3/8) = ?$$

We answer this by taking the area under the approximating normal curve to the left of 3.5/8, which (as reference to Figure 6-5 will show) is nearly the same as the area of the leftmost four bars of the histogram for p:

$$P(\hat{p} \leqslant 3/8) = P(\hat{p} = 0/8 \text{ or } 1/8 \text{ or } 2/8 \text{ or } 3/8)$$

$$\doteq P\left(Z \leqslant \frac{3.5/8 - .5}{.177}\right) = \Phi(-.353) = .362.$$

Table 6-1 shows the correct value to be

$$.004 + .031 + .109 + .219 = .363. \quad \blacktriangle$$

In the last calculation of Example 6-10 it might have occurred to one to use this Z-score:

$$\frac{\hat{p} - E(\hat{p})}{\text{s.d.}(\hat{p})} = \frac{3/8 - .5}{.177} = -.706,$$

and finding the desired probability by looking this up in Table II. But doing this neglects half of the histogram bar corresponding to 3/8, thereby introducing a considerable error—because of the small n. Changing 3/8 to 3.5/8, called a *continuity correction*, gave a much closer approximation to the true probability. However, in cases in which n is large, say 50 or more, the continuity correction is not important (although it cannot hurt).

Example 6-11

The following kind of question will be encountered in analyzing sample results: If a population consists of 40 percent Democrats, what is the probability that a random sample of size 200 will include 50 percent or more (i.e., 100 or more) Democrats?

In mathematical notation, the question calls for an evaluation of $P(\hat{p} \geqslant .50)$ assuming that $p = .40$. This can be accomplished by using the approximate normality of $\hat{p}$. First, express the desired probability in terms of a probability to the *left* of .50 (since the normal table we are using gives areas to the left), and then calculate the Z-score corresponding to $\hat{p} = .50$. For this, we need the mean and s.d. of $\hat{p}$:

$$\text{m.v.}(\hat{p}) = p = .40$$

$$\text{s.d.}(\hat{p}) = \sqrt{\frac{pq}{n}} = \sqrt{\frac{.4 \times .6}{200}} = .0346.$$

And then

$$Z = \frac{.50 - \text{m.v.}(\hat{p})}{\text{s.d.}(\hat{p})} = \frac{.50 - .40}{.0346} = 2.89,$$

from which we find

$$P(\hat{p} \geqslant .50) = 1 - P(\hat{p} \leqslant .50)$$
$$\doteq 1 - \Phi(2.89) = .0019,$$

with the aid of Table II. (Observe that in standardizing we divide by the s.d. rather than the s.e., which is only used when the s.d. is not known. Here we are assuming that $p = .4$, so the s.d. can be computed.)

This computation was just a bit cavalier, inasmuch as the event $\hat{p} \geqslant .5$ is not quite the opposite of $\hat{p} \leqslant .5$. The overlapping point, $\hat{p} = .5$, has a very small probability, since $\hat{p}$ is a nearly continuous variable when n is as large as 200. A more precise approach would employ a continuity correction. Thus, the opposite of $\hat{p} \geqslant .5$ (100 or more Democrats) is 99 or fewer Democrats: $\hat{p} \leqslant 99/200$. And then

$$Z = \frac{99.5/200 - .40}{.0346} = 2.82$$

gives the desired probability as .0024, not appreciably different from the .0019 found earlier. ▲

Example 6-12

Suppose that we plan to toss a fair coin 100 times and ask if you want to bet on the numbers of heads from 45 to 55, or on all the other numbers. How would you bet? Some people feel (and they are not easily dissuaded!) that since there are 11 numbers from 45 to 55 and 90 numbers outside that range, fair odds would be 11 to 90, or about .12 to 1. The corresponding probability is 11/101. As a matter of fact, the range 45 to 55 has considerably more probability than all the other numbers. For, the probability of a number from 45 to 55 is

$$P\left(\hat{p} \leqslant \frac{55}{100}\right) - P\left(\hat{p} \leqslant \frac{44}{100}\right)$$

$$\doteq \Phi\left(\frac{.55 - 50}{.05}\right) - \Phi\left(\frac{.44 - .50}{.05}\right)$$

$$= \Phi(1.00) - \Phi(-1.20) = .7262.$$

[The .05 in the denominators is s.d.$(\hat{p}) = \sqrt{.5 \times .5/100}$.] We have not bothered with a continuity correction because n is large, and because the errors introduced in neglecting it are nearly canceled out anyway, in the process of subtraction.

So fair odds on the range 45 to 55 would be about 73 to 27, or 2.7 to 1—a far cry from .12 to 1! ▲

Example 6-13   *The .400 Hitter*

One often hears talk of the possibility that a major league baseball player will bat better than .400 over a season. This would mean that he must get a hit at least 40 percent of, say, 540 times at bat during the season. Suppose that a player has a 33 percent chance of getting a hit each time he is at bat. This makes him quite an exceptional hitter, for only one active player has a lifetime batting average as large as .330. What is the probability that a .330 hitter gets at least 216 hits in 540 times at bat (or 40 percent)?

This can be calculated if it is assumed that the 540 at-bats are independent trials, with probability .33 of success at each trial. It is approximately the area to the right of .40 under the normal curve with mean

$$m.v.(\hat{p}) = p = .33$$

and standard deviation

$$\text{s.d.}(\hat{p}) = \sqrt{\frac{pq}{n}} = \sqrt{\frac{.33 \times .67}{540}} = .0202.$$

Thus,

$$P(\hat{p} \geqslant .40) = 1 - \Phi\left(\frac{.40 - .33}{.0202}\right) = .0003.$$

Small wonder that none of the presently active players has accomplished this feat.

To be sure, the assumptions of independence of the trials and constancy of the p are questionable. But they may be close enough to reality that the given calculation is of the right order of magnitude. ▲

6.4
CONFIDENCE LIMITS FOR p

In reporting an estimate of a population proportion, it is helpful to include some idea of the accuracy of the estimate. At the very least the sample size should be given, but giving the standard error would provide the reader of the report with a more directly useful measure of error. Sometimes an estimate will be given in the form $\hat{p} \pm$ s.e., that is, the sample proportion as the estimated value of p, plus or minus the standard error. The "$\pm$" suggests, correctly, that the estimate could be in error on either side—too high or too low.

Suppose that in a particular sample it is found (say) that $\hat{p} = .40$, with s.e. = .02. Giving the estimate of p as .40 $\pm$ .02 seems to imply that p is between .38 and .42. This may nor not be so—we cannot know unless we know p! Now suppose that another investigator (independent of the first) took a sample of the same size and found that $\hat{p} = .45$, also with s.e. = .02; his estimate of p would be .45 $\pm$ .02, suggesting that p is between .43 and .47. Which investigator, if either, is right?

Each new sample would yield different "limits," $\hat{p} \pm$ s.e., and whether or not the limits from a given sample include the actual p depends on how close the observed $\hat{p}$ is to p. If the distance from $\hat{p}$ to p is less than the standard error,

$$|\hat{p} - p| < \text{s.e.,}$$

then the limits $\hat{p} \pm$ s.e. *will* include p; otherwise, they will not (see

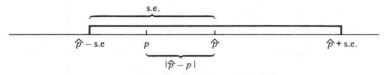

Figure 6-6. Interval centered at $\hat{p}$ that includes p.

Figure 6-6). What are the chances that this will happen with a particular sample?

We can answer this question if the sample size n is at least moderately large—large enough so that the sample proportion $\hat{p}$ is approximately normal. For, the probability that $|\hat{p} - p| <$ s.e. is approximately equal to the probability that $|\hat{p} - p| <$ s.d.$(\hat{p})$, since s.e.$(\hat{p})$ is an approximation for s.d.$(\hat{p})$. And the latter probability is just the probability that an approximately normal variable takes a value within one s.d. of its mean, found by referring to Table II:

$$P[|\hat{p} - p| < \text{s.d.}(\hat{p})] = \Phi(1) - \Phi(-1) = .68.$$

So the answer to the earlier question of which sampler is right is that both of them determined an interval, from a sample, with limits $\hat{p} \pm$ s.e.; and each had a 68 percent chance that his $\hat{p}$ would be close enough to p that the interval from $\hat{p} -$ s.e. to $\hat{p} +$ s.e. would include the true value of p. But we usually will not know which one, if any, managed to succeed in this—indeed, they could both be wrong. But each has a certain degree of confidence that his interval has trapped the value of p, and the 68 percent is a measure of this confidence.

The idea of an interval that has a specified chance of covering the actual population proportion is an attractive one, but most statisticians feel[†] that 68 percent is not large enough. To increase one's confidence, the interval would have to extend more than one s.e. on either side of the point estimate $\hat{p}$. Extending it to *two* s.e.'s on either. side, for instance, increases the chances to over 95 percent:

$$P[\hat{p} - 2(\text{s.e.}) < p < \hat{p} + 2(\text{s.e.})] = P[|\hat{p} - p| < 2(\text{s.e.})]$$

$$= \Phi(2) - \Phi(-2) = .9544.$$

To make this probability exactly equal to .95, a nice round number, the 2 is usually reduced to 1.96, the 97.5 percentile of the standard

[†]The matter is of some practical interest. For example, *TV Guide* (as reported in the article quoted on page 1 of this book) went so far as to contact "a dozen top research companies to ask about their use of the 68 percent level." Not a single one ever used it. Also contacted was the chief of statistics at the Bureau of the Census headquarters in Washington: "In our work we always use the 95 percent level," he said.

normal distribution. The resulting limits are called 95 percent *confidence limits* for *p*.

Large-sample, 95 percent *confidence limits* for a population proportion:

sample proportion ± 1.96 × standard error.

The interval defined by these limits is a 95 percent *confidence interval:* from $\hat{p} - 1.96$ s.e. to $\hat{p} + 1.96$ s.e.

It is possible, naturally, that one might not be content with a 95 percent "level of confidence." To have a higher chance of obtaining a sample whose proportion *p* is close enough to the actual *p* so that $\hat{p} \pm k$(s.e.) *covers p*, the value of the multiplier *k* must be increased—the interval widened. Some values of "confidence level" and corresponding multiplier *k* are as follows [found with the aid of Table II, and repeated as Table II(c)] :

Confidence Level	Multiplier of Standard Error
.80	1.28
.90	1.645
.95	1.96
.99	2.58

Example 6-14   *TV Ratings (more)*

A TV rating of 24 based on a random sample of 993 viewers—that is, a sample proportion of 24 percent—has a standard error

$$\text{s.e.} = \sqrt{\frac{.24 \times .76}{993}} = .01355.$$

The corresponding 95 percent confidence limits for the proportion of the population viewing the program are then

$$\hat{p} \pm 1.96(\text{s.e.}) = .24 \pm 1.96 \times .01355 = .24 \pm .027,$$

and the 95 percent confidence interval (as computed from *this* sample) is

$$.213 < p < .267.$$

Now either the actual *p* *is* in this interval or it is *not*. *No one will ever know!* What we do know is that we used a procedure and formula that have a 95 percent chance of producing an interval that covers the actual *p*. ▲

If a level of confidence is specified along with a desired width of a confidence interval, a sample size can be found that will ensure that these requirements are met. For, the specified level and width determine the standard error, and we have already seen how to choose the sample size to meet a specification for standard error. The next example illustrates the determination of a sample size.

Example 6-15   *How Many to Poll?*

Suppose that we want a 99 percent confidence interval for the proportion of voters favoring a certain candidate that is only 3 percentage points wide—extending 1.5 points on either side of the sample proportion. How large a sample should be polled, assuming random sampling?

The stated criterion for the width of the confidence interval is that

$$.015 = 2.58(\text{s.e.}) = 2.58 \sqrt{\frac{\hat{p}(1 - \hat{p})}{n}}.$$

Square both sides, we have

$$(.015)^2 = (2.58)^2 \frac{\hat{p}(1 - \hat{p})}{n},$$

or [after multiplication by *n* and division by $(.015)^2$]

$$n = \left(\frac{2.58}{.015}\right)^2 \hat{p}(1 - \hat{p}).$$

The right side is largest at $\hat{p} = 1/2$, so a sample of size

$$n = \left(\frac{2.58}{.015}\right)^2 \times \frac{1}{2} \times \frac{1}{2} = 7400$$

will make sure that the confidence interval is no wider than the specified 3 percentage points.

(Usually, national polls are satisfied with half of this accuracy, that is, with twice the confidence interval width. For this, they need only one-fourth as many observations, or 1850, assuming a random sample.)

If something is known about the possible values of *p*, one might get by with a somewhat smaller sample, as we saw in the case of standard error. For instance, if the population proportion of voters favoring candidate A is at least 65 percent, then

$$\text{s.d.} = \sqrt{\frac{pq}{n}} \leqslant \sqrt{\frac{.65 \times .35}{n}}.$$

Setting the rightmost expression equal to .015, we find that

$$n = \left(\frac{2.58}{.015}\right)^2 \times .65 \times .35 = 6730,$$

as compared with the 7400 called for when nothing at all is known about p. This sample size of 6730 is conservative in that if the actual p is greater than 65 percent, the width of the 99 percent confidence interval will not exceed 3 percentage points. ▲

6.5
MORE ON SAMPLE SURVEYS

"Suddenly they are everywhere," states an Associated Press feature story in 1979. "Public opinion polls pervade the nation, affecting its policies and its politics." It goes on to say:

The federal government uses them in its decisionmaking. Political candidates use them in election campaigns. And "what the polls show" on a wide range of issues, from the arms race to abortion, becomes central to public debate.

Yet, the man on the street is suspicious. According to a *Newsweek* article in 1972,

For some Americans polling always has—probably always will— seem a dubious enterprise: They simply cannot believe that by sampling the views of, say, 1200 voters it is possible to discover what 120 million think. As it happens, there are sound statistical reasons why this can be done to a reasonable degree of accuracy.

But despite the "sound statistical reasons" why it *can* be done, it is not always done. Polls are not infallible, and different polls on the same topic do not always agree.

A polling consultant to *The Washington Post* is quoted in the *Associated Press* report as saying, "What really differentiates one poll from another is the . . . artistry." Pollsters talk about sampling error because errors in the art cannot be measured.

Biases, or errors, in polls other than sampling error, can be introduced by nonrandom aspects of the sampling process, as discussed in Chapter 1. Time lag is another source of bias. A poll takes from a couple of days to 2 weeks or more to conduct, and voting preferences change over night; so a poll conducted August 26 and 27, say, although reasonably accurate at that point, may be quite inaccurate as a measure of opinion or preference in October or November. One pollster estimated that 34 percent of the voters switched positions between one poll and the next, in one case.

Another source of error—and a reason opinion polls can differ—lies in the wording of a question. Sometimes changing a single word can reverse the results.

Example 6-16   *To Pray or Not to Pray*

In a widely published news story, George Gallup declared that three out of four Americans favor a constitutional amendment to permit prayers in public schools. The question asked was worded as follows: "Do you favor or oppose an amendment to the Constitution that would permit prayers to be said in the public schools?" The question suggests that at present prayers are not allowed in public schools.

A poll commissioned by Americans United for Separation of Church and State asked this question: "Since the Supreme Court has upheld the right of voluntary prayer, while prohibiting only government sponsored worship activities, should the Constitution be be changed to authorize government sponsored prayer in public schools?" The response to this question was 33.3 percent "yes," and 59 percent "no." No matter which side of this issue one supports, it should be apparent that the wording of the question can make a dramatic difference in the response. ▲

Interview techniques are a factor in the accuracy of a poll. Interviews are done, usually, either by telephone or in person—the latter being five or six times as costly (something like $20 against $3 to $5, in 1981). Telephone calls produce a higher number of "undecided" responses, and as discussed earlier, there may be some selection bias. "Interviews" depending on mailings will generally involve an appreciable response bias, which can be lessened by follow-up calls or visits, or compensated as might be determined by follow-up studies of nonrespondents.

With regard to *sampling* error, which *can* be measured, we have seen how the accuracy of a poll depends on the number of interviews—but *not* on the population size or on the ratio of sample size to population size. Most polls use samples of sizes between 500 and 2000 and (says *Newsweek*) are "considered accurate within three or four percentage points." The formula $.5/\sqrt{n}$ has been seen to provide an upper limit on the standard error of a proportion if the sample is a random sample.

Example 6-17   *Voting for Governor*

A poll was conducted one July to determine voter sentiment in regard to a gubernatorial election the following November. It showed the Democratic candidate leading the Republican candidate by 49 to 45 percent (with 6 percent undecided), overall, and by 53 to 42 percent in union households. It was based on interviews with 610 men and women of voting age. The newspaper account states:

> Results of such surveys are subject to sampling error. For a random sampling this size, it is possible to say that the error will not exceed four percentage points either way. Since this sample is taken only from households with telephones, the error may be larger than for a completely random sampling. For a subsample of the population—for example, the 187 people living in union households—the error could be larger.

What does the statement about "four percentage points" mean? The standard error of a sample proportion for $n = 610$ and an observed sample proportion $\hat{p} = .49$ is

$$\text{s.e.} = \sqrt{\frac{.49 \times .51}{610}} = .02024.$$

Thus, four percentage points would be two standard errors, and the error *can* exceed two standard errors, although only about 5 percent of the time. Notice that with $n = 187$, the standard error will be $\sqrt{610/187} = 1.8$ times as large.

The proportion p being estimated here is the population proportion who favor the Democratic candidate. It is not a good predictor of the final election results, because it does not take into account those who are undecided in July. Sometimes an attempt is made to assess the undecideds to determine how they will divide in November. The 6 percent of undecideds (as estimated by the poll) can clearly throw the election one way or the other. ▲

Example 6-18   *Republicans on the Rise?*

A newspaper headline (*Minneapolis Tribune*, 1978) read "IR Party shows most followers in 4 years." The article thus headlined reported on interviews (obtained by a sampling method not described) of 1225 men and women 18 years old and over, during 4-day periods in May and June. It was said to provide "an approximation of the response that could be expected if all adult Minnesotans had been interviewed." The sampling error was stated as not exceeding four points either way, and we observe that with a random sample of this size, the s.e. will be found as follows:

$$\text{s.e.} \leqslant \frac{.5}{\sqrt{1225}} = .0143.$$

The results of the poll showed 25 percent IR (Independent Republican) in that year, and this was compared with figures for previous years:

1974	20 percent
1975	21 percent
1976	20 percent
1977	23 percent
1978	25 percent

There may be a trend, as the headline suggests; yet it is noteworthy that if the actual proportion had remained *constant* at $p = .22$, then all five observed percentages would be within the "four points either way" stated as limits of accuracy.

The poll also gives percentage breakdowns with respect to other variables—for instance, religion:

	IR	DFL	Independent and Other
Protestants	32	34	34
Catholics	16	51	33

These percentages may involve greater errors (as the report points out), the Protestants in the poll numbering only 668. ▲

We have mentioned, here and in Chapter 1, various possible sources of bias that should be watched for in interpreting a poll. It should be remembered, however, that even if no such biases are evident, the formulas we have given for standard error apply to random sampling, with or without replacement. For practical reasons other methods are often used, methods that involve randomness in part or in stages. For instance, in *cluster sampling* the population would be divided into subgroups—such as households, blocks, counties, and so on—and a random sample of subgroups is chosen. Then, each subgroup is surveyed, either completely, or partially according to some system (e.g., every fourth individual), or partially according to some probabilistic method. The standard error is not as simple as for simple random sampling; in some instances it may be approximately the same, but usually it could be greater, by a factor of as large as 1.5 (or more). You cannot assume that $\sqrt{\hat{p}\hat{q}/n}$ applies automatically.

Example 6–19   *Nielsen Again*

The operational sample for the Nielsen Television Index (NTI) is obtained by sampling a "Master Sample" of some 3000 households. The Master Sample was obtained as follows:

1. The 3070 counties in the United States provided 620 sample counties:
   (a) All 357 "standard" metropolitan counties with a population of at least 240,000.
   (b) 263 groups of counties (averaging 10 counties each), represented by one county for each group.
2. Districts (of a type used in the U.S. Census) were selected systematically within each sample county, with a random start.
3. Blocks or "pseudo-blocks" were chosen from each district, randomly, with probabilities proportional to the numbers of housing units.
4. Sample housing units were selected from the segments obtained in the third stage, generally one per segment. (Additional units were listed as substitutes in the event of refusal to cooperate.)

This is not a random sample. Whether error formulas for random samples apply is beyond our scope. Nielsen asserts that $\sqrt{\hat{p}\hat{q}/n}$ gives, at least approximately, the s.e. for a single telecast.

Figure 6-7 is taken from the *TV Guide* article quoted on page 1.

Figure 6-7. Nielsen ratings, 1977. (From, *TV Guide*, No. 25, 1978.

	NIELSEN AVERAGE RATINGS	

(Top 40 regular shows, Sept. 11 through Dec. 10, 1977)

Rank	Program	Average Rating
1	Laverne & Shirley	31.6
2	Happy Days	31.0
3	Three's Company	26.5
4	Charlie's Angels	25.8
5	All in the Family	25.4
6	Alice	23.8
7	60 Minutes	23.7
8	NBC Monday Movie	23.0
9	On Our Own	22.2
10	Little House on Prairie	22.1
11	ABC Sunday Movie	21.6
12	Eight Is Enough	21.5
12	Rhoda	21.5
12	Soap	21.5
15	Monday Night Football	21.2
16	M*A*S*H	21.1
17	One Day at a Time	21.0
18	Barney Miller	20.9
19	Six Million Dollar Man	20.4
20	The Love Boat	20.3
20	What's Happening!!	20.3
20	Barnaby Jones	20.3
20	Welcome Back, Kotter	20.3
24	The Big Event	20.1
25	Donny & Marie	19.8
26	Family	19.5
27	ABC Friday Movie	19.4
28	Hawaii Five-O	19.3
29	Baretta	19.1
30	Starsky & Hutch	18.9
31	CBS Sunday Movie	18.7
31	Carter Country	18.7
31	The Betty White Show	18.7
34	NBC Saturday Movie	18.6
35	The Waltons	18.3
36	CBS Wednesday Movie	18.1
37	Good Times	17.9
38	World of Disney	17.7
38	Tabitha	17.7
40	Maude	17.6

Using Nielsen's formula, the author has calculated that the sampling error over a 13-week period for a show with a rating of 20 is plus-or-minus 1.8 points. Thus, The Six Million Dollar Man, ranked 19th on the chart, could actually be as high as ninth place or as low as 34th.

The shaded segment is essentially a 95 percent confidence interval: $\hat{p} \pm 1.96$ s.e., based on a standard error that is smaller than $\sqrt{\hat{p}\hat{q}/n}$ by a factor computed from the number of weeks averaged (13) according to a Nielsen formula.

According to *TV Guide*, Nielsen tells its users that the sampling error is 1.3 points for a single telecast with a rating of 20. This is an s.e.—a "typical" error—and is in no sense a limit on the error. (The interval with limits $\hat{p} \pm$ s.e. would have only a 68 percent chance of covering the true p. A Nielsen official, as quoted in *TV Guide*, stated that "the reason we use the 68 percent level is that it makes it very convenient for a user.")

Nielsen does say that studies of response error, nonresponse error, achieved samples, and the like, have been made "from time to time" and supplied to interested clients.

A Ford Foundation executive offered to sponsor a study of audience measurement. Network heads, after discussions, said "What's the matter with the system, anyway? It works fine." (Quoted in *TV Guide*.) They get rating numbers, and advertisers accept them. ▲

Example 6-20 *Slide in Presidential Popularity*

A chart in the July 17, 1978, issue of *Newsweek* (titled "The Long Slide") showed results of three polls concerning President Carter's popularity, month by month over a period of 14 months, starting a few months after he first took office. The Gallup, Harris, and CBS-New York Times polls each reported a percentage of people who approved of the president's performance. The *Newsweek* chart is actually misleading, using 30 percent as a baseline and so accentuating the extent of the "slide." Our Figure 6–8 represents the same data but plotted relative to a baseline of 0 percent.

The CBS-New York Times poll seems the smoothest, and one might speculate that it is the most accurate. However, a closer inspection reveals that the data from this poll are given only quarterly. If we had plotted the other two polls using only the quarterly percentages, they would also look smoother. In any case, we cannot know which polls had the greater errors, since we do not have the corresponding population figures for comparison. Their reliabilities could be compared if we had the sample sizes—as well as some assurance that the samples were taken randomly, so that our standard error formulas could be applied. ▲

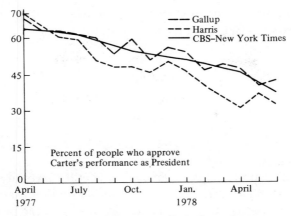

Figure 6-8. Poll results on Carter's performance. (Redrawn from *Newsweek*, July 17, 1978.)

Example 6-21   *The Call-Back Adjustment Caper*

In 1978, a newspaper's polling organization proceeded as follows. In the period September 21 to 24, 613 likely voters were interviewed; between October 16 and 18 they called back as many of the 613 as possible, obtaining responses from 403; and from November 3 to 5 they again called back the original sample, obtaining 371 responses. Percentages in favor of the Democratic candidate for senator were as follows:

September: 43.8 percent.

November: 43.9 percent.

It was found, however, that only 40.8 percent of the 371 persons responding in November had expressed preference for the Democratic candidate in September.

The newspaper editors took the difference 43.9 – 40.8 as an estimate of shift in sentiment toward the Democratic candidate. They then constructed a November estimate of his share by adding 3.1 percent to the September percentage:

$$43.8 + 3.1 = 46.9.$$

This was reported as the November percentage in favor of the Democrat. A similar adjustment was made to the September percentages in the other categories, and the results were reported in the newspaper the day before the election as November figures:

Democrat	47
Republican	46.5
Other	0.5
Undecided	6.

How reliable are these as estimates of population sentiment? Can one compute a standard error?

The method of polling makes analysis difficult. Certainly, one *cannot* apply the standard error formula meant for random samples. The election results were as follows, listed together with the polling organizations unadjusted percentages:

		November Poll	
	Election	Actual	Published (Adjusted)
Democrat	41.5	43.9	47
Republican	57.0	51.4	46.5
Other	1.5	4.7	6.5

Because the actual vote, resulting in victory for the Republicans, differed so markedly from the published figures, which predicted a Democratic victory, there was an outcry (in "Letters to the Editor") charging the newspaper with deliberate distortion. Moreover, the research organization that actually did the survey claimed that the newspaper jeopardized the organization's integrity by doctoring the data. The newspapers lengthy response was, in effect, that the adjustment method had yielded good results in the past. However, the editors hired an independent team of consultants to study the matter, and a report was published at the end of February 1979.

The consultants concluded that the newspaper had made an honest blunder in its use of adjusted double call-back data. They admonished the paper for not keeping up with other polling organizations, after a relatively pioneering start 30 years earlier, and for an unimaginative routineness in the reporting. They recommended that the poll consider "buying a little extra insurance" by going to a sample size of 1000. (This would reduce the standard error from a maximum of .020 to .016—"little," indeed.)

The newspaper published no polls prior to the 1979 fall elec-
tions. ▲

Perhaps an earlier caution is worth repeating here:

Biases—errors other than that encountered in random
sampling—are *not* assessed as part of the standard error.
There is no way of assessing them without information
other than that contained in the sample being used.

KEY VOCABULARY

Population proportion
Sample proportion
Sampling distribution
Estimation
Error of estimation
Absolute error
Standard error
Central limit theorem
Continuity correction
Confidence interval
Confidence level
Sampling error

QUESTIONS

1. In what sense is a "population proportion" a probability?
2. Every statistic defined for a random sample has a "sampling
   distribution." What is it that is "distributed?"
3. Why do we consider it essential that sampling be done at random, for purposes of statistical estimation?
4. What means do we have for finding probabilities for the sample
   proportion $\hat{p}$ as a random variable?
5. If a sample is obtained by methods involving response bias or
   other types of bias, is it possible to correct for these biases?
6. Does the "standard error" give one limits on the sampling error?

7. How is it possible to determine the sample size needed, for a specified degree of accuracy, when the accuracy of estimation depends on the very parameter (p) being estimated, which is unknown?

8. Why do we use the term *confidence interval* rather than *probability interval*?

9. Does widening a confidence interval (for given sample size) make you more or less confident that the interval includes the true parameter value?

10. What can be done about "time-lag" biases in an election poll?

11. Why are polls often content with samples of size 500 to 2000?

12. What kinds of error are included in the notion of "standard error?"

13. Which is more important, the sample size (value of n) or the *relative* sample size (the ratio of n to the population size)?

PROBLEMS

Sections 6.1 and 6.2

*1. A jury panel of 25 persons is selected at random from a large pool of potential jurors which is 20 percent black and 80 percent white.
   (a) Find the mean and s.d. of $\hat{p}$, the proportion of blacks selected for the panel.
   (b) Convert the answer to part (a) to the mean and s.d. of the *number* of blacks among the 25 selected.

2. A roll of 50 pennies is opened and dumped on a table. Assuming this experiment to be equivalent to tossing the 50 coins in independent tosses, determine the mean and s.d. of the proportion of heads among the 50 pennies.

*3. Obtain 50 samples of size five from the chip-in-bowl model for the toss of a coin, by tossing five coins 50 times. After each toss of the five coins record $\hat{p}$, the proportion of heads among the five.
   (a) Did you get the same proportion each time? (Should you?)
   (b) What is the "expected" or mean value of the sample proportion? What is the expected or mean *number* of heads in one toss of the five coins?
   (c) Calculate the average of your 50 $\hat{p}$-values (using the usual

formula for a sample mean); this should be reasonably close to the mean in part (b), which applies to the $\hat{p}$-values in infinitely many samples.

(d) Calculate the s.d. of your 50 p-values; this should be reasonably close to the ideal s.d. of $\hat{p}$ as given by the formula $\sqrt{pq/n}$, which applies to the $\hat{p}$-values from infinitely many samples.

4. Each page of Table VIII of Appendix B has 50 blocks of 50 random digits. It is fairly easy to count the 0's in a block of 50.
   (a) What is the expected or mean value of $\hat{p}$, the proportion of 0's in a block of 50 random digits?
   (b) Count the 0's in each of the 50 blocks on a page and calculate the average of the 50 proportions of zeros.
   (c) Compare the ideal s.d. of $\hat{p}$ and the s.d. of your 50 $\hat{p}$-values.

*5. A random sample of 2400 voters shows 60 percent in favor of a proposal to cut taxes. What is the standard error in this sample proportion as an estimate of the proportion in favor in the population?

6. A 1978 poll of 1000 Florida families divided as follows on a question of casino gambling:

For	46 percent
Against	47 percent
Undecided	7 percent

   (a) Determine the standard error of the proportion in favor, assuming that the sampling was random.
   (b) Would it be safe to predict defeat of the measure to permit casino gambling if a referendum had been held shortly after this poll?

*7. A coin is bent, so that one might suspect (lacking the argument of symmetry) that the probability of heads is not .5. How many tosses would be needed so that the standard deviation of the sample proportion of heads does not exceed .02?

8. A poll is to be taken to determine the proportion of city voters who will vote for a certain candidate for mayor.
   (a) How large a sample would be needed to ensure that the standard deviation of the estimate does not exceed .015?
   (b) A random sample of 750 shows that 42 percent will vote for the candidate. What is the standard error of estimate?

*9. *McCall's* magazine invited its readers in January 1978 to fill out a questionnaire on their religious beliefs, practices, and values. As of May, 60,000 readers had replied. "If there are any doubts that the U.S. is in the midst of a religious revival," states a report in the May issue, "*McCall's* statistics should dispel them. Nine out of ten respondents believe in God, and more than two out of three pray every day."

(a) How accurate are the estimates 9 out of 10 and 2 out of 3?

(b) *Do* the statistics dispel doubts of a revival?

Sections 6.3–6.5

*10. In the experiment of tossing 50 pennies (Problem 2), find the probability that more than 30 heads turn up.

11. Find the probability that there will be more than 72 sixes (or 20 percent) in 360 tosses of a fair die.

*12. A standard normal variable is equally likely to be positive as negative.

(a) Find the probability that six or more of eight observations are positive, using Table I of Appendix B.

(b) Use the approximate normality of $\hat{p}$ to obtain an estimate of the probability in part (a) by use of the normal table (Table II).

13. Use the normal table to approximate the probability of 10 heads among 20 tosses of a coin. (HINT: The bar representing this probability would cover the interval from 9.5 to 10.5.)

*14. Determine a 95 percent confidence interval for the proportion of Floridians, in Problem 6, favoring casino gambling, as of the date of the poll, assuming random sampling. (As it turned out, the election defeated the proposition by 2 to 1.)

15. Determine 95 and 99 percent confidence intervals for the proportion of voters favoring a tax-cut proposition, based on the sample in Problem 5, which showed 1440 out of 2400 to be in favor.

*16. A small box containing 100 tacks was dumped by one of the authors onto a hard tabletop, 25 times. The number of tacks with point up was counted each time, and the total was 1784.

(a) Give a point estimate of p, the probability that a tack lands with point up, together with a standard error.

(b) Give a 99 percent confidence interval for p.

17. To obtain an estimate of voter sentiment that is accurate within 1 percentage point with 95 percent confidence, how large a sample would be required, assuming random sampling?

*18. In estimating the probability of a "six" with an ordinary, but perhaps crudely fashioned, die, how many tosses would be needed to obtain a 99 percent confidence interval of the form $\hat{p} \pm .001$?

Review

19. Seventy-six percent of 21,500 high school student leaders responding to a national survey said they had never had sexual intercourse, according to a 1978 Associated Press report. Questionnaires were mailed to 50,000 students chosen at random by Educational Communications, Inc., from *Who's Who Among American High School Students* (which it publishes), a listing of 318,000 junior and senior high school students recommended for listing by principals, counselors, or youth organizations.

(a) Can you obtain an interval estimate of the proportion of all high school students who have had sexual intercourse? If so, do it, and if not, explain why not.

(b) Could you do it for the proportion of all students listed in the directory? Why or why not?

(c) What inference can be made from the fact that 81 percent of those polled said they are members of an organized religion?

(d) Was the money for the poll wisely spent?

20. The Gallup organization telephoned a national sample of 750 Americans for *Newsweek* magazine. The article asserted: "The margin of error in the survey's findings is plus or minus 4 percentage points." The survey included a number of questions about taxation, and this statement on error was given as applying overall.

(a) Assuming the sampling to be random, determine the maximum standard error of a proportion based on a sample of this size, and determine the probability that the error exceeds ±4 percentage points.

(b) Sixteen percent of the respondents said they objected most to the sales tax (among the different types of tax they paid). Give the standard error for this sample proportion, and a 95 percent confidence interval for the corresponding population proportion.

21. A statistics professor split his class for an examination, asking those with last names beginning with A to L to go to one room, and with M to Z to another. He did not have a class list, but had reason to believe that 120 students would be taking the exam, and prepared 140 copies. The student directory revealed 94 pages of names from A to L and 78.5 pages in the range from M to Z.

(a) How should he divide the 140 papers between the two rooms (which were in separate buildings several blocks apart)?

(b) If he divides them according to part (a), what is the probability that he will run out of exam papers in the A–L room?

22. In Problem 21 of Chapter 4 (which refers to Example 1-16, page 24), the probability of a "yes" response to a question about extramarital affairs or to a question about phone numbers was computed to be

$$P(\text{yes}) = \frac{p}{2} + \frac{1}{4},$$

where p is the proportion in the population who have had an affair.

(a) If 40 percent of the sample answer "yes," what would you estimate p to be?

(b) If the 40 percent figure was obtained for a random sample of size 500, what would be the standard error of your estimate of p?

23. The box of tacks used in Problem 16 was dumped 25 times by the author's son. He obtained a total of 1685 with points up. As in Problem 16, determine point and interval estimates of p based on these trials. (In view of the results, do you think father and son were doing the same experiment?)

24. A research study conducted at the Hebrew University of Jerusalem involved 3658 infants born to women who practiced "middah" (a ritual abstinence at certain times), and who said they conceived during the five days around ovulation.

(a) Of the infants involved, 53 percent were boys. (In the United States about 51 percent are boys.) Construct a 95 percent confidence interval for the proportion of male births among such women.

(b) Of the 145 women who said they became pregnant 2 days after ovulation, 66 percent gave birth to males. Given a random sample of size $n = 145$, determine a 95 percent confidence interval for the population proportion if the corresponding sample proportion is 66 percent. In view of this result, would you believe that the 145 women constitute a random sample from the 3658, or rather that there is something special about conception two days after ovulation?

25. The Harris Poll of July 6, 1980, reported that in a sample of 1492 individuals, 53 percent opposed a tax reduction. Construct a 95 percent confidence interval for the population proportion opposed to a tax reduction, assuming that the sample was a random sample.

26. Continuing the story of Example 6–21: In June 1980, the Minnesota Poll reported on their methods as follows: A computer program was used to select Minnesota phone numbers at random. (The Public Service Commission estimated that 96 to 98 percent of state households have telephones.) Within each household, the particular respondent would be determined in a "statistically unbiased fashion," with no substitutions allowed. Hard-to-reach respondents were called as many as six times, and interviews were arranged at the convenience of the respondent. As to accuracy: "In samples of this size— more than 1200 interviews—the margin of error is estimated at 3.6 percent above or below the figures cited in the poll report."

   (a) Assuming that the sample is a random sample, what would the 3.6 percent figure imply about the meaning of "margin of error"?

   (b) Does choosing one respondent from each household of a random sample of households produce a random sample of respondents? (In particular, does everyone have the same chance of being included in the sample?)

7
Estimating a Mean

IN CHAPTER 6 we studied the problem of estimating a population characteristic—the proportion of individuals in a given category—by using the information in a sample from the population. When it is a *numerical* measure of each individual that is of concern, the *mean* of these numbers over all individuals in the population is a population parameter of common interest. The mean serves to locate the population along the scale of values. The statistical problem we take up now is the estimation of the mean of a population based on the information in a sample drawn from the population.

Example 7-1 *Accounting*

The "book value" of an account receivable of a large company may or may not agree with the actual value, owing to errors in the computerized records. The company would like to know the average error per account, but to determine the error in an individual account requires an audit—a check of original invoices and records of payment. Because this is costly and time-consuming, the company might be willing to settle for an estimate based on a sample of accounts that are subjected to individual audit. Once an estimate of the average error is obtained, an estimate of the total error can be found by multiplying the average by the number of accounts.

If it could be assumed that the company's operations are stable, it might also be reasonable to think of the population of all accounts receivable, present and future, and to consider the sample of present accounts as a sample from this larger population. ▲

The mean is especially useful in contexts where one might be interested in a population total, as in Example 7-1. The total is just the mean value per individual times the number of individuals. However, the mean is useful as a measure of the "middle" of a population of values, even when there is no particular significance to the population total. In studying the effectiveness of a treatment, for example, one often considers the treatment effective if it increases (or decreases, as appropriate) the mean or typical response.

When a population of values is symmetrical, the center value is the population mean, as well as the population median, and is the most natural measure of the middle of the distribution. Measurement errors are often assumed to be symmetrically distributed about zero, in which case the mean of a measured value is the constant being measured.

If a sample is "representative," the *sample mean* is a good estimate of the population mean, according to the meaning of the word "representative." As discussed before, however, unless the sampling involves an element of chance at some stage, there is little reason to expect representativeness. We shall assume throughout that the sampling is random. Using random sampling, with or without replacement, one has a good chance of obtaining a sample whose characteristics resemble those of the population, and in particular, a good chance that $\bar{X}$ will be close to the population mean μ.

The estimation of the mean of a population will be seen to resemble the estimation of a population proportion, discussed in Chapter 6. The reason for this, to be explained in the last section of this chapter, is that estimating a proportion is really a special case of estimating a mean.

7.1
VARIABILITY IN THE SAMPLE MEAN

Different samples will usually have different means, and one can never know (before getting a sample) what value the sample mean will turn out to have. *It is a random variable.*

The sample mean $\bar{X}$ is a *variable* in that its value varies from sample to sample. It is *random* in that its value depends on an experiment of chance—the drawing of a random sample. Drawing a random sample and determining its mean value $\bar{X}$ can be modeled as the random selection of a chip from a population of chips in a bowl. Each chip represents a *sample* whose $\bar{X}$-value is marked on the chip. The bowl of chips is then a *population of possible samples*, as well (because of the $\bar{X}$-value marked on each chip) as a population of $\bar{X}$-values. The $\bar{X}$-values in any specified interval are present in the bowl in number proportional to the probability of that interval. The distribution of $\bar{X}$-values according to these probabilities is the *sampling distribution* of $\bar{X}$.

The sample mean $\bar{X} = \sum X_i/n$ is a random variable whose value depends on the particular sample (X_1, $X_2, \ldots, X_n$) that is drawn. Its distribution is called the *sampling distribution* of $\bar{X}$.

How does one find the sampling distribution of $\bar{X}$? As with any population of chips in a bowl, we can learn about something about it by sampling the chips—in the present case, by looking at a "sample" of sample means.

As a learning exercise we shall obtain a number of samples of given size from a particular population and compute their means. This is not something that one would ordinarily do in practice. For if one actually had, say, 100 independent random samples of size 20, the set of all 2000 observations would be used as a single random sample, one that is much more informative than a sample of 20 for drawing conclusions about the population.

Example 7-2   *Estimating Average Age*

Suppose that we are interested in estimating the average age of the mother among families subscribing to the health organization mentioned in Example 2-1, using a sample of size $n = 5$. We choose this particular population because we happen to have access to the entire population and have calculated (with some effort!) the mean and s.d. of the mothers' ages: $\mu = 28.3$, $\sigma = 6.00$. Ordinarily, we do not know population parameters, but it is easier to see what is going on in the estimation process when nothing is hidden.

The population of 1296 ages can be represented by a bowl of chips—blue chips, say,—one chip for each mother, with that mother's age marked on the chip. The average of the numbers on these chips is then the population mean age, 28.3, and the numbers vary from under 20 to over 40, with an s.d. of 6.00 years. A sample of five is a sample of five blue chips from the bowl, or if we like, it is the five numbers marked on those chips. Their average is the sample mean $\bar{X}$, and we take this as the estimate of the population mean. Its value depends on just which five chips we happen to draw.

In order to judge the reliability of the $\bar{X}$-value of the sample we happen to draw, it helps to consider what could have happened. There are over 30 trillion different samples of size five that we could have drawn. Nevertheless, we can at least imagine that each of these possible samples of five is represented by a white chip, and that the mean age of the five mothers in the sample is marked on the chip. We then have a new "population"—a bowl of over 30 trillion white chips representing the population of possible $\bar{X}$-values. When we draw a sample of five blue chips from the basic population of interest, or from the bowl of blue chips that represent the mothers, we are in effect drawing *one* chip from the bowl of white chips representing the possible samples—and the possible $\bar{X}$-values.

Example 2-1 gave the ages of 30 mothers drawn at random from the population of 1296 mothers. Here we treat the same data as comprising six samples of size 5. They are repeated here in Table 7.1, together with the average age $\bar{X}$ for each sample. The 30 ages are numbers on the 30 blue chips representing the mothers in these six samples. The six average ages are numbers on the 6 white chips representing the six samples. Observe first that, indeed, the means do *vary* from sample to sample—but not as much as do the individual ages. This is perhaps more evident in the plot of Figure 7–1, in which

Table 7-1

Sample Number	Ages	Mean ($\bar{X}$)
1	41, 23, 29, 26, 37	31.2
2	37, 41, 28, 23, 24	30.6
3	24, 27, 29, 25, 22	25.4
4	23, 26, 30, 28, 24	26.2
5	31, 38, 18, 33, 18	27.6
6	21, 42, 23, 34, 33	30.6

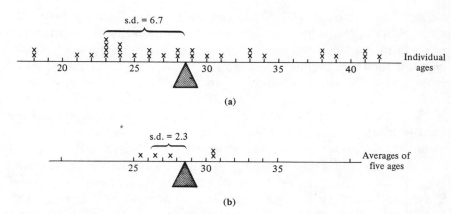

Figure 7-1. Ages, and average ages, Example 7-2.

the 30 individual ages are marked on one scale and the six sample means on a parallel scale. None of the average ages in this set of six samples is equal to the population mean, 28.3; some are greater than 28.3 and some less. Yet, there is less variability in the $\bar{X}$-values, and on the whole they tend to be closer to the population mean than the ages of individual mothers.

To be sure, we have looked at only six of the trillions of possible samples, but the tendency seems clear. Ordinarily, a sample of five ages will have some small and some large ages, so that the average in that sample will not be very far from the average or typical age of all 1296 mothers. (To be continued). ▲

The empirical investigation of the sampling distribution of $\bar{X}$ in a particular case, in Example 7-2, was useful in suggesting some general truths, but is not as satisfactory in this regard as a mathematical derivation. It is possible, in principle, to determine, by mathematical methods, just how the sampling distribution of $\bar{X}$ is related to the basic population—the distribution of X-values. In particular, there is a simple equation relating the mean of the $\bar{X}$-population and the mean of the population of X-values: they are *equal!*

If sampling is random, then

mean of the distribution of $\bar{X}$-values

= mean of the distribution of X-values.

This says that the population of sample means $\bar{X}$ has the same middle value as the basic population of X's being sampled.

The standard deviation of $\bar{X}$, which measures variability in $\bar{X}$ from sample to sample, is related to the s.d. of the population of individual X-values. In the case of a random sample the relation is quite simple:

If sampling is random, and *with* replacement if the population is finite, then

$$\text{s.d. of the distribution of } \bar{X}\text{-values}$$
$$= \frac{\text{s.d. of the } X\text{-value distribution}}{\sqrt{\text{sample size.}}}$$

Observe that, as we noticed in Example 7–2, the variability of $\bar{X}$ from sample to sample is *less* than the variability from one observation X to another if the sample size exceeds 1.

In referring to these facts, it will be convenient to have them in more compact notation. To avoid confusion of the many means and s.d.'s we encounter, we shall use μ to denote only the mean of the population of interest, and σ to denote only the s.d. of that population—the population being sampled. In the following summary we include the case of random sampling without replacement, where the relationship between s.d.'s involves the population size N.

When a sample of size n is drawn at random from a population of X-values with mean and s.d.

$$\text{m.v.}(X) = \mu \qquad \text{s.d.}(X) = \sigma,$$

the sample mean $\bar{X}$ is a random variable whose mean and s.d. are

$$\text{m.v.}(\bar{X}) = \mu$$

$$\text{s.d.}(\bar{X}) = \begin{cases} \dfrac{\sigma}{\sqrt{n}}, & \text{if sampling is with replacement} \\ \dfrac{\sigma}{\sqrt{n}} \sqrt{\dfrac{N-n}{N-1}}, & \text{if sampling is without replacement,} \end{cases}$$

where N is the population size.

(Observe that the extra factor in the formula for the standard deviation in the case of sampling without replacement is exactly the same as it was for estimating a population proportion, in Chapter 6.)

Thus, for random samples of given size n, the values of the sample mean vary from sample to sample, having values that are sometimes greater than the population mean and sometimes less. Their average or mean value is equal to that population mean. But the amount of variability in $\overline{X}$ as measured by its s.d. is less than that of the individual values X by the factor $1/\sqrt{n}$. That sample averages are less variable than individual sample values is a commonly understood phenomenon; averaging, generally, is a "smoothing" operation. In a sample of several observations, the tendency is that some will be large and some small, with the result that the average is neither extremely large nor extremely small. The formulas for mean and s.d. of $\overline{X}$ give numerical expression and precision to a tendency that seems evident from intuition and experience.

Example 7-3   *Estimating Average Age (continued)*

Returning to the population of mothers' ages in Example 7-2, whose mean value is 28.3, we find that the random variable $\overline{X}$ has this same mean:

$$\text{m.v.}(\overline{X}) = \text{m.v.}(X) = 28.3.$$

In the relatively small sample of six sample means given in Example 7-2, the mean value of the means was 28.6. This is not equal to 28.3, but we should not expect that it would be, since the value 28.3 applies to the population of *all* possible samples of size 5 and their means. We had only six.

Using the population s.d., $\sigma = 6.00$, we find the s.d. of the random variable $\overline{X}$ to be

$$\text{s.d.}(\overline{X}) = \frac{\text{s.d.}(X)}{\sqrt{n}} = \frac{6.00}{\sqrt{5}} = 2.68.$$

The s.d. of our six $\overline{X}$-values is 2.3, again not quite equal to the ideal value of 2.68 because we have only six, instead of all possible samples. ▲

7.2
STANDARD ERROR

The *error* in estimating a population mean μ as $\bar{X}$, the mean of a sample drawn from the population, is

$$\text{error} = \bar{X} - \mu.$$

If the sampling is random, the average error is zero, but what counts is the magnitude of the error:

$$\text{absolute error} = |\bar{X} - \mu|,$$

which is just the *distance* between $\bar{X}$ and μ. The r.m.s. average distance or error is the s.d. of $\bar{X}$:

$$\text{r.m.s. error} = \text{s.d.}(\bar{X} - \mu)$$
$$= \text{s.d.}(\bar{X}) = \frac{\sigma}{\sqrt{n}}.$$

Thus, an estimate $\bar{X}$ is reliable if either the population variability is small or the sample is large enough for the operation of averaging to smooth out the population variability.

In practice, one seldom knows the population s.d. σ, so one cannot use it to calculate the standard deviation of $\bar{X}$. However, this s.d. can be *approximated* by using the sample s.d. S in place of σ. The resulting expression[†] will be called the *standard error*, or *standard error of the mean*, abbreviated (as in Chapter 6) *s.e.* It can be computed from sample information.

Standard error of $\bar{X}$, for random samples:

$$\text{s.e.}(\bar{X}) = \frac{S}{\sqrt{n},}$$

where S is the sample s.d.

The standard error is a "typical" amount of error; the actual error in any given case can be smaller or larger than the standard error—and you will never know which, or by how much, as long as the population is inaccessible.

[†]Again, there are those who call $\sigma/\sqrt{n}$ the standard error, and $S/\sqrt{n}$ the *estimated standard error*.

Example 7-4   *Machine Variability*

It has been found that a certain machine that fills "5-pound" sacks of sugar actually fills the sacks with an amount that varies from sack to sack. There is an adjustment for average level, and it has been found from long experience that no matter where this level is set, the standard deviation of filled weights is .040 pound. Suppose now that at a certain level, a sample of 25 sacks has mean weight 5.08 pounds. How reliable is this as an estimate of the mean weight of all sacks produced at this setting?

The standard deviation of $\bar{X}$ is

$$\text{s.d.}(\bar{X}) = \frac{\sigma}{\sqrt{n}} = \frac{.040}{\sqrt{25}} = .0080 \text{ pound,}$$

and because σ is known, this would be used rather than "standard error" in describing the reliability of a value of $\bar{X}$ in estimating μ. On the other hand, if σ were not known, and if the sample of 25 sack weights has an s.d. of $S = .045$ pound, we could compute only the standard error, an approximation to the s.d. of $\bar{X}$:

$$\text{s.e.}(\bar{X}) = \frac{S}{\sqrt{n}} = \frac{.045}{\sqrt{25}} = .0090 \text{ pound.} \quad \blacktriangle$$

Investigators often want to know how large a sample they need in order to achieve a specified level of precision in estimating a mean. Because the population s.d. is a parameter that is usually unknown and is not, like the s.d. of a proportion, a bounded quantity, the question of sample size is not an easy one. If one has some notion of the magnitude of σ, of course, this can be used in the formula for s.d.$(\bar{X})$ to determine a sample size corresponding to a specified s.d. of error. (Statisticians have devised multistage or sequential techniques that use early results in a sampling process to yield information about variability for the purpose of determining how long to keep sampling.)

One thing that can be said about sample size and accuracy is that the standard deviation of $\bar{X}$ is inversely proportional to the square root of the sample size, as it was in estimating a proportion. Thus, a sample of 400 would have a standard deviation of $\bar{X}$ which is half as large as that of the mean of a sample of size 100, and three times as large as that of the mean of a sample of size $9 \times 400 = 3600$.

A word of caution: The computation of a standard error by means of the formula $S/\sqrt{n}$ only makes sense if the sample is a random sample. In particular, it does not make sense if the data at hand do not constitute a sample from some population of interest. If there is no such population, then there is no estimation, no error, and no standard error.

Example 7-5 *Mean Body Temperature*

The average temperature of the 142 student nurses listed in Data Set A (Appendix C) was found in Problem 11 of Chapter 3 to be $\bar{X} = 99.01$, and the s.d. (from Problem 23 of that chapter) is $S = .6227$. The sample size and sample s.d. can be put into the formula for standard error to yield

$$\text{s.e.} = \frac{.6227}{\sqrt{142}} = .052,$$

but does this have a meaning? The 142 nurses were not drawn by a random procedure, but one might consider them as constituting a random sample from a conceptual population of all student nurses in the area—past, present, and future. Whether this is something of interest is doubtful. In particular, it is not clear that it represents a healthy population, since some of the students appear to have been a bit feverish. So the mean of 99.01 would not be a reasonable estimate of the mean of healthy individuals. On the other hand, this mean does describe the "middle" temperature of these 142 student nurses, and possibly this is a population of interest. If so, there is no problem of estimation, and the standard error formula does not apply. ▲

7.3
THE LARGE-SAMPLE DISTRIBUTION OF $\bar{X}$

It is clear from the formula of $\sigma/\sqrt{n}$ for the s.d. of $\bar{X}$ that the variability of $\bar{X}$ decreases as the sample size increases. Indeed, the s.d. tends to 0 as n increases without limit. This means that the larger the sample, the more nearly $\bar{X}$ behaves like a constant from sample to sample—the constant μ. Evidently, $\bar{X}$ will approach μ as a limiting value as more and more data are accumulated. This "law of large numbers," or "law of averages," generalizes the version given for proportions in Chapter 4.

The practical meaning of the law of averages is that when a sample of some given large size is used in calculating a mean, there is a high probability that the calculated sample mean is close to the population mean. But to know how high, and how close, we need to know how to calculate probabilities for *intervals* of $\bar{X}$-values.

The calculation of probabilities of intervals for any random variable requires knowing its distribution. For the population of $\bar{X}$-values, this can be found using theoretical or numerical methods if the distribution of the population being sampled is known or assumed; but it will be different for different populations. However, in the case of *large samples*, there is a most remarkable and fortunate tendency. The ideal histogram for $\bar{X}$-values is *nearly normal* in shape—no matter what population is sampled! (This is reminiscent of the approximate normality of a sample proportion $\hat{p}$. The connection will be explained in Section 7.6.)

The tendency toward normality of $\bar{X}$ is observed in practice, but also has a mathematical formulation, the *central limit theorem*.

Central Limit Theorem:

> In large random samples from a population with mean μ and s.d. σ, the standardized mean (or Z-score)
>
> $$Z = \frac{\bar{X} - \mu}{\sigma / \sqrt{n}}$$
>
> is approximately *standard normal* (m.v. = 0, s.d. = 1).

A more precise statement would assert that the distribution of the standardized mean or Z-score is normal *in the limit*, as n becomes infinite.

Example 7-6

A calculator was programmed to generate 10,000 samples of 10 simulated pointer spins, compute the mean of each sample, and plot a histogram of these means. Figure 7-2 shows this histogram and the smooth histogram for an approximating normal curve. This demonstration does not, of course, prove that the central limit

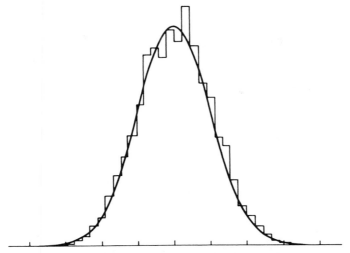

Figure 7-2. Histogram of means of 10,000 samples of 10 pointer spins, with normal curve superposed.

theorem is correct, but it is perhaps useful as empirical evidence that the tendency toward normality asserted by the theorem is a practical reality as well as a mathematical one. ▲

Example 7-7 *ACT Scores*

The composite ACT scores for entering freshmen in a certain liberal arts college have averaged 22.4, over the years, with an s.d. of $\sigma =$ 4.6. The next class will include 400 students, and their average score—if they can be considered to constitute a random sample of 400 from the same population (i.e., with $\mu = 22.4$, $\sigma = 4.6$)—is nearly normally distributed with mean and s.d.

$$\text{m.v.}(\bar{X}) = 22.4, \quad \text{s.d.}(\bar{X}) = \frac{4.6}{\sqrt{400}} = .23$$

The distribution of $\bar{X}$ is shown in Figure 7-3.

The probability that this new sample mean will be less than 22 (say) can be calculated with the aid of Table 2 of Appendix B, after computing the Z-score corresponding to 22:

$$z = \frac{22 - 22.4}{.23} = -1.74,$$

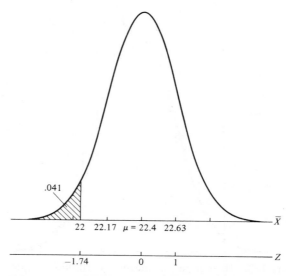

Figure 7-3. Distribution of means of samples of
400 ACT scores (Example 7-7).

whence

$$P(\bar{X} < 22) = \Phi(-1.74) = .041. \quad \blacktriangle$$

7.4
LARGE-SAMPLE CONFIDENCE INTERVALS FOR μ

As in estimating a proportion, so now in estimating a population
mean, an estimate can be combined with its standard error to give
a confidence interval—a range of values that has a specified degree
of assurance of including the true value of μ. We consider here
the case of a large sample.

An interval centered at $\bar{X}$ and extending one s.e. on either side
of $\bar{X}$ is suggested by a commonly used form of reporting an estimate
together with a measure of error, namely, $\bar{X} \pm$ s.e. The interval from
$\bar{X}$ – s.e. to $\bar{X}$ + s.e. will include μ if $\bar{X}$ is within one s.e. of μ; hence,

$$P(\bar{X} - \text{s.e.} < \mu < \bar{X} + \text{s.e.}) = P(|\bar{X} - \mu| < \text{s.e.}).$$

For large samples, this is the probability that a *normal* variable is
within one s.d. of its mean, or about 68 percent, because $\bar{X}$ is then
approximately normal with mean μ and an s.d. given approximately
by the standard error.

For an interval extending a multiple k of the s.e. on either side of $\bar{X}$, the probability is

$$P[\bar{X} - k(\text{s.e.}) < \mu < \bar{X} + k(\text{s.e.})]$$

$$= P\left(\left|\frac{\bar{X} - \mu}{\text{s.e.}}\right| < k\right) = \Phi(k) - \Phi(-k).$$

The relation between this probability, called a *confidence level* and the multiple k is the same as what was given in Chapter 6 for estimating a proportion:

Confidence Level	Multiple of s.e.
.80	1.28
.90	1.64
.95	1.96
.99	2.58

Thus, the 95 percent confidence limits for μ based on a large sample are $\bar{X} \pm 1.96(\text{s.e.})$, the 99 percent limits are $\bar{X} \pm 2.58(\text{s.e.})$, and so on.

Large sample 95 percent confidence limits for a population mean:

$$\bar{X} \pm 1.96(\text{s.e.}),$$

where 1.96 is the 97.5 percentile of the standard normal distribution.

Example 7-8 *Cholesterol Levels*

Table 7-2 gives the cholesterol level of 100 males. The mean and s.d. of this sample are $\bar{X} = 245.7$ and $S = 46.02$. The standard error of $\bar{X}$ is then

$$\text{s.e.}(\bar{X}) = \frac{S}{\sqrt{n}} = \frac{46.02}{\sqrt{100}} = 4.602,$$

and the 95 percent confidence limits for the population mean μ are

$$245.7 \pm (1.96)(4.602),$$

or 236.7 and 254.7. We do not know for sure that μ is between these limits (nor will anyone else!); but if it can be assumed that the sample is a random sample, the procedure we used has a 95 percent chance of success in covering the true population mean. We say that we are "95 percent confident" that μ is between 236.7 and 254.7. ▲

Table 7-2   Cholesterol Level of 100 Males

283	248	280	230	258	204	270	307	368	281
165	234	264	274	229	189	245	220	299	270
171	243	310	289	230	220	389	253	189	295
202	294	226	219	293	229	269	208	218	164
245	312	321	253	208	194	218	280	183	294
261	206	285	306	191	258	245	196	255	272
299	325	269	219	258	295	243	198	270	194
208	260	165	236	196	216	271	227	208	272
220	276	276	249	204	198	246	270	306	232
208	271	199	206	323	148	194	194	198	310

Drawing a sample and using the sample results to construct a confidence interval can be likened to tossing a horseshoe at an invisible stake. The stake is positioned at the unknown population mean μ, and the horseshoe is $2k$(s.e.) units wide. It lands with center at $\overline{X}$. It either does or does not manage to snare the stake, for a "ringer"—but no one can score it because the stake cannot be seen! You can be confident (to the degree of the confidence level corresponding to the multiplier k) that you have a ringer, but occasionally you will miss. But you will not know when you miss, and neither will anyone else.

If the population s.d. is known, at least approximately, the formula for s.e.($\overline{X}$) in terms of n permits a determination of sample size needed to keep the s.e. within a specified bound. The next example illustrates this.

Example 7-9

Suppose that a rather precise method of weighing involves an error that has been found to have an s.d. of about 50 micrograms (μg). How many measurements of a certain weight are needed to yield 95 percent confidence limits for the actual weight that are 15 μg on either side of the sample mean weight?

Assuming a normal distribution for the average weight, we set 1.96 s.d.'s equal to 15:

$$15 = \frac{1.96\sigma}{\sqrt{n}} = \frac{1.96 \times 50}{\sqrt{n}}$$

which means that

$$\sqrt{n} = \frac{1.96 \times 50}{15} = 6.53.$$

So the sample size should be at least $(6.53)^2$, or 43. ▲

Example 7–10

The notion that a 95 percent confidence interval for a population mean μ covers that mean "95 percent of the time" can be demonstrated by an artificial sampling experiment, one that generates lots of samples and computes a confidence interval from each sample mean. We did such an experiment using a population whose mean we know—the spinning pointer on the interval from 0 to 1 (Examples 5–1 and 5–2). For this population, the mean and s.d. are $\mu = .5$ and $\sigma = .289$ (see Example 5–9). A computer generated 200 samples of 25 pointer spins each, that is, with $n = 25$. For each sample so generated, it calculated 95 percent limits:

$$\bar{X} \pm \frac{1.96\sigma}{\sqrt{25,}} \quad \text{or} \quad \bar{X} \pm .113.$$

Thus, a 95 percent confidence interval is .226 unit wide, centered at the sample mean. Figure 7–4 shows a plot of a confidence interval for each of the 200 samples of size 25. A mark ($\times$) is shown at the right if the interval did *not* cover $\mu = .5$. This happens whenever $\bar{X}$ chances to fall farther than .113 unit from $\mu = .5$, and the probability of this is 5 percent. So we "expect" 5 percent of 200, or 10 such misses; in this particular series of trials there happened to be 12 misses. ▲

A few words about the assumption that the sample size is "large:" This assumption was made (1) so that we could use the normal table in evaluating probabilities, when nothing is assumed about the population distribution, and (2) so that we could get by with using the sample s.d. (S) in place of the population s.d. (σ) in finding the Z-score corresponding to an observed $\bar{X}$-value. The size of the sample is usually more crucial for (2) than for (1).

In connection with (1)—use of the normal table—we should point

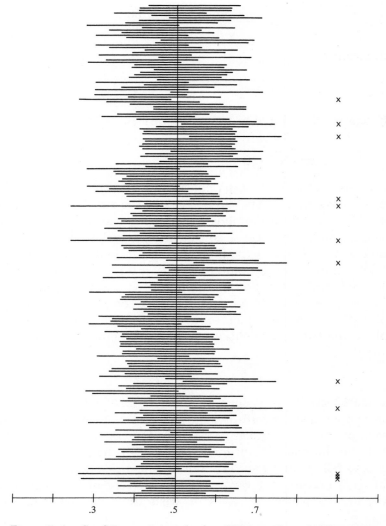

Figure 7-4.   Confidence intervals for 200 samples of 25 pointer
spins.

out that if the population being sampled is normal with known
s.d., then

$$Z = \frac{\bar{X} - \mu}{\sigma/\sqrt{n}}$$

is normally distributed for *any* sample size, down to and including
$n = 1$. Frequently, in practice, the population is close enough to

normal that Z will be approximately normal for samples as small as 10, or even 5. So if the population s.d. (σ) is known, the relation between the confidence level and the constant k used in the limits $\bar{X} \pm k\sigma/\sqrt{n}$ can be based on the normal distribution, even if n is rather small. However, it is rarely the case that σ is known.

When σ is *not known*, the only recourse is to estimate it from the sample. With $S/\sqrt{n}$ in place of $\sigma/\sqrt{n}$, where S is the *sample* s.d., we use the approximate Z-score

$$ Z = \frac{\bar{X} - \mu}{S/\sqrt{n}} . $$

Now, even if n is large enough for $\bar{X}$ to be approximately normal, it may not be large enough for the sample s.d. S to be a reliable estimate of the unknown σ. For instance, if $n = 25$, S can be in error, as an estimate of σ, by 15 percent or so. And the distribution of Z will exhibit a greater variability, from sample to sample, than that of a standard normal variable, owing to the variability in S.

As a rule of thumb, when σ is not known, a sample size of 50 or more will ordinarily be adequate for using "large"-sample results in estimating a mean. If n is between 25 and 50, some hedging may be in order. And if n is less than 25, a procedure designed for small samples should be used. Such a procedure, based on a modification of Z, is taken up next.

7.5
USING THE MEAN OF A SMALL SAMPLE

The distribution of the approximation to a Z-score for $\bar{X}$ which uses $S/\sqrt{n}$ in place of $\sigma/\sqrt{n}$ depends on the population being sampled. The distribution has been derived mathematically and tabled, for the special case of a normal population. And, although one almost never encounters an exactly normal population, it has been found that the distribution derived for the normal case is close enough for approximate results if the population does not deviate too much from normality—if it is not too skewed or heavy-tailed.

The distribution in question, called "Student's t," was discovered in 1908 by Student (the pseudonym for W. S. Gosset, a statistician at the Guinness brewery in Dublin). It finds use in various contexts, and the tabular form that is generally convenient gives percentiles of the variable

$$T = \frac{\bar{X} - \mu}{S/\sqrt{n-1}} = \sqrt{n-1} \times \frac{\bar{X} - \mu}{S},$$

a slight modification of what we called Z.

If $\bar{X}$ is the mean of a random sample from a nearly normal population, the variable

$$T = \frac{\bar{X} - \mu}{S/\sqrt{n-1}}$$

has approximately a t-distribution with $n - 1$ degrees of freedom. (Percentiles are given in Table III of Appendix B for $n \leqslant 31$; for larger n, T is approximately normal.)

The distribution of T depends on the sample size, and the number of "degrees of freedom" $(n - 1)$ may be regarded for the present as simply telling us which row in the table to use.

(If the sample s.d. had been defined to be $\tilde{S}$, with the divisor $n - 1$ instead of n, the denominator in T would be written $\tilde{S}/\sqrt{n}$, which is equal to $S/\sqrt{n-1}$, as we pointed out at the end of Section 3.5.)

In constructing a confidence interval for a population mean μ (if the population is not too far from normal, and if σ is not known), we use the t-table (Table III) rather than the normal table, in finding the proper multiplier for the standard error.

---

95 percent confidence limits for μ:

$$\bar{X} \pm k \, \frac{S}{\sqrt{n-1}}$$

assuming a nearly normal population, where k is the 97.5 percentile of the t-distribution with $n - 1$ degrees of freedom (Table III).

Example 7-11   *Egg-Timer Accuracy*

A "3-minute" egg timer was to be checked to see how long it took to run out after being set and how reliable it was. Twenty trials yielded the following times (read to the nearest second), given as the number of seconds in *excess* of 3 minutes:

18, 17, 12, 8, 18, 15, 6, 8, 7, 17

11, 17, 10, 16, 16, 11, 9, 15, 10, 16.

The mean and s.d. are $\overline{X}$ = 12.85 and S = 3.94, so a 95 percent confidence interval for μ, the "population" mean time in excess of 3 minutes, is given by the limits

$$12.85 \pm 2.09 \times \frac{3.94}{\sqrt{19}} = 12.85 \pm 1.89,$$

where 2.09 is the 97.5 percentile of the t-distribution with 19 (= 20 – 1) degrees of freedom (as found in Table III).

We computed these limits assuming the population to be normal, or close to it, to justify use of a t-percentile. However, a look at Figure 7-5, giving a plot of the data, suggests that this assumption is questionable. Half the data seems centered around 9 and the other half around 17—perhaps a grain of sand gets wedged half the time and slows the flow. Nevertheless, the t-distribution is reasonable in making a crude confidence statement about μ, although it may be that in this situation there are other aspects of the population that are of more interest than μ. Indeed, the phrase "half the data" suggests that possibly the sand ran through faster in one direction than the other. The data do not seem to alternate between high and low values, but we recalled that they were gathered while watching TV, and we often forgot to catch the stopping time. Further trials revealed that the sand did go through faster one way than the other— the glass container was not perfectly symmetrical. ▲

It is to be noted that, as stated in the first "box" of this section, the t-distribution becomes normal as the number of degrees of freedom increases without limit. This is why Table III gives percentiles for degrees of freedom only up to 30. But Table III does include a row of percentiles of the normal distribution labeled "∞," and this is convenient for locating the proper column. For if you know that, for instance, 1.96 is the correct multiplier of the s.e. in the large-sample case, you can simply start at 1.96 in the bottom row and go up in that column until you come to the entry opposite n – 1 degrees of freedom for your sample size n.

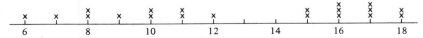

Figure 7-5. Number of seconds in excess of 3 minutes, 20 trials of a 3-minute timer.

7.6
A PROPORTION AS A MEAN

Estimation of a population parameter has been discussed in two contexts—estimation of a proportion and estimation of a mean. Actually, the former can be viewed as a special case of the latter.

A population "proportion," it will be recalled, may be the proportion of individuals of one type in an actual population of individuals, or it may be the probability of some event (the long-run proportion of the time that event occurs) in an experiment with only a conceptual population. In either case we focus on some event to which, if it occurs, we now attach the label $X = 1$; when it does not occur, we say that $X = 0$. The probability p of its occurrence (or the proportion of individuals defining the event) defines the model for the labeling variable X:

Value of X	Probability
1	p
0	$1 - p$

The mean value of X is calculated as the usual weighted sum:

$$\mu = \text{m.v.}(X) = 1 \cdot p + 0 \cdot (1 - p) = p.$$

And since $X = X^2$, for either $X = 0$ or $X = 1$, it follows that $\text{m.v.}(X^2) = p$. Hence,

$$\sigma = \text{s.d.}(X) = \sqrt{\text{m.v.}(X^2) - [\text{m.v.}(X)]^2} = \sqrt{pq},$$

where $q = 1 - p$.

Suppose now that we draw a random sample of size n. The results constitute a sequence of 0's and 1's, summarized by frequency—or by relative frequency (sample proportion):

Value of X	Relative Frequency
1	$\hat{p}$
0	$\hat{q} = 1 - \hat{p}$

The mean value and s.d. of this sample of X-values are now given by the same formulas as in the case of the population, except that p is replaced by $\hat{p}$:

$$\bar{X} = \hat{p} \qquad S = \sqrt{\hat{p}\hat{q}}.$$

The standard error of $\hat{p}$ is then just the standard error of a mean:

$$\text{s.e.}(\hat{p}) = \text{s.e.}(\bar{X}) = \frac{S}{\sqrt{n}} = \frac{\sqrt{\hat{p}\hat{q}}}{\sqrt{n}}$$

Moreover, because $\bar{X}$ is approximately normal for large samples from any population, it is so for large samples from this 0–1 population; therefore, $\hat{p}$ is approximately normal, for large n, with mean p and s.d. $\sqrt{pq/n}$. Also, a confidence interval for μ becomes a confidence interval for p, with limits of the usual form: $\hat{p} \pm k(\text{s.e.})$, where k is again given in Table IIc.

The following summary gives notation and formulas for the general population variable X and the specialization to a 0–1 coding for a proportion:

	General	0–1 Population
Population mean	μ	p
Sample mean	$\bar{X}$	$\hat{p}$
Population standard deviation	σ	$\sqrt{pq}$
Sample standard deviation	S	$\sqrt{\hat{p}\hat{q}}$
Standard deviation of the mean	$\sigma/\sqrt{n}$	$\sqrt{pq/n}$
Standard error of the mean	$S/\sqrt{n}$	$\sqrt{\hat{p}\hat{q}/n}$
95 percent confidence limits for the mean	$\bar{X} \pm 1.96S/\sqrt{n}$	$\hat{p} \pm 1.96\sqrt{\hat{p}\hat{q}/n}$
Central limit theorem	$\bar{X}$ is approximately normal	$\hat{p}$ is approximately normal

KEY VOCABULARY

Sampling distribution
Error of estimation
Standard error of the mean
Central limit theorem
Confidence interval
Student's t-distribution
Degrees of freedom

QUESTIONS

1. What is a *statistical* estimate?
2. Once more: Of all types of samples, why do we use *random* samples as the basis for a statistical estimate?
3. How does the reliability of the sample mean, as an estimate of the true population mean, depend on the sample size?
4. Explain why it is reasonable to expect less variability in the *average* of several observations than in the individual observations themselves?
5. What does the "standard error" tell us about an estimation procedure?
6. There is something especially remarkable about the central limit theorem. What is it?
7. With regard to the problem of determining a sample size to achieve a specified degree of accuracy, why is this harder in the case of estimating a mean than in the case of estimating a proportion?
8. Why is the normal distribution not a good model for the approximate Z-score which uses S in place of σ, when the sample size is small?
9. Investigator A requires 1000 observations to get a 95 percent confidence interval, say, for a population mean. But investigator B requires only 100 observations to get a 95 percent confidence interval of the same width for the mean of his population. How is this possible?

PROBLEMS

Sections 7.1 and 7.2

*1. A measurement X is assumed to be a random variable with mean 10 and s.d. 1. Find the mean and s.d. of $\overline{X}$, the mean of a random sample of 25 such measurements.

2. A measurement error ϵ is random with mean 0 and s.d. 1 millimeter. Measuring a length L yields the value $X = L + \epsilon$. For a random sample of 100 such measurements:
   (a) Find the mean value of the sample mean $\overline{X}$.
   (b) Find the s.d. of $\overline{X}$.

*3. The average systolic blood pressure of the 142 student nurses in Data Set A (Appendix C) is 116.80, and the s.d. is 11.974. Take this group of 142 as a *population*, and using Table VIII of Appendix B, obtain 20 samples of size $n = 6$.† Calculate the mean of each sample. Compare the m.v. and s.d. of your "sample" of 20 means with the ideal values (μ and $\sigma/\sqrt{n}$, respectively).

4. Obtain 50 samples of *three* GPA's from the "population" of 200 students listed in Data Set B (Appendix C), using the random-number table for sampling (Table IX of Appendix B). Calculate the 50 sample means, and find the m.v. and s.d. of your 50 $\bar{X}$-values. Compare these with the ideal m.v. and s.d. of $\bar{X}$.

*5. A random sample of size 100 has mean 42 and s.d. 8.4. Determine the standard error in estimating the population mean to be this sample mean value.

6. Cholesterol readings for a random sample of 81 males have mean $\bar{X} = 245.69$ and s.d. $S = 46.02$. Determine the standard error of the mean.

*7. (a) If you cut a sample size in half (to cut costs), what does this do to the standard error of the mean, other factors remaining unchanged?

(b) If you want to *increase* the accuracy so that the standard error is halved, how would you change the sample size?

Sections 7.3 and 7.4

*8. For a measurement X with mean 10 and standard deviation 1, determine the probability that in a random sample of 25 observations:

(a) $\bar{X}$ exceeds 10.5.

(b) $\bar{X}$ differs from the population mean by more than .5.

9. If a population of platelet counts has mean 234 and s.d. 44 (in 1000's per mm^3), find the probability that a random sample of size 100 would have a mean value less than 230.

*10. A measurement error X is random with mean 0 and s.d. 1. Find the probability that the mean of a random sample of measurements with this error differs from the population mean by more than .05:

†This is done most easily in cooperation with another student or other students.

(a) If the sample size is $n = 25$.

(b) If the sample size is $n = 225$.

*11. Blood samples from 100 individuals have a mean platelet count of 260 (in 1000's per mm^3) with s.d. = 70. Determine 95 percent confidence limits for the mean platelet count in the population from which these 100 were drawn.

12. A 90 percent confidence interval for a population mean was computed to be $10.36 < \mu < 13.64$, based on the mean of a random sample of size 100.

(a) Deduce the value of the sample mean $\bar{X}$.

(b) Deduce the standard error.

(c) Deduce the sample standard deviation.

(d) How large a sample would have been required to obtain a confidence interval only about one-fourth as wide? (Assume that the value of S will be about the same.)

*13. Assuming that the 125 male students in Data Set B (Appendix C), whose mean height and s.d. are $\bar{X} = 70.95$ and $S = 3.165$, can be thought of as a random sample from the campus population of male students, give a confidence interval for the population mean height, with confidence level (a) 90 percent, and (b) 99 percent.

14. What degree of "confidence" would you have in an interval that is one-half as wide as the 95 percent confidence interval you constructed from a given (large) sample?

*15. A research study of dental products (one of those reported in Table 1–1) used a random sample of 294 children from a population of interest. The average number of new cavities per child developed over a 30-month period was 10.88, with standard deviation 6.36. Construct a 95 percent confidence interval for the population mean.

Sections 7.5 and 7.6

*16. A sample of 10 rats of a certain type averaged 38 seconds to solve a certain maze, with a s.d. of 8.5 seconds. Construct a 90 percent confidence interval for the mean time required for this type of rat. (What assumptions are needed?)

17. After the first two runs of the egg-timer experiment described in Example 7–11, with results 17 and 18 (seconds past 3 minutes), we wondered if it was worth continuing, since there seemed to be little variability. Determine 95 percent confi-

dence limits for the mean time based on just these two observations, assuming that the t-distribution is applicable.

*18. Dumping a box of thumbtacks, as described in Problem 16 of Chapter 6, yielded these numbers of points up in the 25 trials:

$$
\begin{array}{ccccc}
71 & 71 & 74 & 74 & 73 \\
82 & 74 & 76 & 77 & 60 \\
72 & 71 & 73 & 67 & 70 \\
69 & 70 & 73 & 69 & 64 \\
67 & 72 & 70 & 72 & 73
\end{array}
$$

(a) Given that the sum of these numbers is 1784, find the average number per trial with points up.

(b) The s.d. of the 25 numbers of points up is 4.1845. Use this to construct a 99 percent confidence interval for the true (or "population") mean number per trial. (Assume that the population from which the 25 numbers were drawn is close enough to normal that the t-distribution can be used.)

(c) Using the fact that the mean number per trial is $100p$, and assuming a probability p for a single tack to land with point up, convert the confidence interval for the mean in part (b) to a confidence interval for p. (It would be interesting to compare this result with the confidence interval for p obtained in Problem 16 of Chapter 6.)

(d) Why is the population of numbers of points up among 100 tacks [as assumed in part (b)] nearly normal?

19. A study of certain teacher training system included a summary of characteristics of the students involved in the study. The tabulation included the following item:

Sex (0 or 1): mean = .71, s.d. = .45, $n = 323$. (Male = 1.)

(a) How many male students were there in this group?

(b) Is the s.d. given consistent with the mean of .71?

Review

20. Suppose that you wanted to estimate the mean height of male students on a large campus.

(a) Among the male students that you have seen, there appears to be a spread of something like 18 inches in heights, so you estimate the population s.d. to be around 3 inches.

Taking this as σ, find the size of sample required to obtain an estimate of the mean with standard error .30 inch.

(b) Taking a random sample of the size determined in part (a), you find $\bar{X} = 70.5$ and $S = 2.4$. What is the standard error?

(c) The average height of the first 100 males in the list in Data Set B is 70.95 inches, with an s.d. of 2.935 inches. Calculate the standard error. (Is this meaningful? That is, can the sample be considered as a random sample of college students? Are there any obvious biases?)

21. The mean score for the 155,623 applicants who took the Graduate Record Examination in 1977 to 1978 was 521, and the s.d., 123. Is there a meaningful standard error here? (If so, find it; if not, explain.)

22. Find the probability that the mean of a random sample of size 64 from a population with $\sigma = 2$ differs from the population mean by more than .50.

23. Experience has shown that the time between arrivals of telephone calls at an exchange (at a certain period of the day) is a random variable with mean .50 seconds and s.d. .707 seconds. Find the probability that the time to the 50th call after a given point in time exceeds one-half minute. [Hint: the average of the times to arrival of the 50 successive calls is just the total elapsed time divided by 50. Hence, the total elapsed time will exceed 30 seconds if the average exceeds 30/50 seconds.]

24. Suppose that you want to estimate the number of bicycle licenses issued in a certain community by checking a random sample of 25 license numbers. Assume that licenses are numbered from 1 to N.

(a) Determine the population mean—the average of the numbers from 1 to N. (This will depend on N.)

(b) If you estimate the average number to be the mean of your sample, how could you then estimate the total number of licenses issued?

To try out this scheme, suppose that the total number issued is actually 850, and use Table VIII to obtain a random sample of 25 numbers in the range of 001–850.

(c) Compute the mean of your sample, and from it derive an estimate of the total number issued (which you happen to know in this artificial situation), using the method of part (b).

(d) Try this estimate:

$$\frac{n}{n-1} \times \text{(largest number in sample)},$$

with $n = 25$, and see whether it comes closer to 850 than the estimate in part (c). When possible, compare the results with those of other members of your class.

25. The mean length of 1000 fish of a certain species caught commercially in a certain coastal area is 24 inches, with standard deviation of 2.0 inches. Compute a 90 percent confidence interval for the population mean length.

26. Any quantity computed from a sample (not just $\overline{X}$) is a random variable and has a sampling distribution. For example, let R denote the range of a set of five observations chosen independently and randomly from the interval 0 to 1 (as in five spins of the "spinning pointer"). Reflecting on what you observed when you obtained such samples in the past, choose the histogram in Figure 7-6 that would best approximate the sampling distribution of R.

27. Describe each statement as true or false (not always true):
    (a) A sample mean is always equal to the population mean.

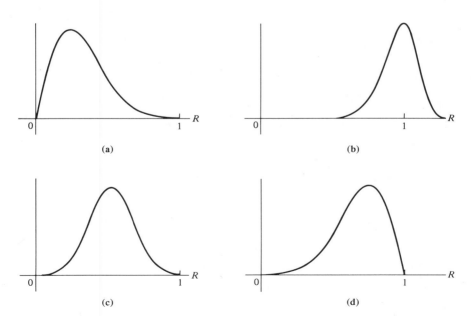

Figure 7-6

(b) The variability in $\bar{X}$ is greater for a sample of size 10 than for one of size 20.

(c) A confidence interval based on 100 observations is narrower than one based on 25 observations (other factors being the same).

(d) Using $\sqrt{\hat{p}\hat{q}/n}$ as a standard error in estimating a proportion p requires that the population size be much larger than the sample size if sampling is done without replacement.

(e) To construct a confidence interval for a mean of a population with unknown characteristics as $\bar{X} \pm k(\text{s.e.})$, with k determined from the normal table, requires a "large" sample.

(f) To construct a confidence interval for a mean as $\bar{X} \pm k(\text{s.e.})$ with k determined from the normal table requires a normal population.

8

Statistical
Tests-One Sample

INVESTIGATORS OFTEN NEED to know whether or not a group of individuals or a group of test results is like some standard population or norm, in some or in all respects. This norm or standard population may be one that is essentially known through past experience; it may be postulated as an ideal that follows from basic "laws"; or it may be specified as the way one wants a population to be. The group of individuals constitutes a sample, and the question is whether or not this can be accepted as a random sample from the standard population. If its characteristics are quite different from what are standard, the only reasonable conclusion may be that it was drawn from some population *other* than the standard one. That is, the differences may be too great to be accounted for as chance variation, or as bad luck in sampling.

Example 8-1 *Thrombocytosis*

Investigating a commonly accepted association of thrombocytosis with malignancy, a team of medical research workers surveyed 153 lung cancer patients.[†] Platelet counts averaged $395,000/mm^3$, with a standard deviation of $S = 170,000/mm^3$. They state that the mean platelet count for healthy males is $235,000/mm^3$ with an s.d. of

[†]S. E. Silvis, N. Turkbas, and A. Doscherholmen, "Thrombocytosis in Patients with Lung Cancer," *J. Am. Med. Assoc.* **211**, 1852 (1970).

44,400/mm^3. Realizing that the mean and s.d. of the platelet count in a sample of size 153 vary from sample to sample, one faces this question: Can the differences between the observed statistics ($\overline{X}$ and S) and the corresponding standard parameters (μ and σ) be accounted for as "sampling error," or should it be concluded that individuals with lung cancer are really not like healthy individuals, with regard to platelet count? ▲

Example 8-2 *Male and Female Births*

Are successive births in a family like independent tosses of a coin as far as the sex of the child is concerned? If so, the distribution of males in families of six children, say, would follow the pattern of probabilities for the number of heads in six tosses of a coin. These probabilities are found in Table I of Appendix B:

Number of Males	Probability
0	.0156
1	.0938
2	.2344
3	.3125
4	.2344
5	.0938
6	.0156

Suppose we were to find, in a sample of 100 families with six children, that the distribution according to number of boys is as follows:

Number of Boys	Frequency	Relative Frequency
0	3	.03
1	12	.12
2	21	.21
3	24	.24
4	23	.23
5	13	.13
6	4	.04

The relative frequencies in this sample are not equal to the corresponding probabilities for numbers of heads in six tosses of a coin. But a lack of agreement between observed proportions and corresponding probabilities is not at all unusual, even when the probabilities are correct for the experiment sampled! The question is this: Is the *extent* of the disparity so great, the "fit" so poor, as to be hard to account for as ordinary chance variation of sample proportions about the corresponding population proportions? If the fit *is* that poor, perhaps there are other ways to account for the discrepancies. For example, it may be that the probability of a male is not 1/2, or is changing with time; it may be that the trials are not independent; and it may be that the sample is not a proper sample for inference. It is because of the existence of such alternative explanations as these that a poor fit is taken as reason to discredit the coin-tossing model as truly representing successive births in a family. ▲

Example 8-3 *Quality Control*

Manufacturers often make regular checks on the output of a production process to see that specifications for their product are adhered to. For instance, a certain dairy has a machine that cuts and packages butter in 1-pound cartons. The actual net weights will vary somewhat from the nominal weight of 1 pound—even when the machine is behaving according to design, so that the process is "in control." But machines wear, knives slip, and the consistency of the butter varies, so the process may go out of control. That is, the mean weight of a package may slip from what it is supposed to be.

Suppose that, to monitor production, 10 packages are taken from the line and weighed. If the average weight $\bar{X}$ turns out to be 1.02 pounds, what does this prove? After all, individual weights vary, and averages will vary from one sample of 10 packages to another. So the observed deviation of .02 pound from the specified weight might be due to chance fluctuation. The question is: How much of a deviation can be attributed to chance fluctuations, and how much would be evidence of a shift in process mean, calling for readjustment? ▲

The hypothesis that the population sampled is the standard population, whether this is a population known from experience, one that is postulated according to some ideal of behavior, or one that has specified characteristics, is usually called a *null hypothesis*. A null hypothesis will be denoted by H_0. (The term *null* means *not different from what is standard*.) The null hypotheses is the three examples considered would be as follows:

Example 8-1 (Thrombocytosis): Patients with lung cancer are no different from healthy patients, with regard to platelet count.

Example 8-2 (Male and Female Births): Successive births in a family are male or female in the same way that successive tosses of a coin are heads or tails.

Example 8-3 (Quality Control): The mean package weight is 1 pound.

8.1
SAMPLE EVIDENCE ABOUT A HYPOTHESIS

To *test* a hypothesis H_0 about a population is to see whether or not a sample from the population has characteristics close to what would be expected when H_0 is true. The process of testing can be thought of as examining a sample for evidence, for or against the null hypothesis.

A test is usually phrased in terms of some statistic, called the *test statistic;* this is chosen to be especially sensitive to variation in the particular population characteristic in terms of which the null hypothesis is specified, as the following example illustrates.

Example 8-4 *ESP*

A subject is asked to name the suit of a card that he does not see, picked at random from a standard deck of playing cards. A null hypothesis about this experiment is that the subject has no extra-sensory powers, and so has one chance in four of naming the correct suit. If p denotes the probability of a correct call, the null hypothesis is

$$H_0: \quad p = \frac{1}{4}.$$

To test this hypothesis, the subject is asked to repeat the experiment a number of times, say n in all, his call at each trial being correct or incorrect. The statistic summarizing the results, one that is naturally related to the parameter p, is the sample proportion of correct calls, $\hat{p}$.

The sampling distribution of the test statistic $\hat{p}$, under the assumption that H_0 is true, will help in deciding whether or not an observed sample proportion lends credibility to H_0. If $n = 10$, for instance, the probabilities (taken from Table I of Appendix B) are as follows:

$\hat{p}$	Probability
0	.0563
.1	.1877
.2	.2816
.3	.2503
.4	.1460
.5	.0584
.6	.0162
.7	.0031
.8	.0004
.9	.0000
1.0	.0000

Something like two or three correct calls in 10 tries (i.e., $\hat{p} = .2$ or .3) would not be unusual at all, when H_0 is true. But 9 or 10 out of 10 would be surprising indeed; and if there were some other, better explanation for such values, their occurrence would cast strong doubt on the null hypothesis, $p = 1/4$. An obvious alternative explanation of 9 or 10 correct calls would be that p is really greater than $1/4$. (To be continued.) ▲

Example 8-5   *Quality Control (continued)*

Consider again the butter-packaging process of Example 8-3. The null hypothesis might be, as suggested there, that the process mean weight is 1 pound. More realistically, it might be that the mean is somewhere between .99 and 1.01, since the extreme precision of the equality $\mu = 1$ is not really required; or it could be that $\mu = 1.01$, set a trifle higher than the nominal package weight so that most packages would weigh at least 1 pound. But assigning some particular value to μ is necessary before probabilities can be calculated. Suppose, for convenience, that we take the null hypothesis to be $\mu = 1$ pound.

The mean $\bar{X}$ of a random sample is clearly related to the population mean μ, so $\bar{X}$ is relevant in testing H_0. A sample mean near 1 pound would make it easy to believe $\mu = 1$, but a sample mean far from 1 pound would suggest that μ is far from 1. This is because for samples of even moderate size, $\bar{X}$ is a fairly good estimate of μ. But how far is "far"? We make this judgment in terms of the sampling distribution of $\bar{X}$ under H_0. This distribution gives the likelihood of the various possible values of $\bar{X}$ when H_0 is true. In particular, the s.d. of this sampling distribution is a measure that is useful in weighing the significance of an observed deviation of $\bar{X}$ from the value of μ that is being tested.

In the case of the 10 package weights, the sampling distribution of their mean $\bar{X}$ is nearly normal. And when the null hypothesis is true, it is centered at $\mu = 1$, with s.d. $\sigma/\sqrt{n}$. (The population s.d. σ is the "process s.d.," describing the variation in weight from package to package.) Values of $\bar{X}$ within $\sigma/\sqrt{n}$ (or one s.d.) of 1 are common when $\mu = 1$; and values of $\bar{X}$ more than two or three s.d.'s from 1 are unusual when $\mu = 1$, tending to cast doubt on H_0. ▲

The distribution of the test statistic when H_0 is true will be referred to as its *null distribution*. In these last examples we have concluded that a value of the test statistic in the "tail" of its null distribution suggests that H_0 may not be the best explanation of the data. These *tail values* (by definition) are those far from the center of the distribution, out where the height of the histogram is close to 0 and values rarely occur. If there are other plausible models that would make a tail value more likely than it is under H_0, observing such a value casts doubt on H_0 and supports these other models as being more plausible than H_0.

The test statistics we shall encounter will have sampling distributions similar to that described by the histogram sketched in Figure 8-1. The values most commonly encountered when H_0 is true are those where the histogram is highest (corresponding to the largest

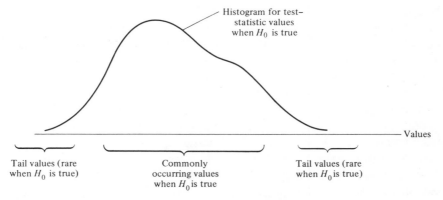

Figure 8-1. Typical null distribution of a test statistic.

concentration of probability); and those rarely encountered are at the tail ends, where the histogram is close to the horizontal axis—regions of very low probability. The sketch is just suggestive; there will generally be possible values of the test statistic between the common and the rare ones, values that neither strongly support H_0 nor cast much doubt on it.

In view of all this, what is the statistician to do? After formulating a null hypothesis, giving some consideration to possible alternatives, deciding on an appropriate test statistic, gathering data, and calculating the value of that statistic—then what? The next step depends on the reasons for conducting the study and on the context of the problem. The next step may not be the same in all cases, and people will often disagree on how to proceed.

A really informative report of an experiment would include the sample size, the value of the test statistic calculated from the data, and the location of this value in reference to the null distribution of the test statistic. But not everyone is able to assimilate all of this information in a useful way.

Some statisticians are fond of reporting test results in the form of confidence intervals. They leave it to the consumer to infer a measure of support for a null hypothesis or for some alternative, or to decide that further sampling is in order.

Some people report the probability under H_0 that the test statistic would take on a value at least as extreme as the observed value. This probability or "tail area" (in the null distribution) is called a *P-value*. It measures the degree of surprise one would feel at obtaining the observed value if H_0 were really true. The smaller the P, the greater the surprise.

Example 8-6   *ESP (continued)*

In the ESP example (8–4), suppose that the subject makes 5 correct calls in 10 tries—a sample proportion of .5. Surely, if this number of successes were reason to believe that the person has ESP, six or more successes would be even more reason to believe it. The probability of 5 or more successes in 10 tries (according to the table given in Example 8–4) is

$$P(\hat{p} = .5, .6, .7, .8, .9, \text{ or } 1) = .0584 + .0162 + .0031 + .0004 \doteq .078.$$

This is the *P*-value corresponding to the observed proportion of correct calls. It is shown as an area (a tail area) in Figure 8-2.

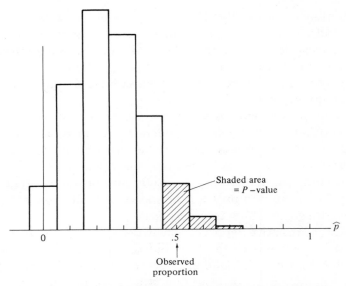

Figure 8-2. Histogram for the sample proportion under H_0:
$p = .25$ $(n = 10)$, for Example 8-6.

Thus, if there is no ESP operating, we are mildly surprised to observe 5 successes in 10 tries. We should be even more surprised if we were to observe 7 successes (the P-value is .0035). ▲

When a test statistic T takes on a value t_0, the *P-value* (for this result) is the probability that T would take on a value at least as far as t_0 (in the same direction) from what is expected when H_0 is true.

The smaller the P-value, the farther out in the tail of its null distribution is the observed value of T. So, with reference to the same sampling experiment, a P-value of .005 is stronger evidence against H_0 than a P-value of .03, say. But in comparing results of different sampling experiments, the P-value is not a good quantitative measure of the credibility of H_0. The next example illustrates this.

Example 8-7

Suppose that the null hypothesis $\mu = 0$ is to be tested, in a population with s.d. $\sigma = 1$. One experimenter uses a sample of size $n = 25$, and another, a sample of size $n = 2500$. Suppose that the first experimenter obtains a sample (of size 25) with mean $\bar{X} = .168$, with corresponding Z-score

$$Z = \frac{.168 - 0}{1/\sqrt{25}} = .84,$$

which yields a P-value of .20 (from Table II). A 95 percent confidence interval for μ is given by $-.22 < \mu < .56$, the limits being calculated as $.168 \pm 1.96/\sqrt{25}$. This interval includes 0, but many other values near 0 as well.

Suppose now that the second experimenter obtains a sample (of size 2500) with mean $\bar{X} = .0168$. The Z-score turns out to be .84, the same as the Z-score found by the first experimenter from his sample, and so the P-value is again .20. The 95 percent confidence interval constructed by the second experimenter would be $-.022 < \mu < .056$, which again includes the value 0, but not so many other values. The evidence in favor of $\mu = 0$ seems different with this larger sample. ▲

The results of an experiment are frequently reported using categories of P-values. The smaller the P-value, the less likely that one would observe as extreme a result, and the more "significant" is the result:

An observed value t_0 of the test statistic T is said to be

     highly significant,   if $P < .01$

     significant,   if $.01 < P < .05$

     not significant,   if $P > .05$.

We have left out the values $P = .01$ and .05 in these inequalities. One reason is that it is unusual for P to turn out to be exactly .01 or exactly .05. Another reason is that the dividing lines ($P = .01$ and $P = .05$) are quite arbitrary. If a P-value is .05 or just under .05, the result might be reported as "barely significant."

A result that is "significant" is signifying that if the null hypothesis is true, the sample we happened to get is not very representative—which is taken as evidence against the null hypothesis. So, the significance here is statistical—it has to do with sampling; and we could be more precise in saying "statistically significant," "highly statistically significant," or "not statistically significant." However, the increase in awkwardness does not seem to pay off in making the concept any more clear or in precluding its misinterpretation, so we (like others) often omit the qualifying adverb "statistical." The difficulties with the language of significance will be discussed further in Section 8.6.

8.2
Z-TESTS

It turns out that, like the sample mean $\bar{X}$, many statistics in common use are approximately normally distributed for large samples. When this is the case, the statistic can be standardized to yield a Z-score, whose significance can then be assessed in terms of the standard normal distribution (Table II of Appendix B). As usual, standardization is accomplished by subtracting the mean value of the statistic from its sample value and then dividing the difference by the standard error. The resulting Z-score measures how much the observed value of a statistic differs from its mean value, expressed as a number of standard errors.

Z-score for testing based on the value of a statistic whose distribution is (at least approximately) normal:

$$Z = \frac{\text{observed value} - \text{mean value}}{\text{standard error}},$$

where the mean value and standard error are evaluated under the null hypothesis.

A word about "standard error" is in order. What is really wanted, in standardizing a random variable, is its standard deviation. When this s.d. is known or computable under the null hypothesis, it can be used as the divisor in forming the Z-score. When it can only be estimated from sample information, the estimate—that is, the standard error—is used as the divisor. For simplicity, we refer to the divisor as "standard error" in either case.

Testing a null hypothesis using the Z-score is based on this idea: When a test statistic deviates from its mean value under H_0 by an amount that is large compared to a "typical" deviation (the standard error), then H_0 should be rejected. This assumes that there is a better way to account for a large deviation than to regard it as merely bad luck in sampling from the model of H_0.

The *P*-value defined by the *Z*-score of a particular value of the test statistic is the tail area in the standard normal distribution beyond that *Z*-score.

In testing a particular value for the population mean μ,

$$H_0: \mu = \mu_0,$$

the natural test statistic is $\bar{X}$, the sample mean. This is approximately normally distributed for large samples—and even for not-so-large samples if the population itself is not far from normal. The *Z*-score corresponding to an observed value of $\bar{X}$ is constructed using the mean and standard deviation of $\bar{X}$:

$$\text{m.v. } (\bar{X}) = \mu_0 \qquad (\text{under } H_0)$$

$$\text{s.d. } (\bar{X}) = \frac{\sigma}{\sqrt{n}},$$

or, when σ is not known (the usual case), using the standard error:

$$\text{s.e. } (\bar{X}) = \frac{S}{\sqrt{n}},$$

in place of the s.d. The *Z*-score is thus

$$Z = \frac{\bar{X} - \text{m.v.}(\bar{X})}{\text{s.d.}(\bar{X})} \doteq \frac{\bar{X} - \mu_0}{S/\sqrt{n}}.$$

The corresponding *P*-value is found in the normal table (Table II) as the probability of getting a *Z*-score at least as far out in the tail of the standard normal distribution as the one observed.

Large-sample *Z*-test for $\mu = \mu_0$, based on the value of $\bar{X}$:
Calculate

$$Z = \frac{\bar{X} - \mu_0}{S/\sqrt{n}},$$

and determine the *P*-value from Table II, as the probability in the tail beyond that *Z*-value.

Example 8–8   *Life of a Light Bulb*

A sample of 50 light bulbs is tested to check the hypothesis that the mean life is (as claimed by the manufacturer) at least 700 hours, against the alternative that it is less than 700 hours. The null hy-

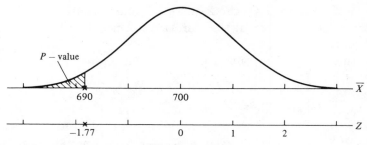

Figure 8-3.   Histogram for $\overline{X}$, Example 8-8.

pothesis is $\mu \geqslant 700$, but we have to pick a single value of μ in forming a Z-score and calculating probabilities. The most crucial value of μ in H_0 is 700, since if our test has a given small probability of rejecting H_0 when $\mu = 700$, it will have an even smaller probability of rejecting H_0 when the mean life exceeds 700 hours. Suppose, now, that the sample mean is 690 hours and the s.d. is 40 hours. Then with $\mu_0 = 700$, the Z-score (see Figure 8-3) is

$$Z = \frac{690 - 700}{40/\sqrt{50}} = -1.77.$$

The corresponding P-value is

$$P(Z < -1.77) = .038,$$

found by entering Table II at -1.77. Thus, the evidence suggests that μ is not 700—and even more, that it is not greater than 700. ▲

In estimating population *proportions* it was found that the sample proportion $\hat{p}$ is approximately normal, with

$$\text{m.v.} (\hat{p}) = p \quad \text{and} \quad \text{s.d.} (\hat{p}) = \sqrt{\frac{pq}{n}},$$

where p is the population proportion and $q = 1 - p$. A Z-statistic for testing a particular value p_0 of p is again constructed by standardization:

Z-test for $p = p_0$, based on the value of $\hat{p}$:
   Calculate

$$Z = \frac{\hat{p} - p_0}{\sqrt{p_0 q_0/n}},$$

and determine the *P*-value from Table II, as the probability in the tail beyond the observed sample proportion $\hat{p}$.

Example 8-9 *ESP (more)*

A test for ESP based on 10 trials was described in Example 8-6. There, the *P*-value corresponding to an observed five successes in 10 trials was found to be .078 (using Table I of Appendix B). Although $n = 10$ is not large, we can use a *Z*-score if we are careful with the "continuity correction" described in Section 6.3. Referring to Figure 8-4, we see that what is wanted is the area of the shaded bars; but since Table II gives areas to the left, we look first for the total area in the unshaded bars. To approximate this, we find the area to the left of 4.5, or on the scale of proportions, to the left of 4.5/10, for which we need this *Z*-score:

$$Z = \frac{4.5/10 - .25}{\sqrt{.25 \times .75/10}} = 1.46,$$

.25 being the value p_0 being tested (corresponding to "no ESP"). The area to the left of $Z = 1.46$ is .9278 (from Table II), so the desired *P*-

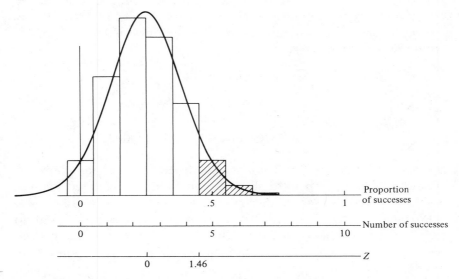

Figure 8-4. Normal approximation of a *P*-value (Example 8-9).

value is approximately .0722. This is not a bad approximation to the actual value of .078, in view of the small sample size. ▲

Example 8-10 *Is Smoky the Culprit?*

An atomic explosion, code-named "Smoky," was conducted in Nevada in the summer of 1957. The U.S. Center for Disease Control has tracked down 450 Smoky participants in a search triggered by a compensation claim for one participant and found 8 cases of leukemia among the 450. The overall rate for men of ages similar to those of the participants is 1/1500. Is the incidence of 8 in 450 statistically significant?

The Z-statistic for testing $p = 1/1500$ in a sample of 450 is

$$Z = \frac{\hat{p} - 1/1500}{\sqrt{(1/1500)\,(1499/1500)/450}},$$

and with $\hat{p} = 8/450$ one obtains $Z = 14.06$—highly significant; or is it?

Our appeal to the normal distribution is *not* really legitimate! The sample size is large, but *not large enough*. Recall that in discussing the approximation of the distribution of a sample proportion by use of the normal distribution, we warned of the need for a large n when p is very small (Section 6.3). In particular, npq should be at least 5 and preferably 10 or more. Here $npq = 450 \times (1499/1500^2) = .3$. However, an exact computation (along the lines of Example 4–21) would show that, for instance, when $p = 1/1500$,

$$P(2 \text{ or more cases in } 450) = .037$$

and that

$$P(8 \text{ or more cases in } 450) = .000000001.$$

So, even though the Z-test is not exactly correct, the sample result (8 out of 450) is so extreme that the P-value calculated from Z would be correct to several decimal places.

Did Smoky cause leukemia? The statistics do not prove this but do point to the possibility, and the CDC proposed to trace other participants in Smoky and to determine what other radiation they may have been exposed to. ▲

8.3
THE T-TEST

Statistics that are nearly normal for large samples may not be so when the sample size is not large. In particular, the "Z-score" for $\bar{X}$ used in testing $\mu = \mu_0$,

$$Z = \frac{\bar{X} - \mu_0}{S/\sqrt{n}},$$

is not well approximated by the standard normal Z, because the denominator is too variable when n is small—as was discussed in Section 7.5. However, as explained there, the modified score

$$T = \frac{\bar{X} - \mu_0}{S/\sqrt{n-1}} = \frac{\bar{X} - \mu_0}{\tilde{S}/\sqrt{n}}$$

(where $\tilde{S}$ is the version of the s.d. with denominator $n - 1$) has a distribution close to the "t-distribution with $n - 1$ degrees of freedom" (Table III of Appendix B)—if the population is not too different from normal. A test based on T is called a "t-test." It is fairly *robust*— that is, relatively insensitive to a moderate departure from the assumption of normality of the population.

Example 8-11 *Conserving Electricity*

The girls in a Brownie troop checked electric power consumption in their homes during a 1-week period—and then again in the following week, during which time they made a special effort to conserve power by turning off lights, using appliances sparingly, and so on. They found decreases in power consumed as follows (in kilowatt-hours):

$$29, \quad 42, \quad -26, \quad 7, \quad -3, \quad 6, \quad 14, \quad 36.$$

The negative decreases (which are increases) may be surprising, but there is considerable variability in consumption from week to week in most homes. For example, if the use of an electric clothes dryer were shifted from just before to just after a child's middle reading, her best efforts may have served only to make an increase smaller than it would have been otherwise.

The mean and s.d. of the eight decreases are $\bar{X} = 13.125$ and $S = 20.88$, so

$$T = \frac{13.125}{20.88/\sqrt{7}} = 1.66.$$

Under the null hypothesis of no change in average consumption, the distribution of T is the t-distribution with 7 degrees of freedom if the sample is a random sample. The P-value is not listed in Table III but is between .05 (for $T = 1.90$) and .10 (for $T = 1.42$). So there is some evidence that the population mean is not 0, even though the P-value is not quite in the range traditionally (and arbitrarily) called "significant." (We gloss over the question of whether the population is close enough to normal for the t-distribution to be applicable. From what we know of the experiment in question, the assumption of near normality seems reasonable.)

Only about half of the Brownies in the troop turned in usable reports (which required three meter readings—one before and after each of the two consecutive 1-week periods), so there is the problem of partial response. The sample may be just a sample from the population of "eager beaver" Brownie families. On the other hand, what is being "tested" perhaps, is the effect of educating children to help conserve electricity. The effort seems worthwhile, at least in the short run, even if only the eager Brownies react to the treatment. ▲

---

t-test for $\mu = \mu_0$ (σ unknown):
Calculate

$$T = \frac{\bar{X} - \mu_0}{S/\sqrt{n-1}},$$

and determine the P-value from Table III (using $n - 1$ degrees of freedom).

The test assumes a normal population but is fairly robust.

It should be noted that if σ can be assumed known, which is not the typical case, then there is no need to approximate σ by S; and a Z-test with $\sigma/\sqrt{n}$ in the denominator of Z can be used if the sample is large enough for $\bar{X}$ to be close to normal.

As the sample size increases, the difference between using the factor $n - 1$ and the factor n becomes negligible. (When $n = 30$, for instance, $(n - 1)/n = .983$.) Moreover, the sample standard deviation becomes a better and better approximation to the population σ, with relatively little variability. So the statistic T will have about the same value as Z, as well as practically the same distribution. The t-test becomes à Z-test.

How does one know whether to use Z (with Table II) or T (with Table III)? The answer is simple: If the population σ is known, use Z, with $\sigma/\sqrt{n}$ in the denominator. If σ is unknown, use T. If in using T the degrees of freedom (d.f.) are beyond those provided for in the t-table, use the row for d.f. $= \infty$, which gives normal percentiles. Thus, you will be automatically led to Z when Z is appropriate. The one thing you have to be alert to, when n is small, is the possibility that the population distribution is radically different in shape from the normal; neither T nor Z is appropriate in such a case.

8.4
A TEST BASED ON RANKS

Hypotheses about the center of a population have been dealt with, thus far, in terms of the population mean, the sample mean being used in constructing tests. A rather different approach for testing the location of a symmetric, continuous distribution, as given by the center of symmetry (which is both mean and median), is one based on ranks. It is applicable to samples of any size and, unlike the t-test, does not require any further assumption about the nature of the population, even in the case of small samples.

The *Wilcoxon signed-rank test* is based on the deviations $X - m_0$ from the value m_0 to be tested as the center of the distribution. If m_0 is indeed the center, some deviations will be positive and some negative, and the positive deviations will tend to be as large and as small the negative ones. If, however, the value m_0 being tested is *not* the center, the deviations will tend to be predominantly of one sign; and what deviations there are of the other sign will be among the smaller ones in magnitude (see Figure 8–5).

The signed-rank statistic is calculated by first arranging the deviations $X - m_0$ according to magnitude, assigning ranks $1, 2, \ldots, n$ to the ordered deviations (the smallest deviation in magnitude has rank 1, the next smallest rank 2, and so on), and then summing the ranks of the deviations of one sign. The sum R_+ of the ranks of the positive deviations and the sum R_- of the negative deviations are related, for

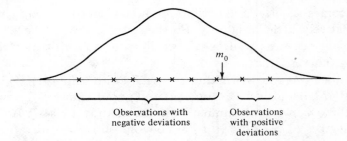

Figure 8-5. Typical sample from a population whose center is not at m_0.

their sum is a constant:

$$R_+ + R_- = 1 + 2 + \cdots + n = \frac{n(n+1)}{2}.$$

So knowing R_+ is equivalent to knowing R_-, and either can be used as the test statistic.

Example 8-12

The times to burnout of 10 light bulbs were given in Problem 34 of Chapter 3 as follows:

390, 705, 640, 882, 940, 480, 564, 690, 728, 1149.

Suppose that this sample is to be used in testing the hypothesis that the median time to burnout is 700 hours, as claimed by the manufacturer. The *deviations* $X - 700$ are

-310, 5, -60, 182, 240, -220, -136, -10, 28, 449,

and in order of *magnitude:*

5, -10, 28, -60, -136, 182, -220, 240, -310, 449.

The sequence of signs is now all that matters:

+, -, +, -, -, +, -, +, -, +,

and the ranks of the +'s in sequence add up to

$$R_+ = 1 + 3 + 6 + 8 + 10 = 28.$$

The -'s have ranks with sum

$$R_- = 2 + 4 + 5 + 7 + 9 = 27.$$

Observe that $R_+ + R_- = 10 \cdot 11/2$. ▲

A basic assumption in calculating probabilities for R_+ is that the population is continuous, an assumption that precludes ties in ranking. It also precludes a deviation of exactly 0, for the probability that X is *equal* to m_0 is zero. In practice, of course, observations are rounded off, so ties and zero deviations can occur. If $X - m_0 = 0$ for any observation X, we ignore this observation, since the deviation is neither positive nor negative and contributes nothing to the analysis. If $X - m_0$ has the same value for two or more observations, the rank assigned to each is the *average* of the ranks they would have if they were slightly different.

Example 8-13 *Median Body Temperature*

The figure 98.6 is used as a "norm," as the typical body temperature (in °F) of a healthy person. The temperatures of the first 20 student nurses listed in Data Set A (Appendix C) are as follows:

$$
\begin{array}{cccc}
98.2 & 98.6 & 98.6 & 100.0 \\
98.6 & 99.4 & 98.2 & 99.6 \\
98.8 & 98.6 & 99.0 & 99.4 \\
98.2 & 98.8 & 98.0 & 99.2 \\
100.0 & 98.0 & 98.8 & 98.6
\end{array}
$$

The corresponding deviations $T - 98.6$ are

$$
\begin{array}{cccc}
-.4 & 0 & 0 & 1.4 \\
0 & .8 & -.4 & 1.0 \\
.2 & 0 & .4 & .8 \\
-.4 & .2 & -.6 & .6 \\
1.4 & -.6 & .2 & 0
\end{array}
$$

Ignoring the five 0's, we rank the remaining 15 deviations according to magnitude:

$$
\begin{array}{cccccc}
.2 & -.4 & -.6 & .8 & 1.0 & 1.4 \\
.2 & -.4 & -.6 & .8 & & 1.4 \\
.2 & -.4 & .6 & & & \\
.4 & & & & &
\end{array}
$$

The rank positions are

$$
\begin{array}{cccccc}
1 & 4 & 8 & 11 & 13 & 14 \\
2 & 5 & 9 & 12 &   & 15 \\
3 & 6 & 10 &   &   &   \\
7 &   &   &   &   &   \\
\end{array}
$$

with averages (used for tied observations):

$$2 \quad 5.5 \quad 9 \quad 11.5 \quad 13 \quad 14.5.$$

The sum of the ranks of the negative deviations is

$$R_- = 5.5 + 5.5 + 5.5 + 9 + 9 = 34.5$$

And R_+ is $15 \times 16/2 - 34.5 = 85.5$. (To be continued.) ▲

To interpret a value of R_+ as evidence for or against the null hypothesis requires a knowledge of the distribution of such values, at least when H_0 is true. Remarkably—and fortuitously—this distribution turns out to depend only on the sample size. It does *not*, in particular, depend on the shape of the population distribution (except that symmetry is assumed). For small samples, left-tail probabilities are given in Table IV of Appendix B. For sample sizes larger than those given in the table, R_+ is nearly normal, so a Z-test (Section 8.2) can be used. Constructing a Z-score from R_+ requires knowledge of its mean and s.d., which are as follows:

$$\text{m.v.}(R_+) = \frac{n(n + 1)}{4}$$

and

$$\text{s.d.}(R_+) = \sqrt{\frac{n(n + 1)(2n + 1)}{24}}.$$

Wilcoxon's signed-rank test for

$$H_0: \text{ median} = m_0$$

in a symmetrical population:

1. Arrange deviations $X - m_0$ according to magnitude.

2. Sum the ranks of the positive deviations to obtain R_+.
3. To interpret R_+:
   (a) When $n \leqslant 15$, consult the null distribution of R_+ in Table IV.
   (b) When $n > 15$, use Table II for the null distribution of the Z-score:

$$Z = \frac{R_+ - n(n+1)/4}{\sqrt{n(n+1)(2n+1)/24}}.$$

(The sum of the ranks of the negative deviations R_- could be used just as well; it has the same null distribution as R_+.)

Example 8-14 *Body Temperature (continued)*

In testing the hypothesis that the median body temperature is 98.6, we use the value of R_- computed in Example 8–13:

$$R_- = 34.5.$$

The P-value corresponding to $R_- = 34$ is found in Table IV in the column for $n = 15$:

$$P(R_- \leqslant 34) = .076.$$

(Although there were originally 20 observations in the sample, five could not be used because their deviations from 98.6 were 0.) The P-value for $R_- = 35$ is .084. Interpolating, we take the P-value for $R_- = 34.5$ to be .080.

Since we have exact probabilities for R_-, we can check the normal approximation to see how close it is at this point ($n = 15$), where Table IV leaves off. We compute the Z-score for $R_- = 34.5$:

$$Z = \frac{34.5 - 15 \times 16/4}{\sqrt{15 \times 16 \times 31/24}} = -1.45,$$

and obtain a P-value (from Table II) of .0735, not far from that found using the exact probabilities. (Presumably, the approximation is even better for samples of more than 15 observations.)

A P-value of around 7 or 8 percent is not "significant" by the usual convention, but there does seem to be some evidence against the hypothesis that the population center is 98.6.

This analysis of temperature data has served to illustrate the use of ranks in a test for location, and the use of Table IV and of a normal approximation. However, researchers are too often content to give the calculations and conclusions as we have just given them, without raising the sorts of questions that show some understanding of their limitations.

Was a rank statistic the thing to use? We could have calculated a T-statistic, now using all 20 observations (with 19 d.f.); it turns out that $T = 1.70$, with a P-value near and only slightly above .05. This is smaller than the P-value for R_- because T uses the magnitudes of the deviations (not just their ranks). (Observe that changing the 100.0's to 99.7's would *not* change R_- but *would* change T.) The t-test assumes normality, and the rank test assumes less—namely, symmetry. A look at the data does not confirm either assumption; neither does it make them unreasonable. Ordinarily, the less one has to assume, the better; but on the other hand, if we had used all the data we had ($n = 142$), the ranking process would have been perhaps even more tedious than finding T (depending on our computational facilities), and T is pretty robust.

Was the sample a random sample? It was not drawn as a random sample, using random numbers or drawing chips from a bowl. It was a "sample of convenience," with the possibility of hidden biases. Yet the way subjects entered the sample would seem to be unrelated to the question at hand, the average of body temperatures, provided that any conclusion is restricted to the population of those who might have entered the sample—the population of all who might be nursing students at the particular institution where the data were collected. Conclusions might also have to be restricted to the time of day when the data were recorded (if there is truth to the common impression that P.M. temperatures are higher than A.M. temperatures).

What, really, was the aim of the investigation? This question should be asked before collecting data, but investigators are often hazy on this point. In the case at hand, if we were really worried about the common belief that a normal or average temperature for healthy people is 98.6, we should have got a lot more data, from a broader population base. If we are only mildly curious as to whether nursing students differ from the general population, the data gathered might suffice, except that one would want to be more careful to see that only healthy persons are included. ▲

8.5
THE CHI-SQUARE TEST OF FIT

The testing of a particular model as H_0 is referred to as testing *goodness of fit*. Tests for $\mu = \mu_0$ based on $\bar{X}$ are actually tests of fit, but they are designed for (and work best for) the case of testing against alternatives expressed in terms of the location parameter μ. When the alternatives are more general or more vague, other tests should be used.

The model for a categorical variable, with k categories, is specified by giving probabilities for those categories. The most general alternative to a particular specification is that some one or more of the specified probabilities are not correct.

Example 8-15 *Testing the Fairness of a Die*

An ordinary die was tossed 100 times to see how often each side would turn up. By symmetry, each face has the same mean frequency if the die is fair: $100/6 = 16.67$. The observed frequencies, the corresponding mean frequencies under the model of fairness, and the amounts by which these differ are shown in Table 8-1. Some observed frequencies are less, and some more, than the mean frequency. The differences must add up to 0, because both the observed and expected or mean frequencies add up to 100. (If the sum is not quite 0, this is because of roundoff.)

Even if the toss is fair, one would be surprised to have the observed frequencies turn out to be equal—especially when the sample size is

Table 8-1

Side	Observed Frequency	Expected Frequency Under H_0: Die Is Fair	Difference
1	14	100/6	−2.67
2	17	100/6	.33
3	20	100/6	3.33
4	20	100/6	3.33
5	18	100/6	1.33
6	11	100/6	−5.67
Sums	100	100	0

not a multiple of the number of categories. Nevertheless, the fit would be considered good if the deviations of observed from mean frequencies under H_0 are small, in some overall or average sense. (To be continued.) ▲

Karl Pearson, in about 1900, introduced a statistic for measuring goodness of fit based on the squared differences between the observed frequencies and the mean frequencies in the model being tested. It is called "Pearson's chi-square":[†]

$$\chi^2 = \sum \frac{(\text{observed frequency} - \text{mean frequency})^2}{\text{mean frequency}},$$

the sum extending over all possible categories. Clearly, if there are large discrepancies between the observed frequencies and the mean frequencies under H_0, then χ^2 will be large. And if there is exact agreement of observed and mean frequencies in all categories, then and *only* then will χ^2 be equal to 0.

Example 8-16 *The Fair Die (continued)*

Table 8-1 gives the differences between observed and mean frequency for the six sides of a die, corresponding to a sequence of 100 tosses. The chi-square statistic is computed from these as follows:

$$\chi^2 = \frac{(-2.67)^2}{16.67} + \frac{(.33)^2}{16.67} + \frac{(3.33)^2}{16.67} + \frac{(3.33)^2}{16.67} + \frac{(1.33)^2}{16.67} + \frac{(-5.67)^2}{16.67}$$

$$= 3.8.$$

The measure χ^2 is not 0, of course, because there are discrepancies between the observed and the mean frequencies under the model of fairness. But the question is: How shall we interpret the value 3.8? Is it "large"—large enough to warrant rejection of fairness? This is the kind of question that we have encountered many times before; it can only be answered if we understand what kinds of values are typical when H_0 (fairness) is true. That is, it is necessary to know the null distribution of the test statistic—of χ^2, in the present example. ▲

The null distribution of χ^2 is quite complicated and depends on the model being tested. However, there is a way out—at least in the

[†]The symbol χ is the lowercase Greek "chi," pronounced "ki" to rhyme with "why."

case of large samples. And this, not just defining χ^2, is Pearson's important contribution: The null distribution of χ^2 in the case of large samples is approximately what is called a "chi-square distribution with $k - 1$ degrees of freedom," where k is the number of categories, no matter what model is being tested! This distribution (for which Pearson gave appropriate mathematical formulas) is tabulated, by percentiles, in Table V of Appendix B, for degrees of freedom from 1 to 30. (This covers the range of degrees of freedom encountered in most goodness-of-fit problems, and for larger numbers a normal approximation is available.)

Example 8-17   *The Fair Die (Still More)*

In testing fairness of a die using the 100 observations summarized in Table 8-1, we found that the value of the chi-square statistic is χ^2 = 3.8. Looking at the row of Table V labeled "5 degrees of freedom" (since $k - 1 = 5$ in this problem), we see that the value 3.8 is just about at the center of the distribution. It is nowhere near the 95th percentile (11.1), so is definitely not "significant." It is, indeed, a rather typical value for χ^2 when the die is fair. (We do not give a *P*-value for it because it does not mean much to say "as far from the center" when the value of a statistic is so close to the center of its distribution (see Figure 8-6). ▲

Chi-square test of a particular model for a categorical variable:
    Calculate

$$\chi^2 = \sum \frac{(f - np)^2}{np},$$

where for each category f is the observed frequency and p is the probability under H_0; find the *P*-value in Table V using $k - 1$ degrees of freedom (k is the number of categories). Large values of χ^2 are significant.

(Because Table V gives percentiles, you will not find a *P*-value directly but can locate it between percentages, corresponding to the percentiles that bracket the calculated χ^2.)

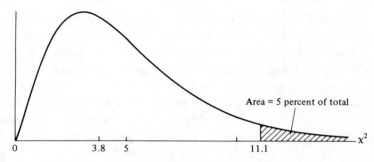

Figure 8-6. Histogram for χ^2, 5 degrees of freedom.

In applying the chi-square test, one should have something like 10 times as many observations as categories (or more), and at most one or two mean frequencies as small as 1. If there are several categories with small mean frequencies, some categories should be combined (in the model and in the data) so that most categories are, say, of mean frequency 5 or more.

If the model to be tested is discrete but has infinitely many categories, or if it is continuous, then the chi-square test can be applied as an approximate test after creating an arbitrary modification of H_0 which has finitely many categories. To do this it is only necessary to specify probabilities for class intervals (or sets of possible values), with a catch-all category of the form "10 or more" if necessary. We dwell no further on such cases. (Better methods exist for testing fit for continuous models, but we do not take them up.)

Pearson's chi-square statistic can be used in testing the fit of a *type* of model—a family of models with an unknown parameter or parameters. All that is required is to obtain a good estimate of the unknown parameters from the sample so that cell probabilities can be estimated for "plugging into" the χ^2-formula; and then it turns out that the large sample null distribution is again of the chi-square type, although with a changed number of degrees of freedom. We shall not go into more detail at this point, although we shall use the method in the next chapter (in Section 9.5) in a specific type of problem.

8.6
MISHANDLING "SIGNIFICANCE"

The methods of statistical testing have been widely disseminated during the past few decades. It is unfortunate that some users, overlooking the rationale of these methods, take them to be something they cannot be. They invoke the phrases "reject the null hypothesis" and "highly significant" in a manner suggesting that an outside authority (Statistics) has settled the matter and proved their point.

The term *significant* is particularly troublesome because it has an everyday meaning as well as a technical meaning in statistical inference. One result is that people sometimes apply a statistical test and deduce "significance" when there is no statistical inference involved.

Example 8-18 *Ages of Blacks and Whites*

The U.S. Census of 1970 showed the median age of black males to be 21.0 and the median age of white males to be 27.6 years. Is the difference of 6.6 years statistically significant?

Even those who have had an explanation of the meaning of statistical significance may fail to appreciate that this is a meaningless question. In a comparison of population characteristics—and the census is supposed to be a survey of the whole population of the United States—there is no question of statistical inference. There are no samples and no sampling fluctuations to explain away observed deviations—no question of statistical significance. The population of black males *is different*, with regard to median age, from the population of white males. Whether this difference has significant social implications, or cultural or medical explanations, is not a statistical matter. ▲

Another problem with the phrase "statistically significant" is that when there *is* a proper sample that yields a statistically significant value of some statistic, this "significance" is often misinterpreted as meaning *practical* significance, or significance for the field of application involved.

Example 8-19 *IQ and Culture*

An Associated Press report in 1978 stated that "IQ levels can be improved significantly by raising children in more intellectual environments," according to a research study by an institute of child development. The study involved 130 black children adopted as babies by white parents with higher than average IQ's—120 as compared with a general average of 100. After some years with the adopting family, the average IQ of the black children was 106, whereas the statewide average for black children was 90.

Whether the difference 106 - 90 is statistically significant or not depends on the population variability, and this was not reported. Apparently, the research study found statistical significance, and the press reporter seized on the word *significant.* The AP report clearly suggests not only that IQ levels can be improved, but also that the amount of improvement is substantial—well worth the effort. But this question is not addressed by a statistical analysis. Actually, since it is generally conceded that IQ tests have a white culture bias, one would be surprised *not* to find that IQ scores can be raised by rearing children in a white household. ▲

Example 8-20

Suppose that we tossed a coin 100,000 times and found 50,316 heads. (There are those who have actually done this much tossing!) The sample proportion of heads $\hat{p}$ is nearly normally distributed, and the observed value $\hat{p}$ = .50316 differs from the ideal proportion p = 1/2 by 2 s.d.'s (The s.d. of $\hat{p}$ is $.5/\sqrt{100,000}$ = .00158.) In testing p = 1/2, this result is significant at the 5 percent level, since the P-value is close to 2 percent.

What has happened, then, is that the sample of 100,000 trials has yielded more heads than would be ordinarily encountered if p were precisely 1/2. But the estimate of p from the sample is about .503, and it would be unusual for this slight deviation from fairness to be of concern. That is, the deviation .003 has little *practical* significance. ▲

A P-value considered "significant" can signify a departure from H_0 which, although real, may be trivial or of no practical significance.

Generally speaking, you should beware of conclusions phrased like these actual quotes:

1. "The training system had measurable and significant influences on performance."[†]
2. "It was found that student performance and instructor ratings both increased significantly."[†]

[†]From the study mentioned in Problem 20, Chapter 1.

3. "Studies show Excedrin to be significantly more effective than aspirin."[†]

4. ". . . our conclusion being that this program leads to a significant increase in monthly income."[‡]

These statements seem to claim practical significance—treatment effects that are worth the cost. In fact, they were given as paraphrases of a conclusion of "statistical significance," which means only that the sample was large enough to detect a departure from the null hypothesis, one that may or may not be significant in the practical sense.

On the other hand, a real and practically meaningful departure from a null hypothesis may not be detected in a statistical test, especially if the sample is not large.

Example 8-21   Checking Instrumentation

A hospital laboratory conducts daily tests of various measuring instruments, using a "standard" blood supply. Each day, a test measurement is made on a small amount of blood from the standard supply. If the hemoglobin level of a standard supply is 15.1, and the lab's measurements have a standard deviation of .40, a measured value of 14.5 would not be considered statistically significant. If the equipment has developed a bias, shifting the mean to 14.5, a single measurement is not precise enough to detect this with assurance. However, a physician might well consider the difference between 15.1 and 14.5 as medically significant if these could be considered as exact levels. Perhaps one reason that the lab can operate at this level of sensitivity is that a patient's hemoglobin level is not a static thing, and the amount of variation in it is on the same order of magnitude as the uncertainty in the measurement process used by the lab. (To be continued.) ▲

If n is small, a sample can produce a large P-value—implying "not significant"—even when the true model is appreciably and significantly different from that of the null hypothesis.

[†]From a TV commercial.
[‡]From an instructor's manual for a textbook in elementary statistics!

Consider, now, some implications in this state of affairs. It is a fact that some scientific journals have an editorial policy of not publishing reports of research that do not have statistically significant results. Suppose that you gather data to test a hypothesis H_0 and find, upon analyzing it, that your test statistic is not significant. You may find it difficult to get this result published—but there is a way! Your null hypothesis model is necessarily precise, so that you can calculate probabilities in this model. However, in making it precise you will almost always make it different from the way things *really* are, even if only slightly different. And a large enough sample will pick up even this slight difference; all you need to do, then, is to keep on sampling and testing until you have enough data to reject H_0.

Any departure from H_0, however slight, will be detected and identified as "statistically significant" if the sample is large enough.

As if this were not enough, suppose that some hypothesis H_0, one of general interest, is really *true*. For every 100 research workers testing this hypothesis, 5 (on the average) will find evidence to reject H_0 (erroneously) at the 5 percent significance level. These 5 can get their erroneous results published, but the other 95 cannot, even though their results are correct!

All of this is not only a criticism of the way people use statistical tests, but also a comment on the notion of hypothesis testing. However, all statistical techniques have similar drawbacks, because they deal with the difficult problem of inference in the face of uncertainty—drawing general conclusions from particular observations. The message for consumers of statistical "demonstrations" is that they must exercise caution. Perhaps our procedures should carry a label: "Possibly harmful if not used with thoughtful care."

When used in conjunction with common sense and caution, statistical procedures can be valuable tools in scientific investigations and in decision making. Distilling a set of data to a single number—the value of some test statistic—can be informative and helpful. But do not ignore the assumptions and the rationale behind the test. In particular:

A statistical procedure is not a substitute for common sense.

8.7
ACCEPTANCE AND REJECTION

So far, we have presented the notion of "testing" a null hypothesis as one of extracting from a set of data evidence for or against that hypothesis, and reporting the result as a *"P*-value." According to the strength of the evidence one may be led to gather more data, or simply to "file" (publish) the report in the appropriate literature— for the information of those interested, and to add to the accumulated knowledge in the field of the investigation.

In the adoption of the 5 percent convention for "significance," we seem to come to a decision, albeit arbitrary, a decision to "reject" the null hypothesis if $P < .05$. But in scientific work, especially, this is seldom a hard and fast decision; for one is always open to further evidence and, possibly, to a revision of previous conclusions. Moreover, failure to "reject" (as when the sample evidence is not "significant") is not usually taken to mean that the null hypothesis is established. Rather, it is simply allowed to stand, pending further investigation.

Sometimes, especially in the areas of business and industry, a statistical analysis of data can help resolve a *decision* problem in such a way that there is a good chance of a correct decision. Whether or not to introduce a new product, to cancel a TV show, to accept a shipment of goods—these are typical decision problems that can make effective use of information in a sample (a sample of the market, the audience, or the shipment, respectively). In such cases a decision must be made—*an action must be taken.* In the simplest case of choosing between *two* actions, we are forced to "put our money" either on the null hypothesis, under which one action is preferred, or on an *alternative hypothesis*, denoted by H_A, under which the *other* course of action is preferred.

Example 8-22   *Accepting a Shipment*

A shipment of disposable oral thermometers is to be accepted for clinic use, or not, on the basis of test results on 10 thermometers chosen at random from the lot. A thermometer will be classified as either *good* or *defective* according as it gives a reasonable reading or not.

Knowing that no manufacturer can guarantee that all its thermometers are good, the clinic is prepared to consider a lot acceptable if it contains no more than 10 percent defectives. Thus, if p denotes the proportion of defectives in the lot, accepting the lot amounts to accepting the hypothesis that $p \leqslant .10$. This is H_0, the null hypothesis. Rejection of the lot amounts to rejecting H_0 and accepting the alternative hypothesis H_A, that $p > .10$.

Because, in a random selection of 10 thermometers from the lot, the quality of the sample will usually reflect the quality of the lot, it seems reasonable to use the sample proportion of defectives as a guide in deciding whether or not to accept the lot. In particular, if there are too many defectives in the sample, it is reasonable to reject the lot—and with it, the null hypothesis. (To be continued.) ▲

In decision problems where the choice is between two courses of action, the "testing" of an hypothesis H_0 against some alternative H_A is carried out as follows:

1. Determine an appropriate test statistic, one whose values will help to make the choice between the two actions more clear.
2. Decide what values of the test statistic will lead you to reject H_0. (The other possible values will then, of course, call for accepting H_0.)
3. Compute the value of the test statistic from the data you gather (or have gathered) and then follow the rule set up in step 2.

Rejection of H_0 in such a test is equivalent to acceptance of the alternative H_A; and, conversely, acceptance of H_0 is equivalent to rejection of H_A.

In testing H_0 against H_A (or if you like, H_A against H_0—the situation is rather symmetric), it is important that the defining of H_0 and H_A, and the setting forth of the rule you will follow in choosing between them, be done *without reference to the data to be used in the decision*. The reason for this is that we are not otherwise able to assess the chances of wrong decisions.

Example 8-23 *Accepting a Shipment (continued)*

One rule the clinic might follow in deciding whether the shipment of thermometers is to be accepted or not is to reject the lot if *any* defectives show up among the 10 thermometers in the sample. For reference, we call this Rule 1:

$$\text{Rule 1:} \quad \text{Reject the lot (and } H_0 \text{) if } \hat{p} \geqslant \frac{1}{10},$$

where (as usual) $\hat{p}$ denotes the sample proportion defective. This rule may be overly protective, in that there is a pretty good chance of getting a defective in the sample even when the lot is acceptable. A somewhat less demanding rule, in this sense, is the following:

$$\text{Rule 2:} \quad \text{Reject the lot (and } H_0 \text{) if } \hat{p} \geqslant \frac{2}{10}.$$

Here, realizing that with $p = .10$ we might "expect" one defective in a sample of 10, we take *more* than one defective in the sample as a signal of a bad lot.

Neither Rule 1 nor Rule 2 is infallible; bad luck in sampling—getting a "nonrepresentative" sample—can lead to the wrong conclusion about the lot. (To be continued.) ▲

In specifying a rule to follow, the values of the test statistic that are taken as reason to reject H_0—or to take the action preferred when H_0 is false—comprise what is called the *critical region* of the test. Ordinarily, these are extreme or tail values of the test statistic; and if a particular value is in the critical region, then any values that are more extreme are also in the critical region, being even stronger evidence against H_0.

As suggested in Example 8-23, almost any conclusion about a population based on the information in a sample can be wrong. A conclusion to "reject H_0" will be wrong if H_0 is true; one will err in this way if a sample happens to be drawn that is not sufficiently representative of H_0, causing the test statistic to fall in the critical region even though H_0 is true. A conclusion to "accept H_0" is wrong if H_0 is actually false; this will happen, when H_0 is false, if one is so unlucky as to draw a sample for which the test statistic does *not* fall in the critical region. These mistakes or errors are of two distinct types (as are the mistakes a jury can make—freeing a guilty person or convicting an innocent one):

Type I error: Rejecting H_0 when H_0 is true.

Type II error: Accepting H_0 when H_0 is false.

We distinguish between these types of error because, in a typical decision problem, their consequences are usually different (as they are in the case of the jury!). Thus, to take an extreme example, forwarding a bad lot of drugs for use may cause deaths, whereas failing to ship a good lot would involve only the waste and disposal of the lot.

If H_0 defines a specific model for the test statistic, and it often does, a probability can then be calculated for an error of type I. If several models are included under H_0, a probability for an error of type I can be calculated for each such model. The alternative H_A is ordinarily quite broad, including many possible models for the test statistic; probabilities for a type II error can be calculated for any specific model included in H_A.

Example 8-24 *Accepting a Shipment (still more)*

In Example 8-23 we suggested two tests—two rules (out of many possible rules)—for choosing between accepting a lot of thermometers and rejecting the lot, based on the sample proportion defective $\hat{p}$:

$$\text{Rule 1:} \quad \text{Reject the lot if } \hat{p} \geqslant \frac{1}{10}.$$

$$\text{Rule 2:} \quad \text{Reject the lot if } \hat{p} \geqslant \frac{2}{10}.$$

Using either rule, there is a possibility of error of one type or the other. These can be assessed using the distribution of $\hat{p}$ given in Table I of Appendix B.

It would seem (and is true) that one has the greatest chance of erroneously rejecting a good lot if the lot is only marginally acceptable, with $p = .10$. For such a lot, the probability that it will be rejected is just the probability of getting a value of $\hat{p}$ in the critical region, calculated for $p = .10$:

$$\text{Using Rule 1:} \quad P(\text{reject lot}) = P(\hat{p} \geqslant 1/10)$$
$$= 1 - P(\hat{p} = 0) = 1 - .3487$$
$$= .6513.$$

$$\text{Using Rule 2:} \quad P(\text{reject lot}) = P(\hat{p} \geqslant 2/10)$$
$$= 1 - P(\hat{p} = 0 \text{ or } 1/10)$$
$$= 1 - .3487 - .3874 = .2639,$$

(See Figure 8-7.) So Rule 2 has a smaller chance of mistakenly rejecting a good lot. However, with regard to the other type of error, Rule 1 is better. Again, probability computations depend on p, so we take a particular value of p among those defining unacceptable lots, say $p = .20$, and calculate the probability of accepting a lot with $p = .20$:

Using Rule 1: $P(\text{accept lot}) = P(\hat{p} = 0) = .1074.$

Using Rule 2: $P(\text{accept lot}) = P(\hat{p} = 0 \text{ or } 1/10)$

$$= .1074 + .2684 = .3758.$$

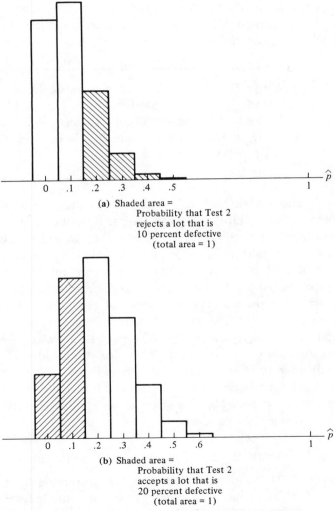

(a) Shaded area =
Probability that Test 2
rejects a lot that is
10 percent defective
(total area = 1)

(b) Shaded area =
Probability that Test 2
accepts a lot that is
20 percent defective
(total area = 1)

Figure 8-7. Some probabilities for Rule 2, Example 8-24.

So when $p = .20$ (and similarly for any $p > .10$), Rule 1 has a better chance of detecting that the lot is unacceptable. ▲

An aspect of this example in the area of acceptance sampling that was not encountered earlier is that the null hypothesis is not specific—is not a single, particular value of p. So the calculation of the probability of erroneously rejecting H_0 depends on which value of p in H_0 one assumes as the population proportion of defectives. The largest such probability, usually for a hypothesis in H_0 that is close to being in H_A (as the value $p = .10$ was close to being in the range $p > .10$, in the last example), is called the *size* of the type I error. It is also thought of as the "significance level" of the test, since we take the occurrence of a value of the test statistic in the critical region as "signifying" that the null hypothesis is to be rejected.

The *significance level* of a test or decision rule, also called the *size of the type I error* and denoted by α, is the maximum probability that H_0 will be rejected when H_0 is true. (If H_0 is specific, there is only one such probability, and this is α.)

You will sometimes hear α referred to as the "probability of a type I error." Although it is this in a sense, the phrase is misleading. It must be remembered that α is a probability in a particular probability model included in H_0. If one asks, "What are the chances that this test will lead to a type I error?" one can only answer conditionally, on the assumption that H_0 is true, or answer with a calculation that takes into account the chances of encountering H_0. For if H_0 is never encountered, one cannot make a type I error!

To relate the accept–reject type of report of a test to the P-value defined earlier, we note that a P-value, for a one-sided test, is just that significance level at which an observed value of the test statistic is on the boundary between significant and not significant—between rejecting and accepting H_0. Sometimes the P-value is referred to as the "observed significance level."

Example 8-25   *Checking Instrumentation (continued)*

In Example 8-21 we described the testing conducted by a hospital laboratory for the purpose of checking its instruments for determining various characteristics of blood samples. Again suppose that a standard supply has a known hemoglobin level of 15.1, and that the laboratory's measurements have a standard deviation of .40. If the measurement made on the standard supply, at the beginning of a day, falls outside the limits 15.1 ± .80—that is, if it differs from 15.1 by more than 2 s.d.'s—the laboratory will assume that the measuring equipment has slipped. Steps are taken to track down the reason.

In essence, the laboratory is testing the null hypothesis that the measurements have mean 15.1 and s.d. .40. The critical region for the laboratory's test is given by $|Z| > 2$ (see Figure 8-8), where Z is the Z-score of the measured value X:

$$Z = \frac{X - 15.1}{.40}.$$

This is a "two-sided" critical region; the null hypothesis is rejected for such Z-values because the *alternative* to $\mu = 15.1$ is two-sided. That is, the laboratory would want to know if the mean shifts *either* way from the value 15.1, and the alternative to H_0 is that $\mu \neq 15.1$.

If it can be assumed that the measurements X are normally distributed, the size of the type I error for the laboratory's test can be found in Table II, using the fact that Z has a standard normal distribution when H_0 is true:

$$\alpha = P(|Z| > 2) = .0456.$$

This is the probability that the laboratory will take time to check over the equipment when it is actually working exactly as specified. Whether the significance level .0456 is reasonable is not for us to say. The laboratory's rule has been found through experience to work well; it seems to be sensitive enough to detect intolerable deteriora-

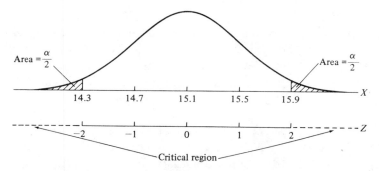

Figure 8-8.   Critical region for Example 8-25.

tion of the measuring equipment without causing too much needless checking.

To be sure, a value of Z in the critical region may be caused by a deterioration in accuracy—by a change in s.d. rather than by a shift in the mean. But this would also warrant checking over the equipment. On the other hand, a change in mean or in standard deviation could occur without the single test measurement's going outside the two s.d. limits; in this case the equipment would not be checked over, and this would be a type II error. ▲

8.8
CONFIDENCE INTERVALS AND TESTS

When hypotheses are formulated in terms of a population parameter (as in Example 8-7), an accept–reject type of test for a particular value of that parameter can be given in terms of a confidence interval. The idea is that if the parameter value being tested is not included within a confidence interval, that value should be rejected; for, the confidence interval should include the "believable" parameter values:

A test for $\mu = \mu_0$:

Reject $\mu = \mu_0$ if the confidence interval for μ, constructed from a sample, does not include μ_0.

The significance level of this test is simply related to the confidence level used in constructing the confidence interval:

$$\alpha = 1 - \text{confidence level}.$$

This is because the probability (when $\mu = \mu_0$) of *including* μ_0 is the confidence level, and the test rejects $\mu = \mu_0$ if the confidence interval does *not* include μ_0. For example, using a 95 percent confidence interval for testing $\mu = \mu_0$ means that the test has significance level

$$\alpha = 1 - .95 = .05.$$

A confidence interval with limits of the form $\bar{X} \pm k(\text{s.e.})$ will include μ_0 if and only if $\bar{X}$ happens to fall within k standard errors of μ_0, on either side. And because k is related to the confidence level by means of the normal table, this means that the test based on whether or not μ_0 is covered by the confidence interval determined from a sample is equivalent to a two-sided Z-test based on $\bar{X}$. It is two-sided because we have constructed confidence intervals to be two-sided, and this is appropriate when the alternative to $\mu = \mu_0$ is two-sided—of the form: $\mu \neq \mu_0$, including μ's on either side of μ_0.

Example 8-26   *Light-Bulb Life (again)*

In Example 8–8 we looked at a test of the hypothesis that the mean life of a light bulb is 700 hours, against the alternative that it is less than 700 hours (the alternative that would be of concern to the consumer). If we had constructed a 95 percent confidence interval for the mean μ using the sample mean of $\bar{X} = 690$ hours and sample s.d. of $S = 40$ hours (for a sample of size $n = 50$), we should have found these limits:

$$\bar{X} \pm \frac{1.96 \times 40}{\sqrt{50}} \quad \text{or} \quad 690 \pm 11.1,$$

which *include* the value 700 being tested. We would not reject $\mu = 700$ at the 5 percent significance level—but this is a two-sided test! In a one-sided test, appropriate for the alternative $\mu < 700$, the Z-value of -1.77 ($P = .038$, as found in Example 8–8) *is* within the critical region for a one-sided 5 percent test: $Z < -1.64$, and μ_0 would be rejected. ▲

8.9
ONE-SIDED VERSUS TWO-SIDED TESTS

When should one use a one-sided, and when a two-sided test? In problems where decisions must be made, and the "accept–reject" approach to testing is appropriate, this question can usually be answered in the light of that context. We have touched on this point in some of the examples earlier in the chapter, but perhaps it bears further discussion.

In Examples 8–22, 8–23, and 8–24, a test was used in deciding to accept or to reject a shipment of thermometers based on the number of defectives among a sample of 10 drawn randomly from the lot. Accepting the lot amounted to accepting $p \leq p_0$ rather than the alternative $H_A : p > p_0$, where p is the (true) lot fraction defective and p_0 is the dividing point between acceptable and unacceptable lots. In this case it seems clear that since large proportions of defectives in the sample point to a large proportion of defectives in the lot, the occurrence of a large value of $\hat{p}$ *should* call for rejection, and a small value of $\hat{p}$ for acceptance of the lot. Thus, a *one*-sided critical region—a one-sided test—is quite in order.

In the problem of testing for ESP (Examples 8-4 and 8-6) where the null hypothesis is a single value, $p = 1/4$, we took values of p *greater* than $1/4$ as the only admissible alternatives. And again we were led to a one-sided test, in which *large* numbers of correct calls suggested the presence of ESP. But one might ask: If p is *less* than $1/4$, does this mean that there is ESP of a "negative" or inverse type? If it does, the test procedure ought to be designed to detect a p that is *either* less than or greater than $1/4$; and sample proportions that are *either* very small or very large would be used as a basis for rejecting $p = 1/4$. However, a person claiming to have ESP would probably be embarrassed by $\hat{p} = 0$, or no successes, and would probably not feel that his claim is substantiated. So, if $p > 1/4$ is the only range of values corresponding to ESP, the test should be one-sided.

The production supervisor on the butter-packaging line (Examples 8-3 and 8-5), on the other hand, would certainly want the average weight to be maintained so that it neither goes above nor dips below the specification. The test used should be sensitive to a slippage either way, so the supervisor should take either an unusually large or an unusually small sample mean weight as evidence that the process needs adjusting. If the process remains in control at the specified level, and if the α of 5 percent is divided so that 2.5 percent is in each tail of the null distribution, then the process would be shut down needlessly about 5 percent of the time—whenever poor luck in sampling gives a sample mean too high or too low. But if the process goes out of control, the supervisor has a good chance of catching it no matter which way it goes.

In reporting a P-value as a measure of evidence for a specific null hypothesis, such as $\mu = \mu_0$, for instance, there is no general agreement as to whether that P-value should be one-sided or two-sided. For example, if a Z-statistic turns out to have the value $Z = 1.645$, the P-value could be defined either as

$$P = P(Z \geqslant 1.645) = .05$$

or as

$$P = P(|Z| \geqslant 1.645) = .10.$$

The latter might be thought appropriate when using a decision rule that rejects H_0 if $|Z| \geqslant K$; but if one is not making a firm decision in the sense of taking some action, then it is not clear which definition is to be preferred. It would seem that the simplest way out is to define the P-value in one way or the other and to state in the report which definition is being used (and with what statistic). What is really wanted is a report of what was done and what the result was.

The confidence interval estimate as a way of reporting an experimental result can also be one-sided or two-sided. Our construction was two-sided, with confidence intervals extending on both sides of point estimates; and if such a confidence interval is used in testing (Section 8.8), the test is automatically two-sided. (It is possible to construct one-sided confidence intervals, and this is sometimes appropriate; but more often than not you will encounter two-sided confidence intervals.)

8.10
TESTING AS INFERENCE

Statistical inference is the drawing of conclusions about a population from the characteristics of a sample drawn from the population. Because a totally accurate inference is generally impossible, it is not surprising that there is some disagreement among statisticians on how best to use sample information for inferences.

Those working in a situation where decisions must be made and actions taken would tend to be "hypothesis testers"—rejecting or accepting some null hypothesis, in effect, by deciding on and taking a course of action as suggested by the data. The rules they adopt for making these decisions are chosen according to the costs of wrong decisions, either by some analysis of the decision procedure that incorporates these costs, or empirically—observing how their rules work in practice and modifying them if necessary.

In scientific work, the usual (and proper) attitude is one of tentative judgment, always being open to the results of further tests, as well as aware of the results of previous experimentation. The language and attitude of "significance testing," and in particular of "P-values," is common in such work. A statistical analysis can point to the need for and areas for further investigation and suggest possible relationships that need study. But statistical analyses will not establish or *prove* scientific hypotheses.

There is a point of contention among statisticians that has tended to polarize them into two schools, the "classical" and the "Bayesian." Consumers of statistics seem to want to know the answer to the question: "What is the probability that my null hypothesis is false?" But it turns out that there is no way of answering this without starting with a predata or *prior* judgment, expressed as a subjective probability, on the part of either the statistician or his client. The Bayesian is willing to use prior probabilities in conjunction with the data, putting them together by means of a mathematical result known as "Bayes' theorem," to obtain postdata or *posterior* probabilities for the falsity of null hypotheses.

The "classical" approach, adopted by those who shun subjective probabilities, is to try to determine what the data in a sample say about the population from which they are drawn, without injecting a personal element. Significance testing, hypothesis testing (in the "accept–reject" approach), and confidence-interval estimation are classical in this sense. Unfortunately, there *is* a personal element in interpreting a *P*-value, in setting arbitrary standards for error sizes, and in setting confidence levels.

Despite the controversy, the two kinds of statistician will usually end up with very similar recommendations in any specific problem. This is especially so in the case of large samples, where the Bayesian's prior judgments about the way things are tend to be overwhelmed by the information in the data; although couched in different terms, his conclusions would be similar to those of the classical statistician.

KEY VOCABULARY

Null hypothesis
Test statistic
Null distribution
Tail values
Tail probabilities
P-value
Statistical significance
Z-test
t-test
Signed-rank statistic
Goodness of fit
Pearson's chi-square

Practical significance
Alternative hypothesis
Critical region
Errors of types I and II
Error size
Significance level
Observed significance level

QUESTIONS

1. What is a statistical hypothesis?
2. What is "null" about a null hypothesis?
3. What information does the null distribution of a test statistic provide?
4. A null hypothesis is often set up in the hope that it will be rejected. What is behind this statement?
5. There are three possible explanations when a test statistic is found to have a value far from what is expected when the null hypothesis is true. What are they?
6. When does one use T and when Z, as the test statistic for hypotheses about a population mean?
7. Why do we judge the significance of a deviation of the sample mean from a hypothesized population mean according to its standard deviation?
8. What assumptions underlie the use of the t-table in finding a P-value corresponding to an observed value of the statistic T?
9. What advantage does the signed-rank test have over the t-test? Does it have any disadvantage?
10. In "goodness of fit," what fits what?
11. How does the large sample distribution of Pearson's chi-square statistic depend on the nature of the population?
12. What does an observed value of 0 for Pearson's chi-square tell you about the data and the model? What does it tell you about the investigator?
13. When a test statistic is judged to be "statistically significant," what is it signifying?
14. What are the two types of error in a procedure used for deciding whether or not to reject a null hypothesis?
15. Why not devise a test so that the type I error size (i.e., the significance level) is 0?

16. Can one make *both* type I and type II error sizes small? If so, how?

17. What is the relationship between the notions of "*P*-value" and "significance level"?

18. Describe the idea behind using a confidence interval for a test procedure concerning a population parameter.

19. In formulating a test statistic and deciding what critical region to use (for a given H_0), is it a good idea to look at the data to see what would work best?

PROBLEMS

Sections 8.1 and 8.2

*1. Consider the ESP testing experiment described in Example 8–4, conducted to test the hypothesis that $p = 1/4$.

    (a) Find the *P*-value corresponding to 4 correct identifications in 8 trials. (Use Table I of Appendix B.)

    (b) Find the *P*-value corresponding to 30 correct identifications in 60 trials.

    (c) About how many correct identifications in 60 trials would give a *P*-value equal (or close) to that in part (a) for 4 correct in 8 trials?

    (d) Suppose that there are 200 subjects to be tested for ESP. Among the 200 results, how many would be expected to turn out to be "significant," if none of the subjects has ESP? How many, on the average, would be "highly significant?"

2. One side of a cube is painted red, the other sides white. Let p denote the probability that the red side turns up in one toss.

    (a) The cube is tossed 5 times to test $p = 1/6$. Find the *P*-value if the red side turns up exactly two times in the 5 tosses.

    (b) The cube is tossed 50 times to test $p = 1/6$. Find the *P*-value if 2/5 of the tosses are red.

*3. The director of graduate studies in a university graduate program observed that the average GRE score in analytical ability for 25 applicants was 600. Test the hypothesis that this group is a random sample from the population of students who took the exam that year, with mean 521 and s.d. 123 (as given in Problem 12 of Chapter 5).

4. The population of all students heading for college in a certain

year averaged 22.4 on the ACT test. The 400 applicants for admission to a certain college that year had a mean score of 21.6, with an s.d. of 4.0. Is the average of this group significantly lower than that of the population? (If so, what does the difference "signify"?)

*5. A panel of 80 prospective jurors is found to be all white. Given that the prospective jurors are chosen at random, would you accept the hypothesis that the population from which this panel of 80 was selected is the population of all adults in a county with 10 percent blacks?

6. The President claims that a majority of the people favor his policy in a certain matter. A public opinion poll of 1500 persons shows 49 percent in favor of his policy. Could he be right, if the sampling is random?

Sections 8.3 and 8.4

*7. Seven skulls found in a certain digging averaged 143.3 millimeters in width, with $S = 5.62$. Might these belong to a race previously found nearby with average 146 millimeters?

8. Ten subjects were tested for increase in reaction time after being given a certain drug. The increases averaged 10 milliseconds with $S = 10.6$. Is this average increase significant (i.e., significantly different from 0)?

*9. An audit of 25 accounts chosen at random from a company's accounts receivable (see Example 7-1) shows that 17 have no error, and the other eight are in error by these amounts:

$$47, \quad -18, \quad 4, \quad 10, \quad -100, \quad 1, \quad 23, \quad -1.$$

Calculate a t-statistic for testing the hypothesis that the average error is 0 and comment on the appropriateness of a t-test.

10. The mean weight of 5-pound bags of sugar (in a certain production facility) is supposed to be 5.10 pounds. Twenty-five bags are taken from the line and weighed, with these results, given as pounds in *excess* of 5.10:

.10	.00	−.05	−.12	−.04
.06	−.02	.00	.08	−.10
−.04	.05	.07	.00	.05
.00	.02	−.08	−.07	−.06
−.11	−.05	−.06	.01	−.14

Test the hypothesis that the mean weight is 5.10, using T.

*11. Do Problem 9 using the signed-rank test in place of the *t*-test.

12. Do Problem 10 using the signed-rank test in place of the *t*-test.

*13. A certain production process is "in control" if a particular dimension X of the produced item is distributed with mean $\mu = 2.7$ and standard deviation $\sigma = .10$. A sample of five parts is checked, and the average dimension is found to be $\bar{X} = 2.8$. It is given that X is normally distributed.

(a) Assuming that σ has not changed, would you say that the process mean has shifted?

(b) If it is *not* assumed that σ is still .10 but it is found that the sample standard deviation is $S = .12$, would you accept the hypothesis of no shift?

(c) If the dimensions X in the sample of five parts are

$$2.60, \quad 2.72, \quad 2.87, \quad 2.90, \quad 2.91,$$

use a rank test for the hypothesis of no shift.

14. To test the hypothesis that (as claimed by the table's compilers) the entries in Table IX of Appendix B are from a population with mean 0, pick 10 consecutive numbers from that table, starting at an arbitrary point in the table and going across or down.

(a) Test the hypothesis that the mean is zero using the *t*-test.

(b) Test the hypothesis that the mean is zero using the signed-rank test.

(c) Pick a second set of 10 numbers from the table and add .50 to each number in the set, and then a third set of 10 numbers from the table, adding 1.00 to each number in this third set. Carry out a *t*-test of the hypothesis that $\mu = 0$, using each of these samples in turn.

(d) Test the hypothesis that $\mu = 0$ using the samples in part (c) in a signed-rank test, and compare the *P*-values with those obtained using the *t*-test in part (c).

Sections 8.5 and 8.6

*15. A coin is tossed 400 times to test the model of equal probabilities for heads and tails. It lands heads 216 times and tails 184 times.

(a) Compute the chi-square statistic for testing $p = q = 1/2$, where p is the probability of heads and q the probability of tails.

(b) Compute the Z-statistic for testing $p = 1/2$.

(c) Is either result significant? (Which statistic would you use?)

16. On a TV game show, prizes are distributed at random behind the numbers 1, 2, 3, . . . , 9, arranged in a square array (3 by 3). Behind one of the numbers there is a "dragon," which, if picked by the contestant, means that the contestant loses at that point. We have been noticing that the numbers in the last row (7, 8, 9) seemed to be occurring more often than they should and decide to test the hypothesis $p = 1/3$, where p is the probability that the dragon is behind 7, 8, or 9. We find in 36 plays of the game† that the 7, 8, or 9 turned up (as the location of the dragon) 21 times. What conclusion can be drawn?

*17. A deck of cards is "cut" 40 times, to reveal a card, with these results according to suit:

Suit	Frequency
Spades	7
Hearts	14
Diamonds	10
Clubs	9

Use the chi-square statistic to test the hypothesis that the suits are equally likely to turn up when the deck is cut.

18. Test the hypothesis that the faces of a die are equally likely, based on 120 tosses that divided as follows among the faces:

Face	Frequency
1	18
2	24
3	12
4	23
5	17
6	26

*19. According to genetic theory, if two recessive genes that give

---

†We confess that we did not watch the show long enough to gather this much data, although the data given reflect a tendency we noticed in the time that we did watch it. Moreover, one contestant, who had been having a long winning streak, remarked that in his 30 chances at the dragon, it had never appeared under number 5.

rise to phenotypes A and B, respectively, are not linked, then the proportions of phenotypes AB, AB^c, A^cB, and A^cB^c will occur in the population in proportions 1/16, 3/16, 3/16, 9/16, respectively. A sample of 80 individuals resulted in the following tabulation:

Type	Frequency
AB	8
AB^c	21
A^cB	12
A^cB^c	39

Carry out a chi-square test of the hypothesis that these recessive genes are not linked.

20. A research team at the Roper Center in Connecticut decided to check the possibility of nonresponse bias in a survey of over 10,000 university and college professors. (Their survey had a response rate of only 53 percent.) They telephoned a sample of 500 nonrespondents and obtained 477 interviews. On one item ("There should be a top limit on income"), the frequencies of answers for nonresponders and the proportions for the 5000 responders were as follows:

	Nonresponders	Responders
Strongly agree	86	.15
Agree with reservations	100	.21
Disagree with reservations	119	.23
Strongly disagree	172	.41

Test the hypothesis that for this item the nonresponders are a random sample from a population with proportions given for responders. What does your test show about nonresponse bias for this item?

*21. Referring to Problem 3, suppose that the director of graduate studies announced that "our applicants are significantly better than the national average." How might this statement be criticized?

22. Referring to Problem 4, would it be correct for the admissions director of the college to conclude that "our applicants are significantly lower than the national average?"

Sections 8.7-8.9

*23. In connection with a certain experiment, involving a sample with mean $\overline{X}$, consider these possible tests for the null hypothesis $\mu = 10$:

Test 1: Reject $\mu = 10$ if $|\overline{X} - 10| > .5$.

Test 2: Reject $\mu = 10$ if $|\overline{X} - 10| > .8$.

(a) Which test has the larger α?

(b) Which test has the greater chance of concluding that $\mu \neq 10$ when this is the case?

24. A coin is to be tossed four times to test the hypothesis that $p = 1/2$, where p is the probability of heads at a single toss. Suppose that you decide to reject this hypothesis if you get all heads or all tails.

(a) What is the size of the type I error (i.e., the significance level α)?

(b) If the coin has two heads (instead of heads on one side and tails on the other), what is the probability that your test accepts $p = 1/2$?

*25. Suppose that the fairness (or "straightness") of a pair of dice is questioned and that a test is conducted involving one toss of the dice. The 36 possible outcomes of the toss can be listed as follows, in the form "(1st die, 2nd die)":

(1, 1) (1, 2) (1, 3) (1, 4) (1, 5) (1, 6)
(2, 1) (2, 2) (2, 3) (2, 4) (2, 5) (2, 6)
(3, 1) (3, 2) (3, 3) (3, 4) (3, 5) (3, 6)
(4, 1) (4, 2) (4, 3) (4, 4) (4, 5) (4, 6)
(5, 1) (5, 2) (5, 3) (5, 4) (5, 5) (5, 6)
(6, 1) (6, 2) (6, 3) (6, 4) (6, 5) (6, 6)

The null hypothesis (fairness) asserts that these 36 outcomes are equally likely.

(a) Suppose that fairness is rejected if the result is "snake-eyes" (1, 1) *or* "boxcars" (6, 6). What is α for this test?

(b) Suppose that fairness is rejected if the point total on the two dice is 11. Find α. [How does this test and that in part (a) compare with regard to type I error?]

(c) Suppose you knew that the person tossing the dice once had in his possession a pair of dice, one with 5's on all six sides and the other with two 6's and four 2's. In light of this, would there be any reason to prefer one of the tests in part (a) or (b) over the other? (Calculate the probability of failing to reject fairness if the altered dice are used.)

(d) Given the situation in part (c), find a good test with $\alpha = 1/9$ based on the point total for the two dice.

26. It is claimed that "tossing" a quarter by spinning it on a smooth surface with the flick of a finger (and letting it slow down and fall) favors heads, with $p = P(\text{heads}) = 2/3$. With this alternative to the null hypothesis $p = 1/2$, and with the aid of Table I:

(a) Find an appropriate test based on 10 spins with $\alpha = .17$.

(b) Determine β, the size of the type II error, for the test in part (a).

(c) Compare the test in part (a) with the test that rejects $p = 1/2$ if 7 or more of the 10 spins result in tails, with regard to α and β.

(d) Compare these two tests with the test in part (a):

(1) Reject $p = 1/2$ if 10 spins yield 3 heads.

(2) Reject $p = 1/2$ if in 10 spins there are fewer than 3 or more than 7 heads.

*27. The laboratory in Example 8-25 uses ± 2 s.d.'s from the normal hemoglobin level of the standard supply as "acceptance limits" in its test. The state requirements for such testing specify ± 3 s.d.'s.

(a) If the laboratory used the state's procedure (rejecting the null hypothesis that the test instruments are operating properly if a measurement falls outside the ± 3 s.d. limits), what would the size of the type I error be?

(b) In view of the smaller α for the state's prescribed procedure, why might the hospital be using a test with a larger α? Would the state consider the laboratory to be in violation?

28. Use the data in Problem 10 (with $\bar{X} = 5.08$) in a two-sided test of $\mu = 5.10$ at $\alpha = .05$, assuming that the population s.d. of $\sigma = .04$ applies. Also, determine the proportion of bags that will weigh less than 5.00 pounds when $\mu = 5.10$, and when $\mu = 5.08$.

*29. Suppose, in Example 8–24, that the lot size is $N = 500$ and that the lot was accepted as the result of a test. After using up the lot, it is found that there were 35 defective thermometers among the 500 in the lot. Is this reason to conclude that the supplier's "p" exceeds .05? [Notice that this takes what was a population in the tests of Example 8-24 (the lot) and treats it as a sample, a sample from the conceptual population of all thermometers produced by the supplier.]

30. In the butter-packaging process described in Examples 8-3 and 8-5, suppose that it is decided to reject the hypothesis that $\mu = 1$ pound if the mean weight of a sample of 10 packages differ from 1 pound by more than .01 lb. Assume that the s.d. of weights is $\sigma = .020$ pound.
    (a) Determine α for this procedure.
    (b) How would you modify the procedure to achieve $\alpha = .05$ (keeping the sample size $n = 10$)? Would this modification have any accompanying adverse effect on the test?

Review

31. The data for Problem 16 (the bonus game for "Tic-Tac-Dough") were as follows:

Dragon Behind	Frequency
1	2
2	2
3	3
4	1
5	4
6	3
7	8
8	6
9	7

(a) Use a chi-square test for the hypothesis that the nine locations are equally likely. [Can you see any reason why this test does not reject the hypothesis of random location while the Z-test (of $p = 1/3$) in Problem 16 *did?*]
(b) Suppose we feel that the mean frequency of four per cell in part (a) is too small, and that we should group some cells together so that the mean frequencies exceed five. Calculate χ^2, using the given data, if the regrouping is into:
    (1) Rows (with groups 123, 456, 789).
    (2) Columns (with groups 147, 258, 369).

32. True or false (not always true)?
    (a) When a test rejects H_0, based on sample results, this means that H_0 is false.
    (b) In testing $\mu = \mu_0$, you never make a type I error if you never encounter a situation in which $\mu = \mu_0$.
    (c) A test for $\mu = \mu_0$ based on the mean of 100 observations has a better chance of detecting that $\mu \neq \mu_0$ than one based on 25 observations.
    (d) If you really want to reject $\mu = \mu_0$ using a test based on $\bar{X}$, all you need is enough data.
    (e) A "statistically significant" result may have inconsequential practical significance.
    (f) If a test results in a P-value of 5 percent, this means that the chances that H_0 is true are 1 in 20.†

33. The hospital laboratory of Examples 8–21 and 8–25 has a standard abnormal blood supply with a white cell count of $21,400/mm^3$. The specification for standard deviation is $1000/mm^3$.
    (a) What is the acceptable range of values for one measurement, in a test with $\alpha = .05$, assuming that $\sigma = 1000$?
    (b) Over a particular 30-day period the laboratory tests average $20,600/mm^3$ with an s.d. of $920/mm^3$. Is this consistent with the hypothesis that the measuring procedure remained "in control" throughout the month?

34. One year before the 1980 New Hampshire presidential primary, columnist George F. Will, writing in *Newsweek* (March 5, 1979), said, "Jerry Brown's challenge will, at a minimum, reveal the extent of the Democratic discontent with Carter. (A poll of 245 New Hampshire Democrats 'likely to vote' gives Carter 41 percent; Brown, 33; no opinion, 26.) That will enable Kennedy to challenge Carter without seeming to *create* disunity." To address the question of whether the quoted figures supports his thesis of "discontent with Carter," assume that Carter and Brown are the only candidates and that the 181 respondents with opinions constitute a random sample, 100 favoring Carter.

†In a 1980 class-action suit concerning sex discrimination, a certain university professor testified for the plaintiffs as an expert on probability. Concerning the significance of a certain proportion she said, in effect, that the P-value was 5 percent. She was then asked, "Does this mean that the probability is 95 percent that there *was* discrimination (i.e., that the null hypothesis is false)?" She replied, "Yes." Do you agree?

(a) Find a 99 percent confidence interval for the proportion of all Democrats who favor Carter.

(b) Suppose that "discontent" is regarded as extensive if the proportion of all Democrats who favor Carter is $< .6$. Test the hypothesis that discontent is not extensive, using $\alpha = .01$.

35. A lot of checking has pretty well established that the average speed on a certain stretch of highway during the summer was 63 mph, a period when the limit was 65 mph. After the establishment of a 55-mph limit to conform to federal regulations, a sample of 100 cars had an average speed of $\bar{X} = 60.4$ mph, with $S = 4.6$ mph.

(a) Is this apparent reduction explainable as chance variation, or has the mean speed been reduced?

(b) Give a 99 percent confidence interval for the new mean speed. Can you say whether the achieved reduction is significant in the practical sense?

36. Medical researchers conducted the following experiment. They perfused the esophagus of each of two dogs with pepsin, simultaneously, one at a low level and one at a high level. They continued until the esophagus of one dog was perforated (and then stopped). They found that in seven trials, the esophagus treated with the high level perforated first. For the hypothesis of no difference between levels of pepsin, determine the P-value corresponding to the observed result. (Use Table I.)

37. Determine whether the value of R_+ in Example 8–12 is significant (in testing $\mu = 700$). Also, given that the sample mean and s.d. are $\bar{X} = 716.8$ and $S = 213.78$, find the P-value corresponding to the appropriate t-statistic.

9
Comparisons

WHENEVER THERE ARE two or more types of individuals, comparisons are inevitable, and we ask such questions as these:

- Do men feel the same way about abortion as women do?
- Are whites less rhythmical than blacks?
- Are left-handed hitters better than right-handed ones?
- Do Holstein cows produce more butterfat than Guernseys do?
- Is the quality of brand A more uniform than that of product B?

Comparisons are also called for when two groups of individuals have been given different treatments, to answer such questions as these:

- Is fluoride toothpaste more effective in fighting cavities than non-fluoride types?
- Is "new math" better or worse than the old in preparing students for advanced work?
- Are smokers more susceptible to disease than nonsmokers?
- Is fertilizer *A* better for growing wheat than fertilizer *B*?

- Does the white cell count among cancer patients vary more erratically than that of healthy individuals?

In some cases it is a new treatment that is to be evaluated in comparison with a standard treatment, or with no treatment at all. Treated individuals are then said to belong to the *treatment group* and the untreated individuals to the *control group*—a standard for comparison.

Individuals used in a comparison test come from a population (usually conceptual) of individuals to whom the conclusions of the test might be applied. Ideally, they would be selected at random and then assigned at random to the treatment or the control group. This amounts to thinking of the whole population as two populations, one treated and one untreated, with a random sample drawn from each. The null hypothesis would be that the populations are identical in character, and the alternative, that they differ—presumably because of the treatment.

Populations can be different in a variety of ways, and one cannot hope to detect every kind of difference between populations with a single statistical test. Perhaps the most obvious kind of difference, and one of the easiest to detect, is a difference in location; and most of this chapter deals with comparing locations. A difference in *average* effectiveness of two treatments, for instance, is often all one can expect and may be sufficient reason to prefer one treatment over the other, even though the one may not outperform the other in every case.

In the case of variables of classification, there is often a need to compare population proportions for the various categories. For example, with regard to smoking history—classified in Example 2-1 as S (smoker), N (nonsmoker), or Q (smoked at one time but quit)—one might ask whether the proportions of S, N, and Q are the same for men as for women. Similarly, voters can be classified as Republican, Democratic, or Other, and one might want to know whether or not the proportions are the same in urban communities as in rural communities.

Whatever comparisons are to be made, it will be assumed that a random sample from each population is available—one from the "treatment" population and one from the "control" population, to use the language of a test for treatment effectiveness. Most of the methods will assume, further, that the samples are independent, although Section 9.7 deals with a particular kind of design in which they are not.

9.1
THE M.V. AND S.D. OF A DIFFERENCE

In comparing locations we shall want to use the difference between the means of independent samples as a test statistic. These sample means are random variables, so their difference is a random variable. The difference varies from one pair of treatment and control samples to another. We need to know how to calculate the mean and standard deviation of the difference in sample means from the m.v.'s and s.d.'s of those individual sample means.

Consider any two random variables U and V. Intuition suggests that the average difference is the difference of the averages, and this is in fact correct:

$$\text{m.v.}(U - V) = \text{m.v.}(U) - \text{m.v.}(V).$$

In particular if U is the mean $\bar{X}_t$ of a treatment sample and V is the mean $\bar{X}_c$ of a control sample, then

$$\text{m.v.}(\bar{X}_t - \bar{X}_c) = \text{m.v.}(\bar{X}_t) - \text{m.v.}(\bar{X}_c) = \mu_t - \mu_c,$$

where μ_t and μ_c are the treatment and control *population* means.

The m.v. of the difference of the means of two random samples is the difference in population means:

$$\text{m.v.}(\bar{X}_t - \bar{X}_c) = \mu_t - \mu_c.$$

Example 9-1 *Comparing Toothpastes*

In a study to evaluate the effectiveness of a fluoride dentifrice, two samples of children were used—208 in a treatment group, and 201 in a control group. The treatment group used the fluoride dentifrice over a period of 3 years, and the control group used a nonprophylactic dentifrice of the same color and flavor. (The "control" in an experiment of this type furnishes a basis of comparison, necessary to see if a "treatment" is better than no treatment.) During the 3-year period the treatment group averaged 19.98 new cavities per child and the control group, 22.39 new cavities per child [S. N. Frankl and J. E. Alman, *J. Oral Therapeutics Pharmacol.* 4, 443–449 (1968)]. These are sample means; but since the samples are large, these means are good estimates of the corresponding population means. The obvious estimate of the *difference* in population means is the corresponding difference in sample means, or 22.39 − 19.98 = 2.41.

Any such estimate based on sample information is subject to error, so this difference 2.41 may be greater than or less than the population mean difference.

The use of random samples guarantees that the *average* error in estimating $\mu_t - \mu_c$ by $\bar{X}_t - \bar{X}_c$ is 0, since the latter difference equals the former *on the average*. (To be continued.) ▲

In judging the reliability of $\bar{X}_t - \bar{X}_c$ as an estimate of $\mu_t - \mu_c$, we shall need an expression for standard error, or to begin with, for the standard deviation of a difference.

The difference $U - V$ of two random variables U and V varies about the mean difference $\mu_U - \mu_V$, and the extent of that variation is indicated by its standard deviation. In the case of *independent* random variables, this is computed from the s.d.'s of U and V as a root-sum-square combination, reminiscent of the Pythagorean theorem:

The standard deviation of the difference $U - V$ of independent random variables U and V is the square root of the sum of the squares of their s.d.'s:

$$\text{s.d.}(U - V) = \sqrt{[\text{s.d.}(U)]^2 + [\text{s.d.}(V)]^2}.$$

Example 9-2 *The Difference of Two Pointer Spins*

Consider a single spin of a pointer with a numerical scale from 0 to 1 (as in Example 5–1). The stopping point is a random variable U with mean and s.d. given by

$$\text{m.v.}(U) = \frac{1}{2}, \quad \text{s.d.}(U) = \frac{1}{\sqrt{12}}.$$

(see Figure 5–15, page 213). The result of a second, independent spin

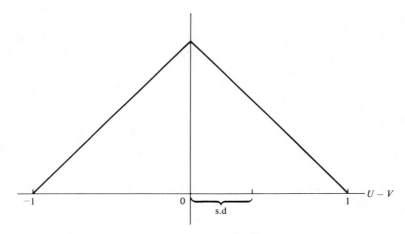

Figure 9-1. Histogram for the difference of two pointer spins.

is a variable V with the same parameters:

$$\text{m.v.}(V) = \frac{1}{2}, \quad \text{s.d.}(V) = \frac{1}{\sqrt{12}}\,.$$

The *difference* $U - V$ between the first and second values can be as small as -1 $(0 - 1 = -1)$ and as large as 1 $(1 - 0 = 1)$; but values near 0 are more common. The histogram for $U - V$ is actually triangular, shown in Figure 9-1. The mean difference is the difference of means:

$$\text{m.v.}(U - V) = \frac{1}{2} - \frac{1}{2} = 0.$$

The s.d. of the difference is the square root of the sum of the squares of $1/\sqrt{12}$ and $1/\sqrt{12}$:

$$\text{s.d.}(U - V) = \sqrt{\frac{1}{12} + \frac{1}{12}} = \frac{1}{\sqrt{6}} = .41,$$

which seems to be of about the right order of magnitude of the "typical" deviation from 0 in Figure 9-1. ▲

9.2
ESTIMATING A MEAN DIFFERENCE

In comparing locations in terms of population means, an obvious measure of discrepancy is the *difference* of those means:

$$\delta = \mu_t - \mu_c,$$

where μ_t and μ_c are the treatment and control population means.[†] To *estimate* this population difference δ it would seem reasonable to use the corresponding difference in sample means, $\bar{X}_t - \bar{X}_c$.

By this time, you should be conditioned to ask immediately: How reliable is $\bar{X}_t - \bar{X}_c$ as an estimate of δ? Or more specifically, what is the *standard error* of the estimate?

The reliability of an estimate can only be determined if the sample is based on probability methods. We need to know, then, how the samples were obtained. If they are *independent* random samples, something can be said about reliability.

We know the s.d.'s of the individual sample means; they are

$$\text{s.d.}(\bar{X}_t) = \frac{\sigma_t}{\sqrt{n_t}}, \quad \text{s.d.}(\bar{X}_c) = \frac{\sigma_c}{\sqrt{n_c}},$$

where n_t and n_c are sample sizes, and σ_t and σ_c are s.d.'s of the treatment and control populations, respectively. If the sample are independent, so that $\bar{X}_t$ and $\bar{X}_c$ are independent, the s.d. of their difference is

$$\text{s.d.}(\bar{X}_t - \bar{X}_c) = \sqrt{[\text{s.d.}(\bar{X}_t)]^2 + [\text{s.d.}(\bar{X}_c)]^2} = \sqrt{\frac{\sigma_t^2}{n_t} + \frac{\sigma_c^2}{n_c}}.$$

The *standard error* of the difference, obtained by replacing the unknown population s.d.'s (σ's) by sample s.d.'s (S's), is then the square root of the sum of the squares of the standard errors of $\bar{X}_t$ and $\bar{X}_c$:

Standard error of the difference of two independent means:

$$\text{s.e.}(\bar{X}_t - \bar{X}_c) = \sqrt{[\text{s.e.}(\bar{X}_t)]^2 + [\text{s.e.}(\bar{X}_c)]^2}$$

$$= \sqrt{\frac{S_t^2}{n_t} + \frac{S_c^2}{n_c}}$$

In interpreting the standard error of any estimator, it helps to know something about the *distribution* of this estimator. One can show mathematically that for large samples the estimator $\bar{X}_t - \bar{X}_c$ is approximately normal, so its s.e. has the kind of interpretation we are used to when using nearly normal estimators. In particular, there is about a 5 percent chance that $\bar{X}_t - \bar{X}_c$ will differ from δ by more

[†]The symbol δ is a lowercase "delta"—the "d" of the Greek alphabet.

than 2 s.e.'s. And so we can construct confidence limits as usual—the estimate plus or minus 1.96 s.e.'s, for a 95 percent interval:

---

95 percent confidence limit for $\delta = \mu_t - \mu_c$:

$$(\bar{X}_t - \bar{X}_c) \pm 1.96 \sqrt{\frac{S_t^2}{n_t} + \frac{S_c^2}{n_c}}$$

(for large independent samples).

Example 9-3   *Comparing Toothpastes (continued)*

Returning to the dentifrice test of Example 9-1, we find that sample s.d.'s are needed to compute the standard error of the mean difference in number of cavities per child. The source article gives these as $S_t = 10.61$, and $S_c = 11.96$. The standard error in the estimate $\bar{X}_t - \bar{X}_c$ is then

$$\text{s.e.} = \sqrt{\frac{10.61^2}{208} + \frac{11.96^2}{201}} = 1.12 \text{ (cavities/child)}.$$

The observed difference (over the period of the study) was 2.41 cavities per child, so the estimate of the population difference δ might be written in the form

$$2.41 \pm 1.12,$$

if it is understood that 1.12 is a "typical" error rather than a limit on the amount of error.

A 95 percent confidence interval for the mean difference in number of cavities (over the 3-year period of the study) extends 1.96 standard errors on either side of the point estimate 2.41, with these limits:

$$2.41 \pm 1.96 \times 1.12.$$

For purposes of testing $\delta = 0$, it would be noteworthy that this confidence interval, $0.21 < \delta < 4.61$, does not cover the value 0. ▲

In working with the formula for standard error given in this section, it is important to remember that its derivation assumes *independent* random samples.

9.3
Z- AND t-TESTS FOR δ = 0

The usual hypothesis to be tested in comparing populations is that there is *no difference* between population means (hence, the "null" hypothesis):

$$H_0: \quad \delta = 0 \quad (\text{or } \mu_t = \mu_c).$$

This null hypothesis says that the treatment is not effective. The alternative would usually be one-sided, saying that the treatment increases the mean response:

$$H_A: \quad \delta > 0 \quad (\text{or } \mu_t > \mu_c).$$

(In some situations, the treatment would be expected to decrease the mean response—as in treating high blood pressure—in which case the alternative would be $\delta < 0$.)

To test the null hypothesis of no difference between population means, we use the Z-score for the difference between sample means. This tells us by how many s.e.'s the sample estimate $\bar{X}_t - \bar{X}_c$ differs from the null value $\delta = 0$:

$$Z = \frac{\bar{X}_t - \bar{X}_c - 0}{\text{s.e.}(\bar{X}_t - \bar{X}_c)}.$$

When $\delta = 0$ and both the sample sizes are reasonably large, this Z-score is approximately standard normal.

Z-test for $\delta = 0$ versus $\delta > 0$ using large, independent samples:
  Calculate

$$Z = \frac{\bar{X}_t - \bar{X}_c}{\sqrt{S_t^2/n_t + S_c^2/n_c}},$$

and find the P-value in Table II of Appendix B.

Example 9-4 *Training Graduate TA's*

A study, some aspects of which are described in Problem 20 in Chapter 1, was carried out to determine the effectiveness of a training program for new graduate Teaching Assistants in Economics. A control group of 323 students was taught in the fall by seven TA's who had not had the training, and a treatment group of 438 students was taught in the winter by the same seven TA's, after systematic exposure to the training system (involving videotaping, seminars, etc.). The students were given pre- and posttests using the "Test of Understanding in College Economics" (TUCE), a nationally validated and normed measure of performance in the introductory economics course. Changes in TUCE scores were summarized as follows:

	Number of Students	Average Change	s.d. of Change
Control (fall)	323	5.94	4.52
Treatment (winter)	438	7.07	4.67

The Z-statistic for testing the hypothesis of no treatment effect (or no difference in mean increase between students taught by trained and students taught by untrained TA's) is

$$Z = \frac{7.07 - 5.94}{\sqrt{4.52^2/323 + 4.67^2/438}} = \frac{1.13}{.336} = 3.36.$$

This is "statistically significant" (being greater than 1.645, or the 95th percentile of Z) in the sense that the sample mean difference is not credible as a chance variation if there is no population difference. Whether the observed difference of 1.13 (an estimate of the difference in *population* means) is of practical significance is not a statistical question.

It should be noticed that since the same TA's were used in the winter quarter as in the fall quarter, it is not really clear whether the improvement was a result of the training system or is simply the improvement that is apt to come with having taught the course once. That is, the two populations involved may be different for reasons other than the difference in treatments, and the confounding of these causes makes it difficult to draw conclusions about the effectiveness of the training system. The study, at the very least, could have divided the group of TA's into one group that would be untreated, and one that would be given the training system, as an attempt to separate the causes. ▲

The case of small samples from normal populations can be treated using the t-distribution. The simplest case is that in which the treatment might change the mean response but not the variability—that is, in which the control and treatment populations have the same s.d.: $\sigma_t = \sigma_c$. We call the common value σ. The s.d. of the difference in sample means then becomes

$$\text{s.d.}(\bar{X}_t - \bar{X}_c) = \sqrt{\frac{\sigma_t^2}{n_t} + \frac{\sigma_c^2}{n_c}} = \sigma\sqrt{\frac{1}{n_t} + \frac{1}{n_c}}.$$

To use this s.d. as the denominator of a standardized score, it is necessary to estimate σ. This can be estimated from the treatment sample as S_t, or from the control sample as S_c—or better yet, as some kind of *average* of S_t and S_c. The larger sample, if one is larger than the other, should provide the more accurate estimate; so in combining S_t and S_c (to exploit the information in both samples), we should weight them according to their sample sizes. A simple weighted average of r.m.s. type is

$$\sqrt{\frac{n_t S_t^2 + n_c S_c^2}{n_t + n_c}}.$$

But to be able to use the t-table without modification, we change the denominator to degrees of freedom:

$$(n_c - 1) + (n_t - 1) = n_t + n_c - 2,$$

to obtain this "pooled" estimate:

$$\hat{\sigma} = \sqrt{\frac{n_t S_c^2 + n_t S_c^2}{n_t + n_c - 2}}.$$

The test statistic for $\delta = 0$ is then constructed as the difference in sample means divided by the s.d. of the difference, with $\hat{\sigma}$ replacing σ in that s.d.:

$$T = \frac{\bar{X}_t - \bar{X}_c}{\hat{\sigma}\sqrt{1/n_t + 1/n_c}}.$$

This has the t-distribution with $n_t + n_c - 2$ degrees of freedom when $\mu_c = \mu_t$. The denominator in this formula for T can be thought of as a standard error, since it is an estimate of s.d.$(\bar{X}_t - \bar{X}_c)$; but in the case of small samples it is not very reliable as an estimate.

A two-sample t-test for $\delta = 0$ (or $\mu_t = \mu_c$), assuming normal populations with equal s.d.'s:

Calculate

$$T = \frac{\bar{X}_t - \bar{X}_c}{\hat{\sigma}\sqrt{1/n_t + 1/n_c}}$$

and determine the P-value using Table III, with d.f. $= n_t + n_c - 2$.

This two-sample t-test is again fairly robust with regard to the assumption of normality.

Example 9-5 *Chest Diameter and Emphysema*

A 1969 study by Kilburn and Asmundsson (*Archiv. Internal Med.* **123**, 379–382) claims to refute a "textbook maxim" that the antero-posterior (AP) chest diameter is increased in patients with pulmonary emphysema. The study included measurements on 16 "normal" males (physicians and technicians without respiratory symptoms) and 25 male patients with emphysema. Measurements were summarized as follows:

	Number	Mean	s.d.
Normal	16	20.2	2.0
With emphysema	25	23.0	2.4

The article states that the difference is "not significant."

The pooled s.d. is calculated[†] as follows:

$$\hat{\sigma} = \sqrt{\frac{15 \times 2.0^2 + 24 \times 2.4^2}{39}} = 2.25$$

The standard error of $\bar{X}_t - \bar{X}_c$ is then

$$\text{s.e.} = \hat{\sigma}\sqrt{\frac{1}{n_t} + \frac{1}{n_c}} = 2.25\sqrt{\frac{1}{16} + \frac{1}{25}} = .72$$

[†]We multiply $(\text{s.d.})^2$ by $n - 1$ because it is likely that the given s.d.'s were computed using an $n - 1$ divisor.

and the observed difference, $\bar{X}_t - \bar{X}_c = 2.8$, is about 3.9 standard errors away from the null mean difference, $\delta = 0$:

$$T = \frac{23.0 - 20.2}{.72} = 3.9 > 2.4 = t_{.99}(39).$$

So the observed difference is "highly significant" and the "maxim" appears to be justified.

In interpreting this result, it would be well to realize that the control sample was taken from a restricted population. This is reflected in the fact that the average age of the control sample was 30.9 years, and in the treatment sample, 56 years. The average weight in the control sample was 74.5 kilograms, and in the treatment sample, 62.1 kilograms. However, taking these differences into account in the interpretation calls for expertise we do not have. ▲

For cases in which one cannot assume that the treatment and control s.d.'s are the same, various modifications of the t-test have been proposed and used.† The inference problem in this case is called the "Behrens–Fisher problem" and has a history of controversy.

9.4
A RANK TEST FOR COMPARING LOCATIONS

We look again at the problem of testing for a shift in location, using now a statistic based on rank order instead of sample means. The null hypothesis is that X and Y have the same distribution, assumed to be continuous; but nothing is assumed about the shape of the histogram. It could be rectangular, triangular, normal, or anything else. This is in contrast to the case of the t-test, where both populations being sampled were assumed normal (or at least, not too far from it).

Figure 9-2 provides a look at typical data from X and from Y, in the three cases of (a) little or no shift, (b) X-shifted to the left of Y, and (c) X shifted to the right of Y. When plotted as marks on a common horizontal scale of sample values, the X's and Y's are or-

†For example, see W. J. Dixon and F. J. Massey, *Introduction to Statistical Analysis*, 3rd ed. (New York: McGraw-Hill Book Company, 1969), p. 119.

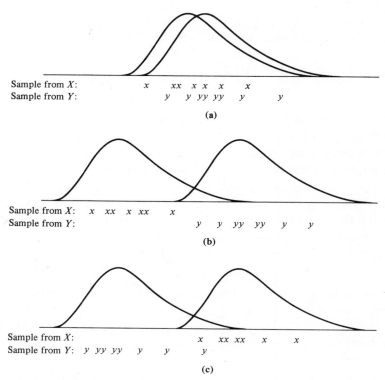

Figure 9-2.   Showing samples from X and from Y, with (a) little shift; (b) Y shifted to the right; (c) Y shifted to the left.

dered together as a single sample. It is evident, in Figure 9–2, that when there is little or no shift, the X's and Y's will be interspersed. But when the X-population is shifted to the right, the ordered observations (including both X's and Y's) will start at the left with a string of Y's and end up at the right with a string of X's, with possibly some intermixing in the middle.

Thus, the relative positions of the X's and Y's in the ordered, combined sample depends upon the relative locations of the X- and Y-populations. To exploit this, we assign ranks to the observations in the combined sample—rank 1 to the leftmost or smallest, and rank 2 to the next smallest, and so on. The ranks of the X's will then tend to be small when the X-population is shifted to the left of the Y-population, and large when X is shifted to the right. The *Wilcoxon rank-sum statistic* is defined as follows:

R_x = sum of the ranks of the X's when ordered together with the Y's (smallest has rank 1, next smallest rank 2, and so on).

The Y-rank sum R_y could be used equally well. However, the sum $R_x + R_y$ is a constant—just the sum of all the available ranks—so the sample results are summarized in one rank sum or the other.

Example 9-6

To illustrate the computation of rank sums, we make up some simple data, temporarily avoiding the complications often present in real data:

$$\text{From } X: \quad 8, 11, 13, 14, 19.$$

$$\text{From } Y: \quad 5, \quad 6, 10, 12.$$

The combined sample of nine observations, in numerical order, is then

$$5, \, 6, \, 8, \, 10, \, 11, \, 12, \, 13, \, 14, \, 19.$$

The Y-values are shown in italics. Indeed, it would suffice to list just the sources of the numbers in this sequence, as follows:

$$Y, \quad Y, \quad X, \quad Y, \quad X, \quad Y, \quad X, \quad X, \quad X.$$

The ranks are $1, 2, \ldots, 9$ (from left to right), so that

$$R_x = 3 + 5 + 7 + 8 + 9 = 32$$

and

$$R_y = 1 + 2 + 4 + 6 = 13.$$

It is not useful to calculate both of these—either one would do. The sum of these two rank sums is $13 + 32 = 9 \times 10/2$, the sum of the integers from 1 to 9. ▲

When X and Y are discrete, there may be observations that are tied. And when X and Y are continuous, ties may occur because of round-off. In either case, if a rank-sum statistic is to be used, we need a procedure for assigning ranks to tied observations. A natural way to do it is suggested by looking at the continuous case. If there are three observations tied at 54, for example, these rounded values would have come from three distinct numbers that could be ranked—say, as 14th, 15th, and 16th—if we only knew what they were. We then assign each of these three tied observations the rank 15—the *average* of the ranks 14, 15, and 16. This treats the tied observations equally and gives them the average of the ranks they would have had if they had not been tied.

With this convention for assigning ranks to tied observations, the rank-sum statistic can be used in the case of discrete populations, and used fairly successfully, even though the tables we shall use were derived under the assumption of continuous populations.

Example 9-7 *Family Ordinal Position*

In a study[†] of differences in attitude between individuals who were first-born, second-born, third-born, and so on, in their families, 120 individuals filled out questionnaires. Scores on nine items dealing with the extent to which the child identifies with and supports the role of the father in the family, for 15 first-born males and 15 second-born males, were as follows:

First born: 40, 41, 44, 49, 53, 53, 54, 54, 56,

61, 62, 64, 65, 67, 67

Second born: 23, 25, 38, 43, 44, 47, 49, 54, 55,

58, 58, 60, 66, 66, 72

These can be arranged in a single-ordered sequence without losing track of their source as first or second born:

1st born

			53 54				67	
40 41	44	49 53 54	56		61 62 64 65	67	
23 25 38	43 44 47 49	54 55	58 60		66	72
			58		66	

2nd born

[†]A. Roost, "A *Q*-sort Analysis of Family Ordinal Position," Ph.D. thesis, University of Minnesota, 1975.

with corresponding ranks:

$$
\begin{array}{cccccccccccccccc}
& & & & & & 7 & & 10\ 12 & \overset{\textstyle 14}{\underset{\textstyle 16}{15\ 17\ 18}} & & 19 & & & & 26\ 28 \\
1 & 2 & 3 & 4 & 5 & 6 & & 9 & & 15\ 17\ 18 & & & 21\ 22\ 23\ 24\ 25 & & & 30 \\
& & & & & & 8 & & 11\ 13 & & & 20 & & & & 27\ 29
\end{array}
$$

The ranks of the second-born individuals' scores can be read from this array, recalling that tied scores are assigned the average of the rank positions they occupy:

$$R_2 = 1 + 2 + 3 + 6 + 7.5 + 9 + 10.5 + 15 + 17 + 19.5$$

$$+ 19.5 + 21 + 26 + 27 + 30$$

$$= 214.$$

The sum of *all* the ranks $(1 + 2 + \cdots + 30)$ is $30 \cdot 31/2 = 465$, so the sum of the ranks of the first born is $465 - 214 = 251$. (To be continued.) ▲

It is clear that a very large or a very small value of R_x (or of R_y) would suggest a difference in locations of X and Y, but there remains the question of how large is "large" and how small is "small." As usual, this can only be answered in terms of the null distribution, which gives the pattern of variation in R_x (or in R_y) when there is no "treatment effect."

In the case of independent samples from continuous populations, the null distribution of R_x depends only on the sample sizes, not on the populations being sampled. For small samples, left-tail probabilities are given in Table VI of Appendix B. For large samples—of sizes n_x from X and n_y from Y—the statistic R_x is approximately *normal*, with

$$\text{m.v.}(R_x) = \frac{1}{2} n_x (n_x + n_y + 1)$$

$$\text{s.d.}(R_x) = \sqrt{\frac{1}{12} n_x n_y (n_x + n_y + 1)}.$$

Because of the way Table VI is constructed, it is convenient to work with the sum of the ranks of the observations in the *smaller* sample. We let X denote the variable in the smaller sample (if one *is* smaller), so that $n_x \leqslant n_y$, and work with R_x.

Wilcoxon rank-sum test for no treatment effect:

H_0: X and Y have the same distribution.

Compute R_x, the sum of the ranks for the smaller sample and find the corresponding P-value in Table VI,† when n_x and n_y do not exceed 10. For larger samples, use a Z-score with Table II:

$$Z = \frac{R_x - n_x(n_x + n_y + 1)/2}{\sqrt{n_x n_y (n_x + n_y + 1)/12}}.$$

(A "large" R_x suggests that X tends to be to the right of Y.)

Example 9–8 *Family Ordinal Position (continued)*

The rank sum R_2 = 214, computed in Example 9–7 for comparing attitudes on the father-role of first-born and second-born males, can be made into a Z-score by "standardization." For this we need

$$\text{m.v.}(R_2) = \frac{15}{2}(15 + 15 + 1) = 232.5$$

and

$$\text{s.d.}(R_2) = \sqrt{\frac{15 \times 15}{12}(15 + 15 + 1)} = 24.11.$$

The Z-score is then

$$Z = \frac{214 - 232.5}{24.11} = -.767,$$

which is not "statistically significant." (On the other hand, there may be a difference between first- and second-born males that is practically significant, even though it is too small to detect with any assurance using samples as small as 15.) ▲

---

†In Table VI, the sample sizes are labeled m and n, with $m \leqslant n$.

Example 9-9   *Fly Mortality*

Suppose fly sprays A and B are tested by applying equal amounts to batches of flies, spray A to nine batches and spray B to seven batches, with these results (in percent mortality):

$$\text{Spray } A: \quad 68, 68, 59, 72, 64, 67, 70, 74, 63$$

$$\text{Spray } B: \quad 60, 67, 61, 62, 67, 56, 58.$$

Plotting these on a common axis, with A's above and B's below, as follows:

$$
\begin{array}{cccccccccccc}
 & & & & & & A & & & & & \\
55 & & A & & A & A & 65 & A & A & A & A & A \quad 75 \\
\hline
 & B & B & & B & B & B & & B & & & \\
 & & & & & & & & B & & &
\end{array}
$$

orders them together so that the B-ranks (the B's make up the smaller sample) can be easily read as

$$1, \quad 2, \quad 4, \quad 5, \quad 6, \quad 10, \quad 10,$$

with sum $R_B = 38$. (The ranks of the 2 B's and 1 A with tied values of 67 each are 9, 10, and 11. The average 10 is assigned to each.) The one-sided P-value, given in Table VI opposite $C = 38$ in the column for $m = 7$, $n = 9$, is .011, almost at the level called "highly significant." ▲

9.5
COMPARING TWO PROPORTIONS

Sometimes the response to a treatment is a categorical variable, and in the simplest case there are just two categories. The model in each case—treatment and control—is defined by the population proportion of one category, or the probability of drawing an individual in that category.

Example 9-10   *Aspirin and Strokes*

Newsweek of July 24, 1978, reported results of a 7-year study, described in the *New England Journal of Medicine*, on the effect of aspirin as a treatment for the prevention of strokes. Of 406 men who had suffered at least one TIA (transient ischemic attack), 200 received aspirin either alone or in combination with sulfinpyrazone, and the rest were given either sulfinpyrazone or an inert placebo. Among the aspirin group, 29 subsequently had strokes or died; in the nonaspirin group, 56 had strokes or died.

Question. Is this, as the *Newsweek* report asserts, "strong evidence that the lowly aspirin may help to prevent strokes?"

The proportion who had strokes or died among those receiving the aspirin is a *sample* proportion:

$$\hat{p}_a = \frac{29}{200} = .145.$$

Similarly, the proportion among those receiving the placebo is a sample proportion:

$$\hat{p}_c = \frac{56}{206} = .272.$$

But our experience with such proportions shows that they are variable, from sample to sample, so that a difference may just be a matter of unlucky draws from populations that are not really different. A test of the null hypothesis of no difference in population proportions: $p_a - p_c = 0$ is in order. (To be continued.) ▲

The difference in proportions of one category between the treatment and control populations, $p_t - p_c$, is estimated by the difference in corresponding *sample* proportions, $\hat{p}_t - \hat{p}_c$. For large samples the test statistic for $\hat{p}_t = \hat{p}_c$ is the Z-score

$$Z = \frac{(\hat{p}_t - \hat{p}_c) - 0}{\text{s.e.}(\hat{p}_t - \hat{p}_c)}.$$

To compute the s.e. it would be proper to combine the s.e.'s of $\hat{p}_t$ and $\hat{p}_c$ (as computed within the treatment and control samples), as was done for the difference of two means generally in Section 9.2, using r.m.s. averaging. However, it is slightly better—because H_0 *assumes* equality of p_t and p_c, and s.d.'s are determined by these proportions—to estimate the common value of p_c and p_t by pooling the samples:

Estimate of the proportion of successes based on independent sample proportions $\hat{p}_t$ and $\hat{p}_c$:

$$\hat{p} = \frac{\text{number of successes in combined sample}}{\text{size of combined sample}}$$

$$= \frac{n_c\hat{p}_c + n_t\hat{p}_t}{n_c + n_t}.$$

This pooled estimate of the common value p is then used in approximating the s.d. of the difference $\hat{p}_t - \hat{p}_c$:

$$\text{s.d.}(\hat{p}_t - \hat{p}_c) = \sqrt{\frac{p_t q_t}{n_t} + \frac{p_c q_c}{n_c}} = \sqrt{pq\left(\frac{1}{n_t} + \frac{1}{n_c}\right)}.$$

Replacing p by $\hat{p}$ and q by $\hat{q} = 1 - \hat{p}$, we obtain an estimate of this s.d., namely, the standard error:

$$\text{s.e.}(\hat{p}_t - \hat{p}_c) = \sqrt{\hat{p}\hat{q}\left(\frac{1}{n_t} + \frac{1}{n_c}\right)},$$

a quantity that can be calculated from the data.

Z-test for the equality of two population proportions, using large, independent samples:

Calculate the Z-score for the difference between sample proportions:

$$Z = \frac{\hat{p}_t - \hat{p}_c}{\sqrt{\hat{p}\hat{q}(1/n_t + 1/n_c)}},$$

where $\hat{p}$ is the pooled estimate of the common population proportion. Determine the P-value from Table II.

Example 9–11 *Aspirin and Strokes (continued)*

We return to the question of Example 9–10: Is the difference in sample proportions:

$$\hat{p}_a - \hat{p}_c = \frac{29}{200} - \frac{56}{206} = -.127,$$

attributable to chance rather than to a difference in population proportions—that is, to a treatment effect?

If the population proportions are equal: $p_a = p_t = p$, the estimate of this common probability or population proportion is the total number who had strokes or died divided by the total number of men in the experiment:

$$\hat{p} = \frac{29 + 56}{200 + 206} = .209.$$

The standard error of the difference between sample proportions is then

$$\text{s.e.}(\hat{p}_a - \hat{p}_c) = \sqrt{.209 \times .791 \left(\frac{1}{200} + \frac{1}{206} \right)} = .040.$$

The Z-score of the difference in proportions is

$$Z = \frac{\hat{p}_a - \hat{p}_c - \text{m.v.}(\hat{p}_a - \hat{p}_c)}{\text{s.e.}(\hat{p}_a - \hat{p}_c)} = \frac{-.127 - 0}{.040} = -3.14.$$

With a one-sided P-value of .0009, this Z is hard to imagine as only a sampling fluctuation. The evidence that "the lowly aspirin may help to prevent strokes—at least in men" does seem strong. (The qualification about men is included because in a group of about 200 women in the study, the observed difference was not statistically significant.) ▲

9.6
COMPARING CATEGORICAL VARIABLES, $k \geqslant 2$

When a classification involves more than two categories, the Z-test of the preceding section cannot be used for comparing two populations, and a different approach is in order.

Example 9-12 *Proposition 13*

After the passage by California of Proposition 13, which reduced the property tax income of local governments by 68 percent, a telephone poll was conducted by *The New York Times* and CBS News to determine sentiment on property taxes in the country as a whole. One question asked dealt with how to make up for lost revenue from property taxes. Results for California and elsewhere in the country were as follows:

	California	Outside California	Combined
Cut services	265	667	932
Add new taxes	108	208	1093
Other, or no opinion	61	218	279
Total	434	1093	1527

Is the feeling on this question the same in California as outside California? The null hypothesis is that it *is* the same.

If the feelings were the same, and if the samples were completely representative, the three columns of frequencies would be proportional. The entries for California and for outside California would be proportional to the combined frequencies, $932:316:279$, and the table of frequencies would look like this:

	California	Outside California	Combined
Cut	265	667	932
Add	90	226	316
Other	79	200	279
Total	434	1093	1527

The 265 in this table is obtained as 932/1527 of 434:

$$265 = \frac{932 \times 434}{1527}.$$

Similarly,

$$226 = \frac{316 \times 1093}{1527}, \quad 79 = \frac{279 \times 434}{1527},$$

and so on. (The easily remembered pattern, to obtain any cell entry, is to multiply the corresponding marginal totals and divide by the overall total.)

In random sampling, the samples are not (usually) exactly representative, and the sample proportions may not agree even when the populations do. But they would on the average, and it is the expected or mean frequencies that we are getting at in the last calculations. The cell entries in the second table are *estimates* of the mean frequencies under the null hypothesis—estimates, because the frequencies in the combined sample, on which they are based, are just estimates of expected frequencies in a sample of 1527. Nevertheless, we take the six estimated cell mean frequencies in forming a *chi-square* statistic with the six observed cell frequencies, as a measure of fit of the data to the null-hypothesis model:

$$\chi^2 = \frac{(265 - 265)^2}{265} + \frac{(108 - 90)^2}{90} + \frac{(61 - 79)^2}{79} + \frac{(667 - 667)^2}{667}$$
$$+ \frac{(208 - 226)^2}{226} + \frac{(218 - 200)^2}{200} = 10.75.$$

Had the observed proportions for California and for outside California turned out to be the same, this statistic would have had the value 0, indicating a perfect fit. But it is not 0, and the question is this: Is the value 10.75 for χ^2 so large as to be hard to account for as chance variation in the null hypothesis model? To answer this, we need the sampling distribution of χ^2 under H_0. ▲

Having used the statistic χ^2 before, you might expect that for large samples its distribution would be approximately of the chi-square type, with degrees of freedom one less than the number of cells (or $6 - 1 = 5$, in Example 9–12). However, the observed frequencies were compared with *estimated* mean frequencies, and fitting the data using estimates obtained from those data tends to reduce the value of χ^2. It can be shown that, whereas the large-sample distribution *is* approximately chi-square under the null hypothesis, the number of degrees of freedom must be reduced—to $(3 - 1) \times (2 - 1) = 2$ in Example 9–12. This corresponds to the "freedom" one has in assigning cell frequencies *given* the marginal totals. In the example, we could specify the 265 and 108; the other cell frequencies are determined by the marginal totals. The *two* cells represent our "degrees of freedom." (In the chi-square test of Chapter 8, with k cells, we were "free" to choose $k - 1$ frequencies, given the total $\sum f = n$.)

The hypothesis that two populations have identical proportions in the categories of some variable of classification is said to be a test of *homogeneity*. The layout of data in a two-way array is called a *contingency table*. The chi-square test just described can be used in comparing not just two, but any number of populations.

Chi-square test for homogeneity in a contingency table with r rows and c columns:

1. Calculate estimated mean cell frequencies from the marginal totals:

$$\text{mean frequency} \doteq \frac{\text{row total} \times \text{column total}}{\text{total number of observations}}.$$

2. Compute

$$\chi^2 = \sum \frac{(\text{observed cell frequency} - \text{mean cell frequency})^2}{\text{mean cell frequency}}.$$

3. Determine the *P*-value from Table V of Appendix B, using $(r - 1) \times (c - 1)$ degrees of freedom.

Example 9–13 *"All the Moos That Are Fit to Print"*

Under the headline above, *The New York Times* reprinted (March 4, 1977) a letter to a British publication, *The Veterinary Record*, reviewing and analyzing a method of sex-ratio control in bovines. A *Times* article in 1974 by a British writer Roald Dahl disclosed that his farmer friend "Rummins" had obtained a 98 percent heifer crop each year from 1916 to 1946 (2516 heifer calves and 56 bull calves) by having his cows inseminated facing the sun.

To test the Rummins method, John P. Bardley of Thatford (as recounted in *The New York Times* of December 9, 1976) faced the inseminating crush south (midday was the customary mating time) and obtained 20 heifers and 4 bull calves. Earlier, in 1975 to 1976, before reading the Dahl article, he had obtained 17 heifers and 33 bull calves with a north-facing crush, and 6 heifers and 11 bull calves in 17 natural matings (uncontrolled as to direction):

	North-Facing	South-Facing	Uncontrolled	Total
Heifers	17	20	6	43
Bull calves	33	4	11	48
Total	50	24	17	91

In this comparison of three "treatments" (directions of mating) the null hypothesis is that the direction of mating is not really a factor. If this were the case, the same proportion of heifers and bulls would be expected in each column as in the "Total" column, or $43:48$.

The table would look more like this:

	North-Facing	South-Facing	Natural	Totals
Heifers	23.6	11.3	8.0	43
Bull calves	26.4	12.7	9.0	48
Total	50	24	17	91

(The entry 23.6 is 50 × 43/91, the 11.3 is 24 × 43/91, and so on.)

Again using Pearson's chi-square statistic to see how well the first table of observed frequencies fits the second table of frequencies expected under the null hypothesis, we compute it for the six cell frequencies:

$$\chi^2 = \frac{(17 - 23.6)^2}{23.6} + \frac{(20 - 11.3)^2}{11.3} + \frac{(6 - 8)^2}{8} + \frac{(33 - 26.4)^2}{26.4}$$
$$+ \frac{(4 - 12.7)^2}{12.7} + \frac{(11 - 9)^2}{9} = 17.03.$$

Whether this seemingly poor fit is evidence of a treatment effect or is explainable as chance variation depends on the null distribution of χ^2, which in this case is approximately chi-square (Table V) with $(2 - 1) \times (3 - 1) = 2$ degrees of freedom. The critical value for a significance level of $\alpha = .01$ is $\chi^2 = 9.21$, so the evidence in favor of an effect appears to be overwhelming. (Of course, one should really take a closer look at the experimental conditions to rule out other factors that might be confounding the results—for example, time. All the south-facing trials were conducted after the others.)

Quite aside from the central issue, Bradley remarked, "I don't expect the bulls mind which way he and the cows are facing—so long as it is the same way." And Dahl's report included the information that just as important to Rummins was the reversal of his technique which permitted him to father four sons because "boys is what you want on a farm."

A final question: Do you believe everything printed in *The New York Times*? ▲

It should be explained that the chi-square test presented here for the case of two or more categories is not really new when two populations are being compared with regard to a classification with just two categories. Indeed, the chi-square statistic for the case of two categories, and the Z-statistic used in comparing proportions (Section 9.5), both of which are appropriate for large samples, are *related:*

$$\chi^2 = Z^2.$$

(This is simply an algebraic identity, not hard to establish from the definitions of the statistics.) Moreover, for any given significance level, the critical value using χ^2 is the square of the critical value in a two-sided Z-test. For instance, the 95th percentile of χ^2 with one degree of freedom is the same as the 97.5 percentile of Z. So the tests are completely equivalent if the alternative is two-sided.

9.7
PAIRED DATA

Sometimes the effectiveness of a treatment can be checked by comparing an individual's response with and without a treatment. Each individual becomes its own "control." The data consist of *pairs* (X_1, Y_1) for individual 1, (X_2, Y_2) for individual 2, and so on. All the X's are obtained under treatment, and all the Y's without it, or vice versa. More generally, the observations (X, Y) may be paired in some other natural way.

Example 9-14 *Carrying Loads*

A study of load-carriage systems[†] reported the results in Table 9-1, giving ratings on a scale (from 6 to 20) of perceived exertion while carrying different types of packs. (This carrying was done on a treadmill, with speed controlled so that the pulse rate was 160 per minute). (The notation M, SD, and SE is that used in the report. One has to get used to differences in notation, for no one system is universal. In particular, "SD" was computed using the divisor $n - 1$.)

It would appear that there are two samples, and this is so, but they are *not independent.* Observations are paired, consisting of an X and a Y for each of seven subjects. The existence of relationship between X and Y seems evident in this example and is supported by the data—the low values of 9 and 11 occur together, as do the high values of 20 and 18.

[†]F. R. Winsmann and R. F. Goldman, *Perceptual and Motor Skills* 43, 1211–1218 (1976).

Table 9-1

Subject	Perceived Exertion	
	Standard Pack, X	Hip Pack, Y
1	11	11
2	9	11
3	12	14
4	11	10
5	13	13
6	20	18
7	16	16
M:	13.1	13.3
SD:	3.72	2.93
SE:	1.40	1.11

It is possible, nevertheless, to test the hypothesis of no difference in exertion for the two types of pack, if we assume that the sample of seven individuals to be randomly drawn, and use the *single sample of differences.*

Subject	Difference, D
1	0
2	-2
3	-2
4	1
5	0
6	2
7	0
	$\bar{D} = -.143$
	$S_D = 1.355$

(These differences and their mean and s.d. were not given in the reference article.)

The t-statistic for testing $\mu_D = 0$ based on the sample of differences D is

$$T = \frac{-.143}{1.355/\sqrt{7-1}} = -.26,$$

If near normality is assumed, the t-distribution with 6 degrees of freedom (Table III of Appendix B) can be used to judge the meaning of this result. (It is far from significant at the usual levels, $-.26$ being near the center of the distribution.)

Alternatively, assuming only the symmetry of the population of differences, one could compute the signed-rank statistic to test the hypothesis that the median difference is 0. For this the nonzero differences are ordered according to magnitude:

$$-2$$
$$1, \quad -2$$
$$2.$$

The sums of the ranks of the positive differences is $R_+ = 1 + 3 = 4$ (and $R_- = 3 + 3 = 6$). The value of R_+ (according to Table IV, with $n = 4$) is not extreme, and H_0 is not rejected.

(It might be noted that the s.d.'s and s.e.'s of the sample of X's and of the sample of Y's as given in the article are not useful in comparing the two types of pack.) ▲

---

t-test for *paired* observations:

Given a pair of scores or responses (X, Y) for each individual in a random sample of n individuals, calculate the n differences $D = X - Y$ and the statistic

$$T = \frac{\bar{D}}{S_D/\sqrt{n-1}}.$$

In testing the hypothesis of no average difference $[\text{m.v.}(D) = 0]$ the P-value is found in Table III, with $n - 1$ d.f., if the differences D are nearly normally distributed.

It may seem at first that the test based on n differences is necessarily less sensitive than one based on twice as much data, namely, on the $2n$ observations in the two samples. If X and Y were independent, this would be true. But when X and Y are positively correlated, there is not as much information in the $2n$ observations as one might think. On the other hand, the variability in the mean difference $\bar{D}$ is reduced; and the less variability in the test statistic, the more sensitive the test.

The reduction in variability in D, and hence in $\bar{D}$, that results when the X and Y in a data pair are positively correlated can be an advantage, significant enough to try to achieve by *introducing* pairing in the experimental design. For instance, suppose that a treatment involves a diet or medicine and is to be tested on human subjects. The response is apt to vary considerably from one individual to another, and this variability in the data may actually mask what could be a worthwhile treatment effect, given samples of a feasible size. To reduce the variability one can pair subjects whose responses are apt to be similar. One way of doing this is to use twins—giving one the standard treatment (which may be no treatment) and the other the treatment being tested, according to the toss of a coin, say. If twins are not available, one can look for pairs of subjects with similar build, age, ethnic origin—whatever identifiable characteristics might contribute to variation in response.

Example 9-15 *Is Preschool Worthwhile?*

Sixteen girls about to enter first grade are given a "readiness test." Eight have been to a popular "preschool" before entering kindergarten, and the others have not. Their scores are as follows:

Preschool: 83, 67, 70, 81, 64, 67, 64, 74 ($\bar{X} = 71.25$, $S = 6.92$).

No preschool: 74, 68, 65, 77, 66, 78, 63, 63 ($\bar{X} = 69.25$, $S = 5.78$).

The two-sample T-statistic for comparing means is

$$T = \frac{2.0 - 0}{6.82\sqrt{1/8 + 1/8}} = .59,$$

which is not significant, assuming independent random samples. (The 6.82 in the denominator is the pooled s.d., computed from the given sample s.d.'s.)

Suppose, however, that the 16 girls had been paired, as closely as possible, according to number of older siblings, socioeconomic class, amount of TV watching, and such other available information as might be relevant—but *not* looking at the test scores themselves, if these are already in hand. The pairs, with their readiness scores (each individual's score as before), turn out to be as follows:

Pair	Preschool	No Preschool	Difference
1	83	78	5
2	74	74	0
3	67	63	4
4	64	66	-2
5	70	68	2
6	67	63	4
7	81	77	4
8	64	65	-1

The means difference, as before, is 2.0. The s.d. is 2.50, and the T-statistic, for the *one* sample of eight pairs, is

$$T = \frac{2.0}{2.50/\sqrt{7}} = 2.12.$$

This *is* significant (7 degrees of freedom).

Pairing the subjects has eliminated, to some extent, sources of variability which, in the two-sample test, were masking the "treatment" effect. Pairing reduced the variability in the difference of mean enough to warrant the sacrifice of 7 degrees of freedom, which increases the critical level from 1.76 to 1.90 (see Table III). Using independent samples would have required considerably larger samples to be able to infer a treatment effect in the midst of individual differences.

But a word of caution is in order. Although a treatment effect seems present, its magnitude (estimated here to be 2.0 ± .945) may or may not have *practical* significance. (A parent might ask: Is a two-point increase in the readiness score worth the tuition for preschool? Indeed, the advantage may even disappear by the end of first grade. Clearly, there is more to be studied.) ▲

Example 9-16   *Smoking and Birthweight*

A newspaper article reported results of a study of babies' birthweights and mothers' smoking. It involved 88 matched pairs of mothers—one smoker and one nonsmoker in each pair. What constituted a match was not explained in the newspaper account, but no doubt mothers were matched in regard to variables that one would expect to be factors in determining a baby's birthweight—body characteristics of mothers and corresponding fathers, length of term, mothers' diet during pregnancy, and so on.

The mean difference in babies' birthweight between smoking and nonsmoking mothers was found to be 1/2 pound. But to judge whether this is statistically significant requires some knowledge of the variability of birthweights. Lacking the data, we can still do something. Birthweights tend to run between 5 and 10 pounds (excluding those prematurely born), so perhaps 1 pound is not a bad estimate of the s.d. If the birthweights in a matched pair were independent, the s.d. of their difference would be $\sqrt{1^2 + 1^2} = 1.4$ pounds, and

$$T = \frac{.5}{1.4/\sqrt{88}} \doteq 3.3.$$

But matching will tend to introduce a positive correlation between the weights in a pair, which will make the s.d. of the difference *less* than 1.4 pounds. This, in turn, suggests that the T-statistic for the actual data would be larger than 3.3—perhaps a lot larger. But even a T-value of 3.3 would lead one to conclude that the sample of differences is *not* a random sample from a population with mean 0. If the sample can be thought of as a random sample from some population, the conclusion would be that smoking mothers have smaller babies (on the average) than do nonsmoking mothers. Whether a mother's smoking inhibits her baby's development is a question of *causal* relationship which this study does not answer. It merely says that there is a relationship. (A stronger case for a causal relationship could be made if the "treatment" of smoking could be assigned randomly to one-half of a group of randomly chosen subjects. Moral considerations rule out this experimental design.) ▲

In applying a one-sample test—either a t-test or a signed-rank test—one may encounter a situation in which all differences are of one sign, or mostly of one sign. It would be natural to wonder if such a phenomenon is not in itself sufficient evidence of a treatment effect, without the calculation of a more sophisticated statistic.

Example 9-17 *Drug Therapy and IQ*

The following are verbal IQ scores on 10 subjects (children ranging from 8 to 14 years of age), with and without a therapy involving the drug ethosuximide:[†]

[†]The data are taken from *Statistics in the Real World*, by R. Larsen and D. Stroup (Macmillan, 1976), which in turn took them from the report of a

Subject	Without Drug	With Drug	Sign of Difference
1	97	113	+
2	106	113	+
3	106	101	−
4	95	119	+
5	102	121	+
6	111	122	+
7	115	121	+
8	104	106	+
9	90	110	+
10	96	126	+

(After each of two consecutive 3-week periods a child was given parts of an IQ test. During one period the child was given a placebo and during another, the drug ethosuximide. The choice between giving the placebo in the first period and the drug in the second, or the other way around was made at random.)

Under the null hypothesis of no difference between the placebo and the drug therapy, the difference (for a given subject) would be positive (+) with probability $1/2$—that is, it would be just like the toss of a coin, if we identify + with heads. The probability of seeing 9 or more heads in 10 tosses of a fair coin is .0108 (found in Table I, with $n = 10$ and $p = .5$.), and this is then the P-value corresponding to the observed 9 +'s among the 10 differences. The outcome of the experiment is thus seen to be statistically significant without our bothering to take the magnitudes of the differences into account. ▲

The test described in Example 9–17 is called a *sign test*. It is essentially a test of the hypothesis $p = 1/2$, where p is the probability that a second measurement is greater than the first in any pair. Tests for hypotheses of the form $p = p_0$ were taken up in Section 8.1 (Examples 8–4 and 8–5) and Section 8.2, for the cases of small and large samples, respectively.

The sign test, being based on only the *signs* of the differences between paired responses, is not as sensitive as the signed-rank and t-tests, which also make use of the *magnitudes* of the differences.

study by W. L. Smith: "Facilitating Verbal-Symbolic Functions in Children with Learning Problems . . . ," in *Drugs and Cerebral Function*, W. Smith, ed. (Thomas, 1970), p. 125.

On the other hand, it is just this aspect of the test that makes it more versatile, for it can be used when numerical scaling is awkward. It is only necessary, for application of the sign test, to judge whether one response in a pair is greater or less than the other.

Sign test for the hypothesis $P(X < Y) = P(X > Y)$, given a random sample of n pairs (X, Y):

Determine the proportion $\hat{p}$ of sample pairs in which $X < Y$. If $n \leqslant 10$, find the P-value in Table I (under "$p = .5$"). If $n > 10$, find the P-value in Table II, using the Z-score:

$$Z = \frac{\hat{p} - .5}{.5/\sqrt{n}}.$$

(Data pairs in which $X = Y$ are excluded from the sample.)

KEY VOCABULARY

Treatment group
Control group
Treatment effect
Rank order
Rank-sum statistic
Pooled variance
Two-sample t-test
Homogeneity
Contingency table
Paired data
Matched pairs
Sign test

QUESTIONS

1. Why should one expect the s.d. of a sum of independent random quantities to be less than the sum of their s.d.'s?

2. What similarity do you find between the construction of a confidence interval for a difference of population means and the construction of a confidence for a single population mean?

3. What similarity do you find between the structure of a Z-statistic for testing the null hypothesis of no difference in population means and that of a Z-statistic for testing the null hypothesis that a population mean has a particular value?

4. In what circumstances is it appropriate to pool two-sample variances?

5. What assumption about the populations makes the use of the t-table for a two-sample t-statistic exactly correct? How critical is it?

6. What assumption about the two samples is needed in using the two-sample t-statistic?

7. What is an advantage of the rank-sum test over the t-test for the hypothesis of no difference in location?

8. How is it that when comparing continuous response variables, we sometimes encounter ties, which have probability 0?

9. Why does our table of rank-sum probabilities give only left-tail probabilities?

10. Why is it that in calculating the standard error of a difference in proportions, we do *not* pool the two-sample proportions, whereas in testing the equality of proportions we *do*?

11. In the case of large samples, the equality of two proportions can be tested using either the chi-square statistic or a Z-statistic. Might you get different results with these two methods?

12. When data come in pairs, why is the standard error of a mean difference not calculable as an r.m.s. average of the s.e.'s of the individual means?

13. When the two-sample t-statistic is applied to paired data, is it apt to be larger or smaller than the one-sample t-statistic applied to the differences if the responses in the pairs are positively correlated?

14. Explain why pairing is sometimes deliberately introduced as part of an experimental design for comparing treatments?

15. In Example 9–16, what aspect of the study makes it impossible to conclude that the mothers' smoking causes smaller birthweights, on the average?

16. What information in the data is not utilized in carrying out a sign test? Is this neglect good or bad?

PROBLEMS

Sections 9.1 and 9.2

*1. Obtain 20 pairs of random digits from Table VII and calculate the s.d. of the (signed) differences. The population s.d. for individual random digits is[†] $\sqrt{8.25}$; compute the population s.d. for the difference of two independently drawn random digits and compare this with the s.d. you computed for the 20 pairs.

2. If in a large group of young couples the s.d. of the husbands' ages is 4.0 and the s.d. of the wives' ages is 3.0, would you expect to find the s.d. of the difference in ages (from couple to couple) to be near 5? (Why or why not?)

*3. An experiment to test the effectiveness of a certain toothpaste ingredient yields the following data:

	Number of Children	Average Number of Cavities per Child	Standard Deviation
Test dentifrice	250	4.0	2.5
Control dentifrice	200	5.0	3.0

Assuming independent random samples, give an estimate of the population mean difference together with a standard error of estimate.

4. A flock of 100 young turkeys was split into two groups of 50 each, and different rations were fed to the two groups for a given period of time. At the end of this period, weight gains were measured in pounds, with these results:

	Number	Mean	s.d.
Ration A	50	15.2	2.2
Ration B	50	14.8	2.0

Give an estimate of the difference in mean weight gains of the populations corresponding to the two rations, together with a standard error of estimate.

[†]The number 8.25 is simply the average of the squares of 0, 1, 2, . . . , 9 minus the square of the average 4.5.

Section 9.3

*5. The 25 individual trials of dumping the box of tacks, for the experimenters of Problems 16 and 23 of Chapter 6, have means and s.d.'s as follows:

	n	Mean	s.d.
Author	25	71.36	4.1845
Son	25	67.40	5.6285

Construct a Z-statistic for testing the hypothesis of no difference in population means (equivalent to no difference in p's, if the tacks in the box fall independently at each trial), and determine a P-value.

6. In Problem 4, is the observed difference in weight gains between the two groups of turkeys statistically significant?

*7. In the United States in 1945, medical care took 4.2 percent of the total of all personal expenditures, or 4.2 percent of 119.7 billion dollars. In 1965, medical care accounted for 6.5 percent of total personal expenditures or 6.5 of 465.9 billion dollars. Is this increase in proportion of expenditures for medical care significant?

8. According to the 1970 census, the median age of blacks in the United States was 22.4, and the median age of whites was 28.9 years. Is the difference in median ages statistically significant?

*9. A group of 28 students in a statistics class was asked to measure their pulse rates. There were 12 smokers and 16 nonsmokers in the group, and their counts (per minute) were as follows:

Smokers: 62, 66, 90, 92, 66, 70, 68, 70, 78, 100, 88, 62.

Nonsmokers: 64, 58, 64, 74, 84, 68, 62, 76, 80,   68, 60, 62,

72, 70, 74, 66.

The means and s.d.'s for smokers are 76.0 and $\sqrt{158.67}$, and for nonsmokers 68.875 and $\sqrt{50.9844}$.

Test the hypothesis of no treatment effect at $\alpha = .10$ against the one-sided alternative $\mu_S > \mu_N$ at the 10 percent significance level, assuming populations with equal s.d.'s (and shapes that would permit a t-test.

10. A research study[†] of female Olympic athletes reports that "runners and swimmers seemed to have significantly broader pelvises" than gymnasts. The data are summarized as follows:

	Mean Bicristal Diameter (cm)	s.d.	n
Runners and swimmers	28.0	.885	15
Gymnasts	25.9	.512	5

   (a) Test the null hypothesis of no difference in mean diameter.
   (b) Criticize the report's statement about significance, and rewrite it in a correct form.
   (Observe also that a "test" here is an inference about conceptual populations. What are they?)
   (c) The study would like to attribute any difference to differences in the use of the body. Is there any other possible explanation?

*11. A group of 32 meditators and a group of 32 nonmeditators[‡] were given 20 trials each on an apparatus called a "mirror-star tracer." Their mean learning scores in terms of reducing errors, and in reducing the required time, were as follows:

	Mean	s.d.	n
Learning—error			
Meditators	24.64	34.80	32
Nonmeditators	18.12	17.47	32
Learning—time			
Meditators	52.15	40.08	32
Nonmeditators	48.27	25.32	32

   Test for the hypothesis of no difference between meditators and nonmeditators in each case.

[†]L. P. Novak, "Working Capacity, Body Composition and Anthropometry of Olympic Female Athletes," *J. Sports Med. and Physical Fitness* **17**, 275–283 (1977).
[‡]I. R. T. Williams, "Transcendental Meditation and Mirror Tracing Skill," *Perceptual and Motor Skills* **46**, 371–378 (1978).

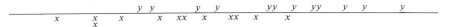

Figure 9-3. Data for Problem 12.

Section 9.4

*12. Independent samples from X and from Y are plotted on a single scale, in Figure 9-3.
(a) Calculate R_x, the sum of the X-ranks.
(b) Assuming approximate normality of R_x, compute the corresponding Z-score and test for significance at the 5 percent level.

*13. The individuals in a random sample of 10 sober subjects and in a random sample of 9 subjects treated with two martinis each were administered a psychological test involving some reasoning and some motor facility. Scores, on a scale from 0 20, were as follows:

    Sober: 16, 13, 13, 17, 11, 16, 14, 16, 17, 12.

    Treated: 16, 12, 8, 10, 14, 12, 9, 12, 13.

Use a rank test for the hypothesis of no treatment effect.

14. Use a rank test to compare the initial pulse rates for smokers and for nonsmokers given in Problem 9.

*15. The group of 28 statistics students who recorded pulses in Problem 9 also recorded them after a 1-minute period of exercise. The 12 smokers and 16 nonsmokers recorded the following *changes* from before to after the exercise:

    Smokers: 14, 12, 4, -8, 16, 2, 8, 24, 26, 5, 22, 26
    Nonsmokers: 24, 22, 16, 10, 0, 4, 13, 42, 16, 8, 16, -4,

                       8, 36, 2, 36.

(a) Apply a rank test for the hypothesis of no treatment difference, where "treatment" refers to smoking.
(b) Apply a t-test for the same purpose as in part (a).

16. At the end of an experiment, blood pressures of a treatment group of eight rats (the treatment being intermittent feeding)

and of a control group of seven rats were as follows:[†]

Pressures		Mean	s.d.
Treatment:	142, 121, 131, 154, 142, 138, 168, 172	146	16.47
Control:	128, 123, 133, 128, 121, 140, 144	131	7.89

(a) Apply the two-sample rank test for the null hypothesis of no treatment effect.

(b) Apply a t-test for the same purpose.

(Assume independent random samples.)

Sections 9.5 and 9.6

*17. A Veterans Administration Medical Center study (reported in the *New England Journal of Medicine*, March 8, 1979) monitored the breathing of normal, healthy men and women, while they slept. Of the 30 men in the study, 20 had episodes of abnormal breathing or low oxygen levels, while of the 19 women, 3 had such incidents. Is this difference between men and women attributable to random sampling fluctuations?

18. In November 1968, spot surveys of parking lots in Nashville were made to see whether supporters of Humphrey and Wallace (in the then forthcoming election) differed in their compliance with a new city ordinance requiring purchase and display of a certain sticker. It was found that 154 of 178 cars with Humphrey bumper stickers displayed the required windshield sticker, and 270 of the 361 cars with Wallace bumper stickers did so.[‡] Calculate a Z-statistic for comparing population proportions. Is it statistically significant?

*19. In 1941, Ted Williams won the American League batting championship with an average of .406 (185 hits in 456 at-bats) and Joe DiMaggio was second with an average of .357 (193 hits in 541 at-bats).

(a) Is there a statistically significant difference between these two proportions?

[†] J. Falk, M. Tank, and S. Forman, "Schedule Induced Hypertension," *Psychosomat. Med.* 39, 252–263 (1977).

[‡] From "Wallace Supporters and Adherence to 'Law and Order,'" in R. A. Baron and R. M. Liebert, eds., *Human Social Behaviors* (Homewood, Ill.: Dorsey Press, 1971), pp. 217–225.

(b) Their lifetime statistics are as follows:

	At-bats	Hits	Average
TW	7706	2654	.3444
JD	6821	2214	.3246

Their lifetime averages differ by only about .020, but they are based on many more at-bats. Is this difference significant?

(The assumptions of independence and time-invariant probability of getting a hit do not hold here, but experience suggests that they are not grossly inaccurate.)

20. Repeat the analysis of the preceding problem for the *home run* probabilities of Ted Williams versus Joe DiMaggio based on number of lifetime home runs of 521 for TW and 361 for JD. [Their lifetime home run averages differ by only about .015, but now there is more accuracy available, since $p(1 - p)$ is smaller.]

*21. The Gallup Poll once sampled 1520 individuals in the United States and asked each if he or she favored a constitutional amendment prohibiting abortions except when the mother's life is in danger. The following table gives the results (rounded to make the computation easier):

	Favor	Oppose	No Opinion
Men	320	380	60
Women	360	360	40

In the report it was stated that "analysis of the findings reveals that men tend to be more pro-abortion than women." Analyze these data and see if you agree.

22. Opinions on the usefulness of current theories of behavior modification were solicited from 200 psychologists and 80 psychiatrists; out of 138 responses, 110 were usable. Responses were cross-classified as follows:[†]

[†]G. P. Koocher and B. M. Pedulla, "Current Practices in Child Psychotherapy," *Profess. Psychol.* 8, 275–286 (1977).

	Usefulness			
	Never	Occasionally	Often	Always
Psychologists	12	26	29	18
Psychiatrists	5	13	7	0

What conclusion can be **drawn**?

*23. Referring to the check of nonresponse bias in Problem 20 of Chapter 8, we find that the original respondents divided as shown in the following two-way table. Also given are the results for the sample of original nonresponders:

	Original Sample	Nonresponders
Strongly agree	750	86
Agree with reservations	1050	100
Disagree with reservations	1150	119
Strongly disagree	2050	172

Test for homogeneity.

24. Using the data and χ^2-value in Example 9–12, test for homogeneity of in-state and out-of-state opinions on California's Proposition 13.

Section 9.7

*25. In each of seven pairs of overweight twins, one is picked at random and assigned a standard reducing diet A, and the other is given new diet B. Weight losses over a given period are as follows:

Pair Number	Diet A	Diet B
1	10	15
2	6	10
3	19	21
4	13	12
5	8	14
6	7	6
7	3	6

Test the hypothesis of no difference in effects of the two diets against the alternative that diet B is more effective, at the 10 percent significance level:

(a) Using a signed-rank test.

(b) Using a t-test.

26. A report[†] of a study on "rational-emotive" therapy gives the following data—mean scores of 11 subjects on anxiety tests taken before and after therapy:

	Pretest	Posttest
Mean	29.82	32.54
s.d.	5.13	7.10

The value $T = 3.48$ is given in the report and termed "significant."

(a) Is the given T-value a two-sample t-statistic? Should it be?

(b) If the given T-value is a one-sample statistic (for 11 pairs of scores), deduce the s.d. of the differences in pre- and posttest scores. (Is this calculable from the given s.d.'s—which is all that the report says about variability?)

*27. Increases in normal heart rates were measured for eight subjects who watched a pornographic movie and again after they watched an Alfred Hitchcock suspense film:[‡]

Subject Number	1	2	3	4	5	6	7	8
"Adult" movie	1.2	10.0	5.6	16.8	3.2	5.6	16.4	-4.0
Hitchcock	7.6	1.6	4.0	6.8	-.8	5.6	-1.2	8.0

Calculate an appropriate rank statistic. Is it significant?

28. The article referred to in Example 9-14 also gave data on the treadmill speed required to maintain a pulse rate of 130 per minute, for each of eight subjects using each of two types of pack, as follows:

[†]S. P. Hymen and R. Warren, "An Evaluation of Rational-Emotive Therapy," *Perceptual and Motor Skills* 46, 847–853 (1978).

[‡]From N. Bernick, A. Kling, and G. Borowitz, "Physiological Differentiation of Sexual Arousal and Anxiety," *Psychosomat. Med.* 33, 341–352 (1971).

Subject	Standard Pack	Hip Pack
1	4.65	5.71
2	3.31	3.75
3	4.96	4.63
4	5.89	5.97
5	5.36	4.70
6	4.38	4.31
7	4.78	5.97
8	5.65	5.81
Mean:	4.87	5.11
SD:	.76	.81
SE:	.1	.3

(a) Are the SE's (taken from the article) correct? Are they of any use?

(b) Test the null hypothesis of no difference in type of pack.

*29. Find the P-value corresponding to the sign-test statistic,

(a) For the data of Problem 25.

(b) For the data of Problem 27.

(c) For the data of Problem 28.

30. Twelve matched pairs of plants, six at each of two experiment stations, were used for comparing two types of fertilizer, fertilizers A and B being assigned to the two plants in a pair by the toss of a coin. Analyze the response data using the sign test:

Station 1		Station 2	
Type A	Type B	Type A	Type B
6.9	7.5	5.4	7.0
7.8	8.3	5.3	7.1
9.0	8.1	6.8	6.8
9.4	8.6	5.6	6.4
9.4	9.8	6.2	6.0
9.8	9.2	6.4	6.7

Review

31. An elementary statistics text gives the following data (presumably artificial) for weight losses by five individuals who used diet A and five who used diet B:

$$\text{Diet } A: \quad 0, \quad 0, \quad 0, \quad 1, 49$$

$$\text{Diet } B: \quad 9, 10, 10, 10, 11$$

These data were presented in a problem intended to point out that the *t*-test is not always a good tool.

(a) If you did blindly use a two-sample *t*-test, what conclusion would it call for?

(b) Why do you think the two-sample *t*-test might not be appropriate?

(c) Apply a rank test to these data. (Comment on the appropriateness of the rank test as a substitute for the *t*-test.)

(d) Which diet would you prefer if you wanted to lose a lot of weight?

32. A school administrator notes that in one city there are 10 high schools; these have an average enrollment of 1800 (s.d. = 300). In a nearby city, there are eight high schools, with an average enrollment of 1300 (s.d. = 250). The administrator computes a two-sample *t*-statistic to be $T = 3.56$ and reports a statistically significant difference in average enrollments. What do you think of this report?

33. Each child among six pairs of twins in a children's home received a cup of milk at 9 o'clock, and one of the twins in each pair received in addition 1 tablespoon of honey dissolved in the milk. After 6 weeks, increases in hemoglobin were recorded:

Pair	Honey	No Honey
1	19	14
2	12	8
3	9	4
4	17	4
5	24	11
6	22	15

Use an appropriate test for the null hypothesis of no treatment effect against the alternative that the honey increases the hemoglobin level.

34. The 1960 census gave these numbers of people in various types of occupation, in thousands:

	Male	Female	Total
White collar	26,607	12,119	28,726
Blue collar	20,571	3,640	24,211
Service	2,911	5,438	8,349
Farm	4,396	999	5,395

Are the differences in proportions between men and women statistically significant? (Think before you compute.)

35. In a study[†] involving life-styles, questionnaires were distributed to 1400 couples; 339 were returned, of which 107 were usable. Results concerning the religion of the respondents were as follows:

	Background	Current
Protestant	132	60
Catholic	55	11
Jewish	11	9
None	9	120
Other	7	14

Compute χ^2 for this contingency table. Is this computation a useful exercise? (Why or why not?)

36. Students in a course in business administration were split into several small groups, and the groups were then assigned at random to be part of a treatment or part of a control sample. Those students in the control sample worked as individuals in identifying the problems of a certain company, from a lengthy case study, and wrote individual reports. Those in the treatment sample discussed and analyzed the case within groups, but subsequently also wrote individual reports. Reports of both control and treatment individuals were graded without the grader's knowing the student's name or whether the student was in the control or in the treatment sample. The letter grades were coded according to the usual scale (A = 4, B = 3, etc.), except that .3 was added for a plus and .3 subtracted for a minus:[‡]

Control: 3.7, 2.7, 3.0, 2.3, 2.0, 2.3, 2.7, 1.3, 2.3, 3.3.

Treatment: 2.0, 1.7, 2.7, 1.3, 3.3, 2.3, 2.3, 3.0, 4.0, 2.0, 3.0, 4.0, 3.3.

Test the hypothesis that the "treatment" of group discussion has no effect.

37. Suppose that a new drug has been discovered for treating leu-

[†]G. A. Thoen, Ph.D. thesis, University of Minnesota, 1977.
[‡]P. H. Anderson, "The Effect of Case Study Method on Problem Solving," Ph.D. thesis, University of Minnesota, 1975.

kemia. Its proponents suggest that if taken when the disease reaches a certain stage, it will increase the patient's expected life. Half of 400 patients are selected at random, for treatment with the drug. The other 200 are not given the treatment, and the experiment is "double-blind." The average number of months of life in the control group turns out to be 20.1 with an s.d. of 5, while for the treated group these statistics are 24.0 and 12, respectively.

(a) Is there a significant difference?

(b) Suppose that you had leukemia and wanted to be sure as possible of living 12 months; would you want the treatment? Explain, making a reasonable assumption about the populations if need be.

(c) Answer part (b) for 36 months in place of 12.

38. The two experimenters (author and son) who dumped the box of thumb tacks 25 times each, as described in Chapter 6, Problems 16 and 23, obtained 1784 and 1684, respectively, as the total number of tacks (out of 2500) with points up. Test the hypothesis that the probability of a tack's falling with point up is the same for both experimenters. (This gives a test that is different from the Z-test in Problem 5. *If* the 2500 tosses can be considered in each case to be independent tosses with a fixed p, this test is more appropriate to the problem. If there is any question about the independence of tacks in dumping, the Z-test in Problem 5 is the appropriate one. But in that case the test deals not with a "p" but with the average number of points up in dumping a box of 100.)

39. In October 1978, the number of years of professional football experience of the 40 players on the Dallas Cowboy's roster had mean 5.7 and s.d. 3.71. The number of years of experience of the 39 players on the Minnesota Vikings' team had mean 7.31 and s.d. 4.99. Were the Cowboys significantly less experienced, on the average, than the Vikings?

40. A study[†] reports results of tests on attitudes toward feminism, the scores (totaled over 18 items) being given on a scale from 18 to 90, high scores indicating pro-feminism. Subjects were classified as "high," "medium," and "low" by scores on preliminary tests to determine degree of authoritarianism. The

[†]G. Sarup, "Gender, Authoritarianism, and Attitude Toward Feminism," *Soc. Behav. Personality* **4**, 57–64 (1976).

results for the "high" and "low" categories are as follows:

Authoritarianism Rating	Mean	s.d.	n
Low	67.7	11.8	30
High	52.4	13.0	31

Calculate the value of an appropriate statistic for testing the null hypothesis of no difference between those with low and those with high authoritarianism ratings, and determine a *P*-value, assuming independent random samples.

(The original sample included 53 students at a midwestern university, 6 high school students attending an NSF program at the university, and 14 persons working and living in neighboring communities. However, since not enough cases of *really* high authoritarianism ratings were found among these subjects, the researchers added 55 more respondents—19 belonging to two fundamentalist churches and 36 employees of business establishments in nearby towns. Perhaps this could be called a "scrounged sample.")

41. It is asserted in Example 9-14 that the X and Y measurements of exertion for an individual are *related*. Calculate the coefficient of correlation r for the seven pairs (X, Y) given in that example.

10
Correlation
and Independence

MANY IMPORTANT STATISTICAL PROBLEMS involve several variables and their interrelationships. For example, a sociologist would be interested in relationships among the rates of inflation, unemployment, birth, and suicide. An anthropologist would want to understand the relationships among various skull measurements—such as breadth, basialveolar length, nasal height, and basibregmatic height. Botanists, for purposes of classification, would want to understand the relationships among petal length, petal width, sepal length, and sepal width for various varieties of iris. Educators deal with individual students, who are measured or classified according to achievement, aptitude, and treatment (instructional methods). And so on.

A first step in dealing with such problems is to consider the case of two variables. A "relationship" between two variables exists when changes in one variable tend to accompany changes in the other. Our basic concern is the extent and nature of such relationships. In Chapter 2 we introduced some ways of summarizing bivariate data, and in Chapter 3 we explained how one can measure a certain type of relationship between numerical variables. Here the question will be one of inference: Given a sample of bivariate data, what conclusions can be drawn about a relationship between the population variables?

10.1
RELATIONSHIP AND CAUSE

When random variables A and B are related, there are three possible explanations:

1. A change in variable A causes a change in B.
2. A change in variable B causes a change in A.
3. Neither variable causes the other; their relationship is the result of other factors that affect them both.

When variable A deals with a phenomenon preceding the phenomenon of variable B in time, its variation could not be caused by that of B. However, even in such a case, the conclusion that then changes in A must cause changes in B—although tempting—is unwarranted, because of the possibility of explanation 3.

Example 10-1 *Storks and Babies*

It was once noticed that there existed a strong relationship between the number of stork nests and the number of babies born in English villages. This does not mean that storks bring babies, nor that babies attract storks. Both variables were determined in part by the size of the village.

This somewhat frivolous example is given to show how silly it can be to take the existence of a relationship between two variables as evidence of cause. ▲

Example 10-2 *Smoking and Disease*

A strong correlation has been rather firmly established between cigarette smoking and lung cancer and other diseases. The smoking usually precedes the disease; so if one causes the other, it would have to be smoking that tends to induce the disease. The possibility that other factors, possibly genetic, are "causing" both smoking and disease has made it hard to resolve the issue. The Surgeon General has a warning of a health hazard on cigarette packages, and in 1979 released still another report on the association of smoking and disease; but the American Tobacco Institute has yet to concede defeat. ▲

When the relationship between variables is one of cause and effect, it is useful to know this so that one can achieve certain effects by adjusting the cause. This is why people tend to go beyond a justifiable inference of a relationship to one of cause and effect that is not justified by statistical evidence alone.

To establish causality, it is essential to *control* the variable thought to be a "cause." We have already encountered statistical problems in which the cause variable is a *treatment*. Analyzing response data gathered for treated and for untreated individuals, or for samples of individuals given each of several types of treatment, will give evidence for or against the hypothesis of cause–effect. Methods of statistical comparison, used for such inferences, were introduced in Chapter 9 and will be developed further in Chapters 11 and 12.

The existence of a relationship may indeed be useful in suggesting the possibility that it is a causal relationship. However, without control of the causing variable, such a possibility must be investigated by a study of the mechanisms involved.

All of this is not to say that unless one can establish one variable as a cause, the existence of a relationship is of no interest. When variables are related, whether or not one is a "cause" and the other its "effect," the relationship can be exploited to *predict* one variable from a knowledge of the other. Thus, if college performance is related to aptitude test scores, the test score of an individual can be used in predicting his or her performance in college. Again, if physiological variables A and B are related, and the more difficult-to-measure B is more relevant to diagnosis than A, one can use the easier measurement A to "predict" the value of B.

10.2
INDEPENDENCE OF CATEGORICAL VARIABLES

In Chapter 4 we defined independence of random variables. For a population of individuals that are cross-classified according to two variables, independence is manifest as proportionality of rows and proportionality of columns.

Example 10-3 *Suit and Denomination*

A deck of 52 playing cards can be thought of as a population. Each card has a *suit* and a *denomination*. To simplify the discussion, we consider just these categories for denomination: *ace* (A), *face* (K, Q,

J), and *other* (2, 3, . . . , 10). The population numbers in each combination of a suit with a denomination are shown in the following two-way table:

	S	H	D	C	
Ace	1	1	1	1	4
Face	3	3	3	3	12
Other	9	9	9	9	36
	13	13	13	13	52

Observe that the rows are all in proportion, and that the columns are all in proportion.

The proportionality of the columns is telling us that the odds on ace, face, and other are the same in the subpopulation of spades as in the subpopulation of hearts, and so on, and the same as in the population as a whole. That is, the subpopulations are (in the language of Section 9.6) "homogeneous." By the same token, the subpopulations of ace, face, and other are homogeneous as regards the variable "suit"—that is, the proportions of spades, hearts, diamonds, and clubs are the same in the subpopulations according to denomination. ▲

From Example 10–3 we see that the hypothesis of *independence* of two categorical variables is equivalent to the hypothesis that the subpopulations determined by the categories by one of the variables are homogeneous with respect to the other variable. So we can test for independence by means of the chi-square statistic presented in Chapter 9 for testing for homogeneity.

Example 10-4 *Therapeutic Pets*

A study of 92 patients hospitalized for a heart attack or other serious heart disease was reported at a 1978 meeting of the American Heart Association. These 92 patients were classified according to whether or not they were alive a year after their hospitalization, and also according to whether or not they owned a pet:

	Lived	Died	
Pet	50	3	53
No Pet	28	11	39
	78	14	92

This contingency table looks just like those analyzed in Section 9.6; but it is the result of sampling 92 individuals and classifying each according to two variables, so we speak of independence (or lack of independence) rather than a comparison of pet owners and nonpet owners (among heart patients) with respect to survival.

Using the marginal totals as in Section 9.6, we compute estimates of the mean cell frequencies:

$$\frac{78 \times 53}{92} = 45,$$

and so on,[†] and put them in a two-way table:

	Lived	Died	
Pet	45	8	53
No Pet	33	6	39
	78	14	92

(The numbers involved in the computation of the mean 45 are in italics to show the pattern.) In this table of mean frequencies, rows and columns are in proportion—as they would be in the population, under the null hypothesis of independence. The test statistic is

$$\chi^2 = \frac{(50-45)^2}{45} + \frac{(3-8)^2}{8} + \frac{(28-33)^2}{33} + \frac{(11-6)^2}{6}$$

$$= 8.60,$$

which defines a P-value of less than 1 percent, found in Table V of Appendix B, using $(2-1)(2-1) = 1$ degree of freedom. (Degrees of freedom are computed as in Section 9.6: number of rows minus 1 times number of columns minus 1.) The result is "significant." That is, the data seem to say that the sample is not from a population in which having a pet is independent of surviving a year after hospitalization for a heart ailment.

If these 92 patients can be considered a random sample from some larger population of interest, the conclusion is that the variables of interest are related in that population. Even so, it may be, as the

[†]These estimated mean frequencies have been rounded because their actual values are so close to integers, the 45, for instance, being 44.935 to three decimal places. This round-off will affect the computed value of the test statistic. If we had kept one decimal place (as we have done in examples in Chapter 9), the computed χ^2 would be 8.95 instead of the 8.60 that results from using frequencies rounded to the nearest integer.

researcher acknowledged, "that a specific kind of person who owns a pet is more likely to survive." We cannot conclude that having a pet is what *causes* survival.[†]

The sample, in this retrospective study, is a sample of convenience. Whether it can be thought of as a random sample of some conceptual population, including future heart patients whom we might want to treat by advising acquisition of a pet, is arguable. ▲

Chi-square test for independence, given the contingency table summary of n observations in a random sample, with r rows and c columns:

1. Calculate row and column totals.
2. Calculate estimates of cell means:

$$\frac{\text{row total} \times \text{column total}}{n}.$$

3. Calculate the test statistic:

$$\chi^2 = \sum_{\text{all cells}} \frac{(\text{cell frequency} - \text{mean cell frequency})^2}{\text{mean cell frequency}},$$

using the estimates from (2) as mean cell frequencies.
4. Compare χ^2 with percentiles of the chi-square distribution with $(r - 1)(c - 1)$ degrees of freedom.

In testing for homogeneity in Chapter 9, where subpopulations were defined by different kinds or levels of treatment, we assumed that sample sizes would be specified for the sample from each subpopulation. When neither variable is controlled, there are two ways of sampling. One is to take a random sample of given size from the overall population of interest and determine the value of each variable for each individual in the sample. When tabulated in a two-way array, the results define a contingency table that looks just like the contingency tables of Chapter 9. The summary in Example 10–4 is of this type because the basic sample was that of 92 heart patients, each classified according to the two variables of interest.

[†]That is, we cannot conclude that the relationship is a causal one, on the basis of the *statistical* analysis. Nevertheless, as one physician stated in a TV interview, older people might do well to have a pet—for the exercise in walking it, and to keep their minds occupied with something other than "bad thoughts." (At any rate, he recommends this to his patients in the belief that the relationship *is* causal.)

The other method is natural when the subpopulations defined by the categories of one variable are conveniently available. One can then take a sample of specified size from each subpopulation, classifying the individuals drawn according to the other variable. The resulting summary will again look like a two-way contingency table. But whichever sampling method is used, the test for independence is the same.

Example 10-5

Suppose that we wanted to investigate the possibility of a relationship between smoking habits and a student's college of enrollment. A contingency table summarizing data on 200 students might look like this:

	Arts	Technology	Agriculture	Education
Smoker	23	16	9	12
Nonsmoker	27	34	41	38

To obtain such data, we could have taken a sample of size 50 from each college and surveyed the individuals drawn as to smoking. Or, we could have taken a sample of 200 from the combined population of the four colleges, asking each individual about smoking and college of enrollment. The table would look the same, except that it would be unusual for the number in each college to have turned out to be 50, if the sample of 200 were drawn from the population as a whole. That is, in taking a sample of 50 from each college, we fix the lower marginal totals at 50; in taking a sample of 200 from the population as a whole, the lower marginal totals would be unpredictable.

One could also imagine drawing a sample of 60 from the population of smokers and one of 140 from the population of nonsmokers—if lists of these populations were available.

We can ask any of these three questions: Are the colleges of enrollment homogeneous with regard to smoking? Are the populations of smokers and of nonsmokers the same with regard to college? Are the variables of "college" and "smoking habits" independent in the population of students as a whole? But we have seen that an affirmative answer to any one implies an affirmative answer to the other. Moreover, the statistical method to be used will be the same no matter which question we have in mind and no matter which of the sampling methods described we use. ▲

10.3
TESTING FOR ZERO CORRELATION

The coefficient of correlation between too numerical variables associated with each member of a *population* is usually denoted by ρ
("rho," the Greek version of r). It describes the degree of *linear*
association between the variables in the population. A random sample from the population is a set of pairs $(X_1, Y_1), \ldots, (X_n, Y_n)$,
which we represent graphically in a scatter diagram. The *sample*
correlation coefficient r, calculated for these n pairs, is a natural
estimate of the population correlation coefficient ρ. As in every
estimation problem, there is here a question of reliability of the
estimate—the calculation of a standard error and the construction of
confidence intervals. We shall not take up these matters, examining
only the question of whether or not there is any correlation at all in
a given population.

A correlation or zero between two numerical variables means that
there is not even a hint of a linear relationship between them, although they may be related in other ways. In many applications,
however, the bivariate *normal* model mentioned in Section 5.5 is a
fairly good model, and in this kind of model a correlation of zero
actually implies complete independence of the two variables.

Even if the correlation in a bivariate normal population is zero, a
sample will generally exhibit a correlation that is *not* exactly zero,
owing to random sampling fluctuations—the luck of the draw. The
obvious question, then, in this: How far from 0 can a sample correlation be when the population correlation is 0? The answer, of course,
depends on the null distribution of r—on the distribution of r under
the null hypothesis of zero correlation in the population.

It has been found that a certain function of r turns out to have a
familiar distribution, when $\rho = 0$, namely, the variable

$$T = \sqrt{n - 2}\, \frac{r}{\sqrt{1 - r^2}}.$$

The statistic T has a t-distribution with $n - 2$ degrees of freedom, if
the data constitute a random sample from a bivariate normal popula-
tion with $\rho = 0$. A large value of $|T|$ is taken as evidence against the
hypothesis $\rho = 0$, and the P-value for a given value of T is found
using Table III of Appendix B.

Real populations are seldom (if ever) exactly bivariate normal, but
fortunately the t-distribution works rather well in using T as long as
the population is not too far from bivariate normal.

Table 10-1

Infant	Glucose (mg/100 ml)	Lactate (mM)
1	53	5.79
2	58	4.60
3	62	4.20
4	47	1.65
5	48	2.38
6	31	5.67
7	3	12.60
8	41	3.40
9	30	7.57
10	37	2.48
11	28	4.36

Example 10-6 *Glucose and Lactate*

In a study of infants' blood components, the glucose and lactate content 2 hours after birth were given,[†] for 11 infants who were small for their length of gestation. They are shown in Table 10-1.

The correlation coefficient in this sample is $r = -.70$, and the corresponding value of T is

$$T = \sqrt{11 - 2} \frac{-.70}{\sqrt{1 - .70^2}} = -2.94$$

This exceeds 2.82, the 99th percentile of t with 9 degrees of freedom—"highly significant." That is, if $\rho = 0$, such a large sample correlation is very hard to account for as random sampling variation in a bivariate normal population with $\rho = 0$.

However, a scatter diagram—and it never hurts to look at a graphical representation of data—shows clearly (Figure 10-1) that the high sample correlation is caused mainly by infant number 7. Without that one pair of readings, the correlation would be -.25, yielding a T of -.734. This is not at all significant ($P > .20$). Further investigation would seem to be in order—perhaps more data, and a check of the "outlier," number 7.

Whether the basic assumption in the t-test of normality is fulfilled or not would be hard to say on the basis of the available data. However, the test is known to work fairly well even when the condition is not met. Indeed, it may be easier to accept the assumption of nor-

[†]M. W. Haymond et al., "Increased Gluconeogenic Substrates in the Small-for-Gestational-Age-Infant," *New England J. Med.* **291**, 322–328 (1974).

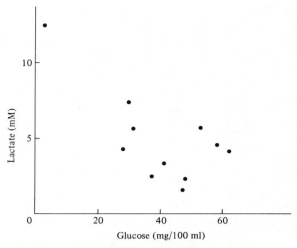

Figure 10.1.   Scatter diagram for Example 10-6.

mality than that of a random sample. The sample was probably a
"sample of convenience," and one should consider the possibility of
hidden bias. ▲

Example 10-7 *Early and Later Snowfall*

A TV weather announcer asserted in early December that because
the snowfall through November had been light, we could expect a lot
of snow during the rest of the season. The weather almanac for the
area gives the snowfall figures, year by year since 1920, and these are
given here in Table 10-2. (The scatter diagram for these data was
given in Figure 3-4.)

The correlation coefficient for these 54 pairs of snowfall amounts
is $r = -.119$, which in turn leads to

$$T = \sqrt{54 - 2}\, \frac{-.119}{\sqrt{1 - .119^2}} = -.864.$$

The number of degrees of freedom is 52, for whose percentiles we
refer to the standard normal table. The P-value is about 20 percent,
and one would not reject $\rho = 0$, or zero correlation in the population,
at the usual levels.

These last conclusions are based on the assumption of a bivariate
normal distribution, whereas the cloud of data points in Figure 3-4
bears little resemblance to those in Figures 3-13 to 3-17, which are

Table 10-2 Snowfall Data

Oct.–Nov.	Dec.–May	Oct.–Nov.	Dec.–May	Oct.–Nov.	Dec.–May
1.1	19.5	2.4	39.2	6.8	32.3
16.6	26.7	.6	44.5	10.3	10.9
1.2	41.0	26.3	26.2	5.7	13.4
1.4	31.0	3.8	20.1	10.6	21.2
3.2	20.0	2.9	31.5	2.4	37.8
6.9	30.1	10.3	16.6	2.5	78.8
5.0	25.1	2.5	31.4	5.6	28.9
5.3	51.8	4.5	35.4	0	28.9
3.6	43.8	4.1	20.9	4.3	69.4
4.6	22.9	21.8	27.3	1.6	34.5
5.2	9.0	2.5	35.8	3.6	73.8
0.7	45.9	2.7	48.9	1.1	16.4
2.4	32.3	5.6	83.3	4.9	63.2
6.0	19.6	11.6	67.4	6.2	57.2
8.5	43.1	10.1	32.8	6.3	48.4
2.6	56.2	1.9	23.8	13.2	51.0
1.8	42.8	6.8	27.1	1.1	40.6
1.2	28.3	8.5	36.7	18.0	33.2

typical of data from a bivariate normal distribution. The distributions of X by itself and Y by itself, which should be normal if (X, Y) is bivariate normal, are actually quite skewed:

X (Oct.–Nov.)		Y (Dec.–May)	
Depth	Frequency	Depth	Frequency
0–2.9	21	0–10	1
3–5.9	15	10–20	5
6–8.9	8	20–30	15
9–11.9	5	30–40	14
12–14.9	1	40–50	9
15–17.9	1	50–60	4
18–20.9	1	60–70	3
21–23.9	1	70–80	2
24–26.9	1	80–90	1

Nevertheless, because the t-test used is fairly robust with regard to the nature of the population distribution (as in Section 8.3), the use of the normal table has some justification—but caution is advisable. There remains the question of randomness—that is, are the snowfall figures like independent observations from a population, or are there trends and/or cycles present? We leave this open. ▲

10.4
THE CORNER TEST FOR ASSOCIATION

A simple test for association of two variables, based on counting, has been devised by Olmstead and Tukey, who call it the *corner test*. It does not assume any more about the population of pairs than that the variables are of the continuous type.

This freedom from assumptions about the population is an advantage, but the test is sometimes called "quick and dirty." This means that it is easy to use, but, by virtue of its crudeness, apt not to be as sensitive as a test that takes more effort to apply. The test is perhaps easiest to explain by example.

Example 10-8   *Diastolic Versus Systolic Pressure*

The first 14 subjects in Data Set A (Appendix C) were recorded as having the following blood pressures:

122/78	110/68
130/78	104/76
132/74	108/70
102/68	110/70
108/74	100/60
130/76	90/60
100/76	102/76

These pairs are plotted in Figure 10–2 as points in the plane. The median diastolic pressure is 108, and the median systolic pressure, 74. The first step is to draw a set of axes centered at the median point (108, 74), and this has been done in the figure. Quadrants defined by these axes are numbered (as before) I, II, III, and IV. Quadrants I and III are thought of as positive quadrants, II and IV as negative.

Starting at the right of the data, a vertical line is moved to the left over to the point where further movement would have passed by points in *both* quadrants I and IV; the points passed by (these are circled in the figure) are counted, with + assigned to the count if

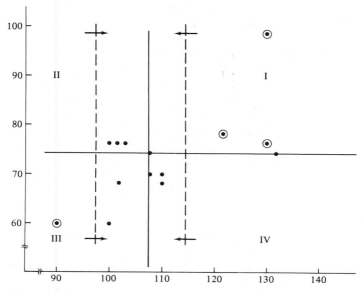

Figure 10-2

they are in quadrant I, and − if they are in quadrant IV. Thus, the three circled points contribute +3.

Next, starting to the left of the data, a vertical line is moved to the right over to the point where further movement would pass by points in both quadrants II and III. One point in quadrant III, circled in the figure, would have been passed by, and the count is +1. This same procedure is then followed for a line moving down from the top, and a line moving up from the bottom. The counts, evident in the figure but not marked there, are +2 for the line moving down (for two points in quadrant I), and +2 for the line moving up (for two points in quadrant III). The total count is

$$C = 3 + 1 + 2 + 2 = 8.$$

The test statistic is the absolute value of this total count. (In some problems the count will be a negative number.)

The idea of the test is that if there is little association, the moving lines will not encounter many points exclusively in one quadrant. But if there is a high degree of association, the points passed by first will lie in diagonally opposite quadrants, and C will be large in magnitude. The distribution of C under the null hypothesis of no association is needed to tell whether an observed value is small enough to

Table 10-3   Tail Probabilities for the Olmstead-Tukey Statistic $P(C \geqslant k)$

k	\multicolumn{5}{c}{Sample Size}
	6	8	10	14	∞
0	1.000	1.000	1.000	1.000	1.000
1	.933	.904	.911	.911	.912
2	.756	.754	.757	.758	.755
3	.600	.600	.601	.604	.600
4	.467	.462	.466	. .469	.462
5	.311	.351	.352	.355	.347
6	.222	.262	.259	.261	.252
7	.156	.182	.187	.188	.178
8	.111	.126	.133	.132	.122
9	.100	.084	.093	.092	.081
10	.100	.054	.064	.063	.053
11	. .100	.037	.044	.043	.034
12	.100	.030	.029	.030	.022
13	.00	.029	.019	.020	.013
14		.029	.013	.014	.008
15		.029	.010	.010	.005
16		.029	.008	.007	.003
17		.000	.008	.005	.002
18			.008	.003	.001
19			.008	.002	.001

accept that null hypothesis, and tail probabilities are given in Table 10-3 for various small, even sample sizes. Referring to this table shows that the *P*-value corresponding to $C = 8$, when $n = 14$, is about 13 percent:

$$P(|C| \geqslant 8) = .132.$$

This is not "significant." There is some indication of association in the data, but not enough with this crude test (if they constitute a random sample) to rule out the hypothesis of no population association between the variables.

(Of course, in a serious study of this question one could easily get enough data to resolve it beyond reasonable doubt.)

The data can, of course, be analyzed using the correlation coefficient. One finds the sample correlation to be $r = .606$, with a corresponding value of T:

$$T = \sqrt{14 - 2} \; \frac{.606}{\sqrt{1 - (.606)^2}} = 2.64.$$

This *is* significant, and the T-test has detected an association where the corner test did not. ▲

This instance of a "quick and dirty" test has been included to show the kind of preliminary analysis that is often convenient and useful at an early stage of an investigation.

KEY VOCABULARY

Causal relationship
Association
Independent variables
Population correlation coefficient
Bivariate normal population
Linear relation

QUESTIONS

1. What is meant by saying that two variables are "related?"
2. When two variables are related, does it necessarily follow that changes in one *causes* changes in the other? Explain.
3. Under what regimen of data collection can an observed association of two variables imply a cause/effect type of relation?
4. What kind of association is measured by the correlation coefficient?
5. Suppose that a population is cross-classified on two categorical variables A and B, and that A and B are independent. What value of χ^2 would we get if we calculated it for this cross-classification?
6. Suppose that we look at a random *sample* from a population classified on independent variables A and B. Will the chi-square statistic calculated from the sample contingency table have the value 0?
7. Is the correlation coefficient of a random sample from a bivariate population equal to the population correlation coefficient?
8. If, as is the case, the "corner test" is not very sensitive in detecting an association between numerical variables, what is the point of it?

9. The chi-square distribution is not an exact sampling distribution for Pearson's chi-square statistic. In what circumstances is it a good approximation?

10. Suppose that a contingency table is constructed from a sample of given size, and it turns out that there are several cells with expected frequencies (as estimated from the sample) less than 1. What can be done so that we can still use the chi-square table for this sample?

PROBLEMS

Sections 10.1 and 10.2

*1. An article in *Psychology Today* (January 1974) reported on a comparison of 57 students classed as "colaholics" (40 to 111 ounces of cola per day) and 112 students, the "control group," who averaged only 8 ounces per day. Twice as many of the colaholics reported sleeping difficulties as in the control group, and 58 percent of the colaholics as compared with 41 percent of the control group said they sometimes felt jittery. Does drinking cola cause sleeping difficulties and jitters?

2. Census data show an appreciable correlation between years of education and income—say, for men in a given age bracket and in a certain class of occupation. Does this tell you that (for such men) the higher incomes are the result of more education?

*3. In a certain population of 120 males, 45 wear glasses. In a population of 80 females, 30 wear glasses. In the combined population of these 200 individuals,
   (a) What proportion wear glasses?
   (b) Are wearing glasses and sex-independent variables?
   (c) What proportion of all individuals are male?
   (d) What proportion of glasses wearers are male?
   (e) What proportion of all individuals are female *and* wear glasses?
   (f) If one took a random sample of 25 individuals from the population, would males and females be present in the sample in the same proportion as in the population?
   (g) If the random sample of 25 [in part (f)] were cross-classified on sex and wearing of glasses, would rows be proportional (as they are in the population)?

4. A (hypothetical) population of students on an "Ag" campus is cross-classified on the college of enrollment and according to sex. The proportion of the population in each combination of categories is as follows:

	College		
	Bio	HomeEc	Ag
Female	1/8	1/4	1/8
Male	1/8	0	3/8

(Thus, e.g., 3/8 of the students are males in the College of Agriculture. So the *probability* is 3/8 that a student selected at random is a male from that college.)

(a) Given that the population consists of 400 students in all, make a table showing the *frequencies* in each cell (i.e., for each combination of college and sex).

(b) Are sex and college of enrollment independent in *this* population?

(c) What proportion of the subpopulation of males is in the College of Agriculture?

(d) What proportion of the biological science students are male?

(e) What proportion of the students not in Home Economics is male?

(f) Construct a cross-classification of 400 individuals in which the row and column totals are the same as in (a), but in which the columns are proportional.

*5. In a certain city with a 10 percent black population, the unemployment rate for blacks is 15 percent. For whites it is 5 percent.

(a) What percent of the total population is unemployed?

(b) What fraction of the unemployed is black?

(c) Make a probability table like that in Problem 4, for the two variables race and employment status, based on the given information. Identify the fractions in (a) and (b) in this table.

(d) Are race and employment status independent in this population?

6. Suppose that two-thirds of the drivers encountering a certain stop sign are males. If three-tenths of the males and three-fifths of the females come to a full stop:

(a) What proportion of all drivers come to a full stop?

(b) What proportion of those who stop are females?

(c) Is coming to a full stop independent of sex?

*7. Suppose that you observe 100 drivers at the stop sign in Problem 6, and note whether or not they come to a full stop, and whether the driver is male or female—with these results:

	Full Stop	Not Full Stop
Male	18	42
Female	22	18

On the basis of these data, would you say that coming to a full stop is independent of sex?

8. Students in a statistics class were surveyed as to smoking and use of corrective lenses, with results as follows:

	Never Smoked	Smoke Now	Did but Quit
Use glasses	15	5	6
No glasses	39	5	7

(a) Would you expect to find a relationship?

(b) Carry out a chi-square test of independence.

(c) What do you think of this as a sample, and of the conclusion in part (b)?

*9. A 1978 newspaper poll (The *Minneapolis Tribune's* "Minnesota Poll") asked 611 people this question: "If you were an employee, who do you think would generally make a better boss for you personally—a man or a woman?" Results were as follows:

	Men	Women
Prefer men	188	160
Prefer woman	19	39
No difference or not sure	107	98

Is the preference for a man or woman as boss independent of the sex of the responder (in Minnesota)?

10. Cross-classify the 200 students in Data Set B (Appendix C)

according to sex and the number of math courses taken. Are these independent variables in the population?

*11. Cross-classify the 200 students in Data Set B (Appendix C) according to class and to number of math courses taken. Would you expect these variables to be related? Calculate the chi-square statistic for testing independence.

Sections 10.3 and 10.4

*12. A random sample of 27 male students showed a correlation of $r = -.0366$ between height and pulse rate. Assuming a bivariate normal distribution, test the hypothesis of zero correlation.

13. A sample of 41 female students showed a correlation of $r = .335$ between height and weight.
    (a) Assuming a bivariate normal distribution, what conclusion would you draw about the hypothesis of zero correlation?
    (b) Why is the correlation coefficient apt to be an inappropriate measure of a relationship between height and weight?

*14. The correlation coefficient between weight and life satisfaction index among the 75 female students listed in Data Set B (Appendix C) is $r = -.139$. Calculate the t-statistic for testing the hypothesis that the correlation is 0 in the population from which this sample is drawn. Assuming random sampling, determine the corresponding P-value. (Some things to think about: *Is* this a random sample from some population? Would you expect a correlation? If so, would you expect it to be positive or negative? Is the observed correlation "significant"? Does either variable "cause" the other?)

15. Calculate the correlation coefficient of the radial pulse and respiration of the first 15 individuals in Data set A (Appendix C). Is this significantly different from 0?

*16. Plot the data points (X, Y), where X = height and Y = satisfaction with life, for the first 20 individuals on the list of "main campus males" in Data Set B (Appendix C). Apply the corner test for association. What does the result tell you?

17. Plot the data points (X, Y), where X = GPA and Y = satisfaction with life, for the first 20 individuals on the list of "main campus males" in Data Set B, Appendix. Apply the corner test for association. What does the result tell you?

18. Plot the data points (X, Y), where X = weight and Y = satisfaction with life, for the first 14 individuals on the list of "main

campus females" in Data Set B, Appendix C. Apply the corner test for association. What does the result tell you?

Review

19. Fifty patients in an Air Force hospital[†], with active duodenal ulcerations, were assigned at random to either a treatment regimen involving an antacid or to one involving a placebo (made of starch, lactose, sodium, saccharine, and flavoring). After 30 days of treatment, "healing" was recorded if the ulcer was reduced in size by two-thirds. Results were as follows:

	Healing	No Healing
Antacid	24	3
Placebo	17	6

What do you conclude from these data?

20. The 50 states of the United States are cross-classified in the following table according to the party of the governor in 1975 and according to the leading party in the 1976 presidential election:

		Governor (1975)	
		D/I	R
Presidential	D	18	5
preference			
(1976)	R	20	7

(a) What proportion of states with Republican governors in 1975 went Republican for President in 1976?

(b) What proportion of all states went Republican for President in 1976?

(c) What proportion of states that went Democratic in the 1976 Presidential election had Democratic governors in 1975?

(d) What proportion of all states had Democratic governors in 1975?

[†]Reported in an article by D. Hollander and J. Harlan in *J. Am. Med. Assn.*, **226**, 1181–1185 (1973).

(e) Is Presidential preference in 1976 independent of the party of the governor in 1975?

21. In connection with a move to change the type of school calendar being followed at a certain large university, a sample of 346 faculty members (out of several thousand) was surveyed and cross-classified according to the type of appointment and preference, as follows:

| | Type of Appointment | | |
	12-Month	9-Month	Other
Traditional quarter	85	54	1
Early start-finish quarter	63	30	1
Traditional semester	23	35	1
Early semester	15	24	2
4-1-4 semester	4	8	0

The committee report to the Faculty Assembly included this footnote: "Chi-square test significant at $\leq$ .01 level." Check this statement. (NOTE: To avoid too many cells with small expected frequencies, you should group "other" together with "9-month.")

22. A climatologist, referring to the prospects for an approaching winter in the Twin Cities, was quoted in the newspaper as saying, "Warm Julys have been statistically linked to cold winters." Table 10-4 gives average July temperatures and average temperatures in the following winter months of December, January, and February over the period 1964 to 1974.

Table 10-4

Year	Average, July	Average, Dec.–Jan.–Feb.
1964–1965	75.0	11.6
1965–1966	69.2	14.7
1966–1967	74.0	14.1
1967–1968	68.7	16.9
1968–1969	69.3	14.4
1969–1970	71.2	12.6
1970–1971	72.7	12.9
1971–1972	66.0	10.5
1972–1973	67.5	14.8
1973–1974	70.2	11.1

(a) Determine the correlation coefficient. Does its algebraic sign agree with the climatologists claim?

(b) Assuming the 10 pairs to have come from a "population" of such pairs as a random sample, what conclusion could you draw about a correlation in the population?

23. The data represented in Figure 3–4 (page 130) are shown here in Table 10–2.

(a) Make a stem-leaf diagram for X and one for Y, and find the median values. (For Y, round-off snowfall amounts to the nearest inch before making the stem-leaf diagram.)

(b) Draw a vertical line through the median X-value and a horizontal through the median Y-value, and determine the value of the statistic used in the corner test. Conclusion?

24. There are 53 points in the data of Figure 3–5 (page 130). Find the median X and the median Y, and then apply the corner test for association. (Is the result surprising, in view of the figure?)

25. Suppose that eight data points happen to fall on a straight line.

(a) Find the value of r, the sample correlation coefficient, and the corresponding P-value for testing $\rho = 0$.

(b) Find the value of C, the corner-test statistic, and the corresponding P-value for testing the null hypothesis of no association.

11
Linear Regression

LIKE PAIRS OF CATEGORICAL OBSERVATIONS, pairs of continuous observations arise in two types of sampling situations. In one, two characteristics of each individual in a sample are measured or observed, and the result is a sample of bivariate data of the type studied in Chapter 10 with regard to the existence and extent of an association between the two variables. In the other type, a variable x is set or controlled at various values—perhaps by the setting of a dial or the application of known dosages, and a *response* variable y is measured for each setting (i.e., at each value of x). With the controlled variable x set at a particular value, the measurement of y typically involves random errors, so the measured value Y is one of a "population" of values determined or indexed by x.

When, for given x, the error component of the measured Y has mean 0, the "true" response y is then the mean value of Y. In a regression analysis we are concerned with the relationship between this mean value of Y and the controlled variable x. The relationship is linear in the simplest cases of interest, and it is this type of relationship that we deal with here. We digress slightly in Section 11.1 to review the algebraic notion of the "equation" of a line.

11.1
THE EQUATION OF A LINE

When a straight line (usually referred to simply as a *line*) is drawn on a set of coordinate axes, the x- and y-coordinates of each of its points satisfy a simple type of algebraic relationship, called *linear*

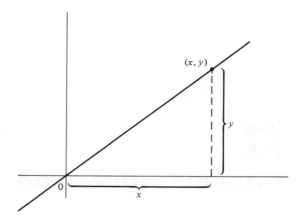

Figure 11-1

for the obvious reason. The purpose of this section is to review some facts about linear relationships and their graphs.

The coordinates of a point on a line through the origin, as in Figure 11-1, are always in a fixed ratio—according to the facts about similar triangles. Thus, the ratio y/x in Figure 11-1 is always the same (unless both x and y are 0), for the given line. This ratio—call its value b—is the same for every point on the line and is termed the *slope* of the line. It represents the amount of vertical rise per unit of horizontal travel in moving on the line from (0, 0) to any point (x, y). In algebraic form, the statement that the ratio is constant is the *equation* of the line:

$$\frac{y}{x} = b \quad \text{or} \quad y = bx.$$

Thus, a line with slope 2 is represented by the equation $y = 2x$ and rises 2 vertical units for each unit of horizontal displacement from left to right. A line with slope 1/3, represented by $y = x/3$, rises 1 unit for each 3 units of horizontal displacement from left to right. A line with slope 1 increases 1 unit in the y-direction for each unit of increase in the x-direction; and if the "unit" on each axis is represented by the same length, the line makes an angle of 45° with the horizontal.

A line parallel to $y = bx$ but passing through the point (0, a), as in Figure 11-2, is characterized by the fact that the y-coordinate of

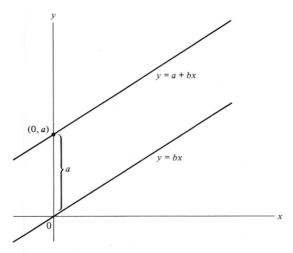

Figure 11-2

each of its points is greater by the amount a than the y-coordinate on $y = bx$ at the same x-value:

$$y = a + bx.$$

The ordinate a of the point where the line intersects the y-axis is called the *y-intercept* of the line.

We see then that a line with slope b and y-intercept a has the equation $y = a + bx$. Conversely, if an equation of this type is given, say $y = -2 + 3x$, its graph is a line, drawn by starting at the y-intercept $(0, -2)$ and drawing a line through it with slope $b = 3$ (going up 3 units for each horizontal unit to the right), as in Figure 11-3.

The equation of a line through a given point (x_0, y_0) with slope b is easily found by first writing its equation in coordinates (x', y') measured from (x_0, y_0):

$$x' = x - x_0$$
$$y' = y - y_0.$$

The equation satisfied by these new coordinates is $y' = bx'$, so the equation satisfied by the xy-coordinates is

$$y - y_0 = b(x - x_0).$$

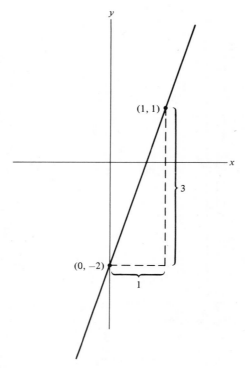

Figure 11-3

Reasoning in reverse, one finds that the equation

$$y - 4 = 2(x + 1),$$

for example, represents or characterizes a line through the point $(-1, 4)$ with slope 2.

Equation of the line with slope b and y-intercept a:

$$y = a + bx.$$

Equation of the line through (x_0, y_0) with slope b:

$$y = y_0 + b(x - x_0).$$

11.2
A CONTROLLED VARIABLE

Some examples of a response y to a controlled variable x are these:

x	y
Dose of a drug	Change in blood pressure
Temperature	Yield of a reactor
Amount of advertising	Amount of sales
Amount of carbon	Hardness of an alloy

If there is a relationship in such a case, it is clearly a causal relationship; we say that y is a *function* of x. It is this function, called the *regression function* (for historical reasons to be explained later), that is the target of our investigation.

Measuring y for various selected values of x will often show that a measurement has a random component—a random "error," referred to as such whether it be measurement error, or the result of unaccountable factors corrupting the otherwise simple dependence of y on x. We use the capital letter Y to denote the *measured* response, which includes these errors.

Example 11-1 *Iron Content and Corrosion*

Thirteen specimens of Cu–Ni alloys, each with a given content of iron (x), were tested for corrosion (in salt water for 60 days). Corrosion was measured by weight loss (Y) in mg/decimeter^2/day, as follows:[†]

Iron, x	Loss, Y
.01	127.6, 130.1, 128.0
.48	124.0, 122.0
.71	113.1, 110.8
.95	103.9
1.19	101.5
1.44	91.4, 92.3
1.96	83.7, 86.2

[†]Given in N. R. Draper and W. L. Smith, *Applied Regression Analysis* (New York: John Wiley & Sons, Inc., 1966), p. 37.

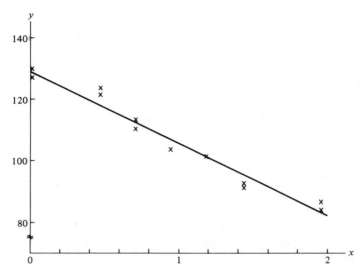

Figure 11-4. Data plot for Example 11-1.

Examination of the plot in Figure 11-4 shows that the data are not far from falling on a straight line. It may be that the true relationship is linear, but that the data points are not exactly collinear because of randomness (measurement errors and variations in experimental conditions and experimental material). The object of a statistical analysis of the data would be to extract this linear relationship as accurately as possible. ▲

Using the simplest form of functional relationship, a linear one, we conceive of measurements Y as being constructed like this:

$$Y = \text{regression function} + \text{error} = (\alpha + \beta x) + \epsilon,$$

where the slope β and intercept α of the linear regression function $\alpha + \beta x$ are population parameters, usually unknown, and ϵ is an error component. The errors are regarded as observations on a random variable with mean 0 and variance σ^2, so the error distribution is the same for all x. Because ϵ is assumed to have mean 0, the mean of Y is the linear function $\alpha + \beta x$. This is the regression function.

When a number of observations on Y are made for various values of x, it is assumed (in the simplest model) that the error components are independent random quantities, each distributed *normally* about the regression function:

Linear regression model:

$$Y = \alpha + \beta x + \epsilon,$$

where ϵ is normal with mean 0 and s.d. σ, and repeated measurements Y are independent.

[The assumption that the relationship between m.v.(Y) and the controlled variable x is a *linear* one is restrictive, although not quite as restrictive as it may seem at first. For, it is often possible to transform one or both of the variables x and Y—for example, by taking logarithms—so that what was not linear becomes linear. On the other hand, this may require assuming that the normally distributed error adds on to the transformed variable. But it turns out that the methods and theory that apply when the regression function is linear are in fact adaptable to more general types of relationship.]

As in Figure 11-4, the random-error component will usually scatter data points so that they do not fall on any line, let alone the regression line. Just as we have tried in previous chapters to estimate unknown quantities from data, we shall try to find a line that in some sense "fits" the data points and is a good approximation to the regression line. Of course, it is usually not hard to draw a line on a scatter diagram that seems about right, fitting it by eye. However, it is more desirable to have a less subjective means for finding a line that fits, and the method of the next section provides such a means.

11.3
LEAST SQUARES

Given a set of data points (x, y), the question is how to fit a straight line as an estimate of the regression line. The *method of least squares* provides a means for doing this, one that goes back to Karl Friedrich Gauss (1809). The method actually can be used for a general problem of "curve fitting," but it turns out to be particularly simple when the "curve" is a straight line.

Given any trial line, $y = a + bx$, its degree of fit to the given data points can be measured by the r.m.s. average deviation of those points from the line, where the deviations are measured *vertically*. Using vertical deviations seems natural in view of the assumptions that the data points deviate in the vertical direction from the regression line—by the amount of error component in the y-measurement. The vertical distance of (x, y) from the line $y = a + bx$ is

$$y - (a + bx),$$

where y is the distance up to (x, y) from the x-axis, and $a + bx$ the distance up to the line (see Figure 11-5). This vertical distance, a deviation of the point (x, y) from the line, is termed a *residual*.

"Fitting a line" is equivalent to choosing two numbers, a and b, the intercept and slope of the line. The method of least squares is to choose a and b so that the r.m.s. residual is as small as possible. Minimizing this quantity, or equivalently, its square:

$$\frac{1}{n} \sum (y - a - bx)^2,$$

is accomplished (as can be shown algebraically) by choosing slope and intercept as follows.

The line that minimizes the r.m.s. residual has slope

$$b = \frac{S_{xy}}{S_x^2} = \frac{\sum xy - n\bar{x}\bar{y}}{\sum x^2 - n(\bar{x})^2},$$

and passes through the point $(\bar{x}, \bar{y})$:

$$a = \bar{y} - b\bar{x}.$$

(As in Section 3.8, S_{xy} is the sample covariance and S_x^2 the variance of the x's.) Since the least square line is determined by the data as an estimate of the regression line, it is also called the *empirical regression line*.

The minimum r.m.s. residual, achieved for the line whose slope and intercept are computed in this way, will be found useful. It is given by

$$\text{minimum r.m.s. residual} = S_y \sqrt{1 - r^2},$$

where r is the correlation coefficient of the set of data points.

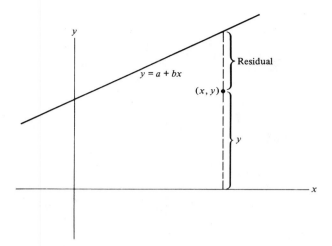

Figure 11-5

Example 11-2

To illustrate the computation of a least-squares line without excessive numerical complication, we consider these four data points: $(0, 2)$, $(1, 1)$, $(2, 4)$, $(5, 5)$. A tabular array with columns for x^2, y^2, and xy facilitates the computation:

	x	y	x^2	xy	y^2
	0	2	0	0	4
	1	1	1	1	1
	2	4	4	8	16
	5	5	25	25	25
Sums:	8	12	30	34	46
Means:	2	3	30/4	34/4	46/4

From these means we find

$$S_x^2 = \frac{30}{4} - 2^2 = \frac{14}{4}$$

$$S_{xy} = \frac{34}{4} - 2 \times 3 = \frac{10}{4},$$

whence

$$b = \frac{S_{xy}}{S_x^2} = \frac{10}{14} = .714$$

and

$$a = \bar{y} - b\bar{x} = 3 - \frac{10}{14} \cdot 2 = \frac{11}{7} = 1.57.$$

The least-squares line is thus

$$y = 1.57 + .714x.$$

The correlation coefficient, which is a measure of closeness to linearity, can be calculated using $S_y^2 = 46/4 - 3^2 = 2.5$:

$$r = \frac{S_{xy}}{S_x S_y} = \frac{2.5}{\sqrt{3.5 \times 2.5}} = .845.$$

The r.m.s. average residual of the data about the least-squares or best-fitting line is then

$$\sqrt{\frac{1}{n} \sum (y - a - bx)^2} = S_y \sqrt{1 - r^2} = .845.^\dagger \quad \blacktriangle$$

Example 11–3

Example 2–29 gave data from an experiment in ophthalmology, data whose scatter diagram, shown in Figure 2–16, suggested a linear relationship between the reciprocals of substrate concentration and reaction rate. The relevant sample statistics are found to be

$$\bar{x} = 12.52, \qquad S_x^2 = 53.39,$$
$$\bar{Y} = .254, \qquad S_y^2 = .00720,$$
$$S_{xy} = .562.$$

The slope and intercept of the least-squares line are computed from these, as follows:

$$b = \frac{S_{xy}}{S_x^2} = .0105$$

---

†Although the r.m.s. residual here happens to be equal to the coefficient of correlation, they would usually differ.

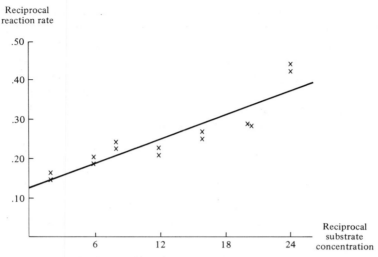

Figure 11-6. Data for Examples 2-29 and 11-3, showing the least-squares line.

$$a = .254 - .0105 \times 12.52 = .1225.$$

The equation of the line is thus

$$y = .1225 + .0105x.$$

This line is drawn on the scatter diagram in Figure 11-6. ▲

11.4
INFERENCE ABOUT THE SLOPE

When a regression function is linear, of the form $\alpha + \beta x$, the intercept α and the slope β are *population parameters* (as is also the s.d. σ of the deviations about the regression line); so in place of a we write $\hat{\alpha}$ and in place of b we write $\hat{\beta}$. This circumflex notation emphasizes our use of these quantities as estimators of the population intercept and slope, respectively. In this notation, the least-squares line is

$$y = \hat{\alpha} + \hat{\beta}x,$$

where (as in the preceding section)

$$\hat{\beta} = \frac{S_{xy}}{S_x^2}, \quad \hat{\alpha} = \bar{Y} - \hat{\beta}\bar{x}.$$

The slope $\hat{\beta}$ of the least-squares line is an estimator of the true slope β. Like all estimators, it has a sampling distribution. In a *normal* regression model defined in Section 11.2, the distribution of $\hat{\beta}$ is normal, with mean β and s.d. $\sigma/\sqrt{nS_x^2}$. The parameter σ, the error s.d., is usually not known, but it can be estimated from the sample to obtain a standard error for $\hat{\beta}$. Since the residuals $Y - \hat{\alpha} - \hat{\beta}x$ can be thought of as constituting a sample of errors, their r.m.s. average provides an estimate of σ, the error s.d.:

$$\hat{\sigma} = \sqrt{\frac{1}{n} \sum (Y - \hat{\alpha} - \hat{\beta}x)^2} = S_y \sqrt{1 - r^2}.$$

Estimating the slope β by the slope $\hat{\beta}$ of least-squares line:

$$\text{m.v.}(\hat{\beta}) = \beta$$

$$\text{s.d.}(\hat{\beta}) = \frac{\sigma}{\sqrt{nS_x^2}}$$

$$\text{s.e.}(\hat{\beta}) = \frac{\hat{\sigma}}{\sqrt{nS_x^2}},$$

where

$$\hat{\sigma} = S_y \sqrt{1 - r^2}.$$

An interval estimate for β is given as usual by limits which are a multiple of the s.e. on either side of the point estimate $\hat{\beta}$:

$$\hat{\beta} \pm k(\text{s.e.}).$$

The multiplier k can be found in Table IIc of Appendix B, in the case of large samples, since then $[\hat{\beta} - \text{m.v.}(\hat{\beta})]/\text{s.e.}(\hat{\beta})$ has a nearly normal distribution.

For small samples, the quantity

$$T = \sqrt{n - 2} \; \frac{\hat{\beta} - \hat{\beta}}{\sigma/S_x},$$

has a *t*-distribution with $n - 2$ degrees of freedom. The 95 percent confidence limits in this small-sample case are

$$\hat{\beta} \pm k \frac{\hat{\sigma}}{\sqrt{n - 2} \, S_x},$$

where $k = t_{.975}(n - 2)$ is the 97.5 percentile of t with $n - 2$ degrees of freedom.

To test the hypothesis $\beta = \beta_0$, we use the t-statistic

$$T = \sqrt{n - 2} \; \frac{\hat{\beta} - \beta_0}{\hat{\sigma}/S_x},$$

which has the $t(n - 2)$-distribution if $\beta = \beta_0$. In particular, to test the hypothesis $\beta = 0$ (corresponding to no change in response when the controlled variable is varied), we use

$$T = \sqrt{n - 2} \; \frac{\hat{\beta}}{\hat{\sigma}/S_x} = \sqrt{n - 2} \; \frac{r}{\sqrt{1 - r^2}} \; ,$$

the same statistic as proposed in Chapter 10 for testing $\rho = 0$ in a bivariate normal model.

Example 11-4

Consider again the ophthalmology experiment of Example 11-3 (and the earlier Example 2-29). There were 14 data points, with

$$\bar{x} = 12.52, \qquad S_x^2 = 53.39,$$

$$\bar{y} = .254, \qquad S_y^2 = .00720,$$

$$S_{xy} = .562,$$

where x is the reciprocal substrate concentration and y the reciprocal reaction rate. The estimate of slope was calculated in Example 11-3:

$$\hat{\beta} = \frac{S_{xy}}{S_x^2} = .0105,$$

and we now compute its standard error. The estimate of error variance is

$$\hat{\sigma}^2 = S_y^2(1 - r^2) = S_y^2 - \frac{S_{xy}^2}{S_x^2} = .00128,$$

so we find that

$$\text{s.e.}(\hat{\beta}) = \frac{\hat{\sigma}}{\sqrt{nS_x^2}} = \sqrt{\frac{.00128}{14 \times 53.39}} = .00131.$$

And 95 percent confidence limits for β are given by

$$\hat{\beta} \pm 2.18 \; \text{s.e.}(\hat{\beta}) = .0105 \pm .00286,$$

where 2.18 is the 97.5 percentile of t with 12 (or $n - 2$) degrees of freedom. ▲

11.5
PREDICTION

When neither X nor Y is a controlled variable, but one variable, say X, is observed before or in place of the other, a common problem is that of predicting the variable Y *given* the value of X. The variable X is then called the *predictor variable*. Some typical pairs of variables of interest are these:

Predictor, X	To Be Predicted, Y
ACT score	College GPA
Height of father	Height of son
Snowfall in Oct.-Nov.	Snowfall in Dec.-May
Frequency of a cricket's chirps	Temperature

In these examples there is not much question that there *is* a relationship between X and Y. The problem is to exploit the relationship to obtain a better prediction of Y using the value of X than could be made without the knowledge of X.

We consider first the simpler problem of predicting a random variable Y *without* the help of some related variable. Such a prediction would be a sheer guess without at least some rudimentary knowledge of the distribution of values of Y. We suppose that the mean value of Y is known. This mean value is in the midst of the possible values of Y, and we have thought of it as "typical." It is not unreasonable, then, to predict that the value of Y will turn out to be its mean value, μ_Y. This predicted value will usually be in error, and the amount of error is the deviation of Y from the predicted value, $Y - \mu_Y$.

Prediction of a random variable Y with mean value μ_Y:

Predict that Y will be μ_Y.

The r.m.s. prediction error is σ_Y.

Note that the r.m.s. prediction error given here is the s.d. of the error—which is equal to the s.d. of Y because the error differs from Y by a constant, μ_Y.

Example 11-5

Suppose that X and Y are scores on two standard exams. These are given to thousands of individuals, so that we have come to know (just about, at any rate) the population. Suppose that each score has mean 70 and s.d. 10, and that the coefficient of correlation between them is $\rho = .90$. To predict the Y-score for a student without knowing that student's X-score, one would predict Y to be 70, the mean of the Y-scores. The r.m.s. error of this prediction would be 10, the s.d. of the Y-scores.

On the other hand, if we know the student's X-score, we ought to be able to make a more accurate prediction—one with a smaller r.m.s. prediction error, inasmuch as there is a high degree of association between X and Y. (To be continued.)   ▲

Although we have been describing prediction as though the population characteristics were known, they are seldom known in practice. When the population of all Y-values is not known, but we have a *sample* of values $Y_1, \ldots, Y_n$, the obvious substitute for the prediction μ_Y is the sample mean $\bar{Y}$. With this prediction we have not only the error inherent in the variability of Y, but also the error in the estimation of μ_Y by $\bar{Y}$. To avoid this complication, for the present, we look at the problem of predicting Y given the value of X in the case where the relevant characteristics of the whole population of interest are known.

As a reminder, we point out that the relevant population characteristics—mean, s.d., correlation—are not conceptually different for a population of values than for a sample of values. It is just that in the case of a population, there is no uncertainty. We use different notations for sample and for population:

Quantity	Sample	Population
Means	$\bar{X}, \bar{Y}$	μ_X, μ_Y
s.d.'s	S_X, S_Y	σ_X, σ_Y
Correlation	r	ρ

We have in mind, then, a population of pairs (X, Y)—that is, a population of individuals, each with two measurements, X and Y. And we assume that we know the population means, s.d.'s, and correlation.

To see how knowledge of an X-value might help in predicting the corresponding Y-value, we consider the population of possible pairs, represented approximately by a "cloud" of data points, such as that shown in Figure 11-7. Although this cloud is not an exact model for the population, it serves to suggest the model. Suppose, then, that a prediction of Y is needed for an individual whose X-value is the one marked on the x-axis in Figure 11-7. Because the picture is incomplete (the population has more points than are shown in the cloud), there may be no points with precisely the marked X-value; but in a thin strip with X-value close to the one marked there are many points. The Y-values of these points constitute a population whose *mean* can be used as the predicted value of Y, for the given X-value.

When we take the mean of the Y-values in the strip at a given $X = x$ as the predicted value of Y, the r.m.s. prediction error is just the s.d. of the Y-values of points in the strip. And when there is a relationship between X and Y, like that in Figure 11-7, the r.m.s

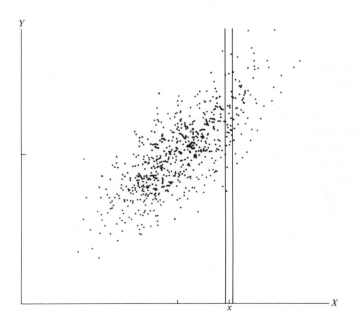

Figure 11-7. Bivariate normal data, showing a conditional distribution.

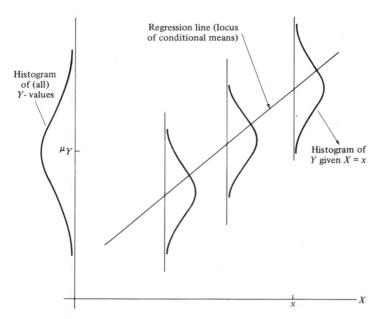

Figure 11-8.   Conditional histograms of Y at various X-values.

prediction error will be smaller than what it would be if the prediction did not make use the given X-value. For if X were *not* given, we should have to use the mean of the Y-values of *all* of the points in the cloud, as the predicted value of Y, and the r.m.s. prediction error would be the s.d. of *all* Y-values. In Figure 11-7, the s.d. of the Y-values for points in the strip at $X = x$ appears to be about $1/2$ to $2/3$ of the s.d. of the Y-values for all points in the cloud. And the prediction that makes use of the knowledge of X is that much more reliable.

In one of the simplest useful models for (X, Y), that called *bivariate normal* (mentioned first in Section 5.5), the mean values of Y for given X all lie on a straight line—the *regression line*, a line that passes through the point (μ_X, μ_Y) with slope $\rho\sigma_Y/\sigma_X$. (If $\sigma_X = \sigma_Y$, the slope is just ρ, the correlation coefficient between X and Y.) The distribution of Y-values at any given X is normal, and is the same normal distribution at each x. These conditional distributions are shown for a typical case in Figure 11-8 as histograms, for several values of X. (The cloud in Figure 11-7 was actually from a bivariate normal distribution, and the histograms in Figure 11-8 are approximately the histograms for the distributions of Y in vertical strips at the indicated X-values.) The histogram of *all*

Y-values is shown at the left; it, too, is a normal curve. The s.d. for *Y*'s at a given $X = x$ is

$$\sigma_Y \sqrt{1 - \rho^2}.$$

Thus, we see that by using *X* judiciously, we can cut the r.m.s. error of prediction from σ_Y to $\sigma_Y \sqrt{1 - \rho^2}$, which is smaller than σ_Y by a factor that is close to 0 when *Y* is highly correlated with the predictor variable *X*.

Now, suppose that the population is *not* bivariate normal. If there is a high degree of linear correlation—that is, if ρ^2 is near 1—it would be only natural to try a straight line as a tool for prediction. It can be shown that the linear function with the smallest r.m.s. prediction error, in any case, is again

$$y = \mu_Y + \rho \frac{\sigma_Y}{\sigma_X} (x - \mu_X),$$

and that the corresponding r.m.s. prediction error is $\sigma_Y \sqrt{1 - \rho^2}$.

Linear prediction of *Y*, given *X*:

Given $X = x$, predict *Y* to be

$$y = \mu_Y + \rho \frac{\sigma_Y}{\sigma_X} (x - \mu_X).$$

The r.m.s. prediction error is $\sigma_Y \sqrt{1 - \rho^2}$.

We have denoted by *Y* the variable that is to be predicted, given a value of the predictor variable—also called the *regressor*. In a problem involving variables not yet so labeled, one speaks of "the regression of income on age," for example; this language means that age is taken to be the regressor, or predictor variable, and income is the variable to be predicted. If variables have been labeled *X* and *Y* according to our usual convention, our interest may occasionally reverse, to focus on the regression of *X* on *Y;* in this case the *X* and *Y* are interchanged in the various formulas we have given. (The regression line of *X* on *Y* is different from the regression line of *Y* on *X*, unless $\rho^2 = 1$.)

Example 11-6   (continuing Example 11-5)

If test scores (X, Y) have a bivariate distribution with means $\mu_X = \mu_Y$ = 70, s.d.'s $\sigma_X = \sigma_Y = 10$, and correlation $\rho = .90$, we predict the Y-score of an individual with $X = x$ to be

$$y = 70 + .90(x - 70).$$

(The slope here is ρ, because $\sigma_Y/\sigma_X = 1$.) That is, we take the height of the regression line (above the x-axis) at $X = x$ as the predicted value of Y. The r.m.s. prediction error in such a prediction is

$$\sigma_Y \sqrt{1 - \rho^2} = 10 \times \sqrt{1 - .9^2} = 4.4,$$

or about 44 percent of what it was in predicting Y to be μ_Y, without knowing the value of X.

In particular, then, some predicted Y's corresponding to given X-values are these:

x	Predicted Y
85	$70 + .9(85 - 70) = 83.5$
70	70
40	43

One curious phenomenon here is that an individual with X-score *above* the mean of 70 is predicted to have a Y-score somewhat *less* than the X, and one with an X-score *below* the mean of 70 is predicted to have a Y-score somewhat *more* than X. ▲

The term *regression* in the name for the line used in prediction refers to the phenomenon noticed at the end of the last example— to the "regression effect." To explain this, we shall refer to diagrams in which a bivariate normal distribution is represented by an ellipse, suggesting the elliptical nature of the cloud of data points from such a population (seen, for example, in Figures 3–11 to 3–17, and 11–7). With the variables X and Y scaled (at least in the graph) so that $\sigma_X = \sigma_Y$, the percentiles of X and Y match up according to the 45° line, or major axis of the ellipse in the schematic representation of the distribution by means of an ellipse. Thus, for example, the 80th percentiles are so matched in Figure 11–9. However, it is apparent that the *average* Y for an individual with an X-value at the 80th per-

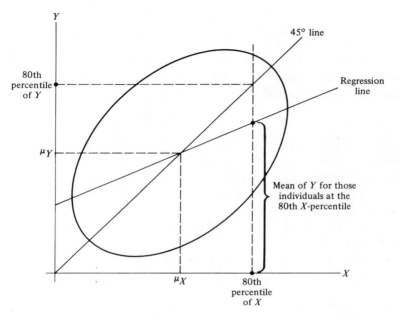

Figure 11-9. The "regression effect."

centile will be *below* the 80th percentile of Y. That is, we would predict a Y-value somewhat lower than the 80th percentile for an individual at the 80th percentile of X. There is an apparent "regression" back toward the mean. The geneticist Francis Galton used the word *regression* for this effect, when he noticed that the son of a very tall man tended to be not quite as tall as the father.

The regression effect becomes more pronounced as the magnitude of the correlation decreases. If $\rho = 1$, there is no regression effect, since the regression line is then the 45° line (in the standard plot). But as the magnitude of ρ decreases, the regression line rotates away from the 45° toward the horizontal, and the predicted value of Y gets closer and closer to μ_Y, for any given $X = x$. Figures 11-10, 11-11, and 11-12 show ellipses corresponding to correlations of .85, .42, and .07, respectively; and the increased regression effect, going to smaller correlations, is evident in these graphs.

Suppose now that the population parameters relevant to linear prediction (means, s.d.'s, and correlation) are *not* known, as has been assumed in the discussion up to this point. Then, of course, the regression line used for predicting Y from a knowledge of X is not known. However, given *data* $(X_1, Y_1), \ldots, (X_n, Y_n)$, one can

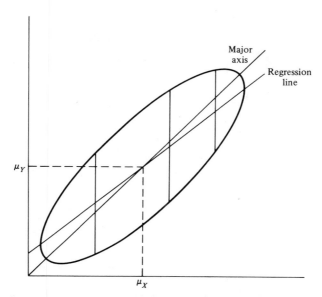

Figure 11-10. Schematic representation of bivariate normal distribution with $\rho = .85$.

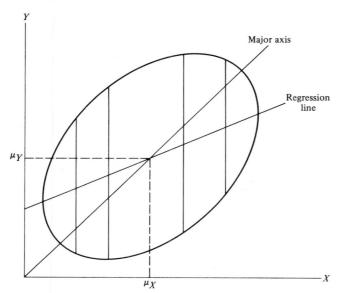

Figure 11-11. Schematic representation of bivariate normal distribution with $\rho = .42$.

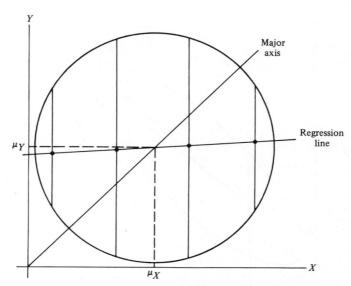

Figure 11-12.   Schematic representation of bivariate normal distribution with $\rho = .07$.

estimate the unknown parameters by the corresponding sample statistics, to obtain the empirical regression line:

$$y = \bar{Y} + r \frac{S_Y}{S_X} (x - \bar{X}).$$

(Notice that this is, indeed, the least-squares line for the given data—the line used in estimating the locus of means of Y when X is a controlled variable.)

If the empirical regression line is used for predicting Y for an individual with a given X-value, instead of the population regression line, the prediction will be in error in part because of the variability in Y at the given X-value, and in part because the estimated conditional mean of Y at that X-value (as given by the least-squares line) will be in error. These two sources of error show up in a mathematically derived formula for r.m.s. prediction error given in the following summary.

Predicting Y for given X, based on data $(X_1, Y_1), \ldots,$ (X_n, Y_n):

Given $X = x$, predict the corresponding Y to be

$$y = \bar{Y} + r\frac{S_Y}{S_X}(x - \bar{X}).$$

The r.m.s. prediction error is given approximately by

$$S_Y\sqrt{1 - r^2}\;\sqrt{1 + \frac{1}{n} + \frac{(x - \bar{X})^2}{nS_X^2}}$$

when the population regression function is linear.

The $1/n$ in the second radical stems from the estimation error in locating the line vertically at x. The third term in that radical, involving $(x - \bar{X})^2$, increases as the X-value for which the prediction is desired moves away from the center of the data. This is because when the slope is inaccurate, the effect of this inaccuracy increases as one moves away from $\bar{X}$ (see Figure 11-13). Observe that the first term in the second radical does not disappear as the number of points in the data increases. That is, even with perfect knowledge of the regression line, one is stuck with the inherent variability measured by the s.d. of Y at the given x.

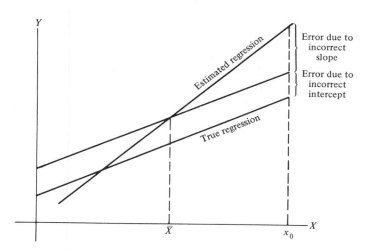

Figure 11-13. Effect of incorrect slope on prediction error.

Example 11-7 *Assessed Value and Selling Price*

A sample of 31 properties sold in a California community yielded the following statistics[†] on X, the assessed value, and Y, the selling price (in dollars):

Means: $\quad \bar{X} = 46{,}700, \quad \bar{Y} = 50{,}950,$

S.d.'s: $\quad S_X = 37{,}100, \quad S_Y = 39{,}900,$

Correlation: $\quad r = .998.$

Putting these values in the equation of the least squares line, one obtains

$$y = 50{,}950 + .998 \times \frac{39{,}900}{37{,}100} (x - 46{,}700)$$

or

$$y = 826 + 1.073x.$$

Using this, one would predict the selling price of a piece of property assessed at $80,000 to be

$$826 + 1.073 \times 80{,}000 \doteq 86{,}700.$$

An estimate of the r.m.s. error involved in predicting Y for an X-value of 80,000 is as follows:

$$39{,}900 \times \sqrt{1 - (.998)^2} \times \sqrt{1 + \frac{1}{31} + \frac{(80{,}000 - 46{,}700)^2}{31 \times (37{,}100)^2}} = 2590.$$

This assumes a linear regression function, which is reasonable in view of the high degree of linear correlation. However, the computations would only be applicable in the locality from which the sample is taken, and at a time not far removed from the period of the data, since the relationship between assessed value and selling price would surely change with time and location. ▲

KEY VOCABULARY

Regression function
Regression of Y on X
Residual

[†]Reported by D. Friedman, R. Pisani, and R. Purvis in *Statistics* (New York: W. W. Norton & Co., Inc., 1978). Data consisted of the properties sold during a 2-week period in 1975.

Error component
Method of least squares
Predictor variable
Prediction error
Regression effect

QUESTIONS

1. Why do we use *vertical* deviations as a basis for measuring the closeness of fit to a set of data points by a straight line?
2. Are regression functions always linear?
3. What are the population parameters in the normal, linear regression model?
4. Given a value of the predictor variable X, why do we take the y-value on the regression line at that X-value as the predicted value of Y?
5. Suppose, in a bivariate population, that the correlation is -1. How much prediction error would you expect to make using the regression line for prediction?
6. Using an empirical regression line based on a certain data set, why is it risky to predict Y if the given X-value is much outside the range of X-values of the data points?
7. A least-squares regression line goes through the "middle" of the data points on which it is based. What does "middle" mean here?
8. Why is the hypothesis $\beta = 0$ sometimes of special interest?

PROBLEMS

Sections 11.1–11.3

*1. (a) Find the slope and y-intercept of these lines:
   (1) $x + y = 4$.
   (2) $2x - 5y = 10$.
   (3) $x = 4y + 7$.
  (b) Write the equation of the line:
   (1) With slope $1/3$ and y-intercept 6.
   (2) With slope -2 and passing through $(1, 1)$.
   (3) With slope 4 and x-intercept 3.
*2. Given the data points in the table,
  (a) Determine the least-squares regression line of y on x.

x	y
1	1
2	5
3	7
4	7

(b) Determine the average squared deviation about the least-squares line.

(c) The points constituting the above data all fall on the curve $y = -x^2 + 7x - 5$ (a parabola). But if this relationship fits the data perfectly, how do you account for the fact that the correlation coefficient is less than 1, suggesting a less-than-perfect relationship?

3. Three observations on (X, Y) are obtained, as follows: $(0, 3)$, $(6, 0)$, and $(2, 2)$.

(a) Determine the least-squares regression line of Y on X.

(b) Determine the correlation coefficient r.

(c) Determine the r.m.s. deviation about the least-squares line.

(d) Plot the data, and then comment on your answers to parts (a)–(c).

*4. An article in *The Farm Quarterly*† reports an experiment conducted to determine which angle of horn elevation‡ would result in the greatest milk production:

Angle of Lift (deg)	Pounds Produced
32	76
78	128
82	156
101	167
127	202
156	313
180	582

†Winter 1946: "A study of cumulative results in dairy improvement on Maine farms by horn elevation," by John Gould, New England writer and humorist.

‡A scrub Ayrshire on a farm near Peppermint Corner, Maine, knocked off her horns trying to get out of the barn. When the horns were replaced, the angle of adjustment was 13.5° in excess of the original angle and subsequently a radical increase in production was noted. This suggested the experiment of altering the horn elevation in steps to discover the relationship.

Determine the least-squares line for the regression of production on angle of elevation. (This example illustrates the danger of assuming that linearity extends much beyond the range of the data, for the report notes that after the seventh trial, the cow's horns were green and sticking straight up, and the cow was dead—presumably from the strain of overproduction.)

5. An experiment was performed to investigate the damage to residential structures caused by blasting. For 13 blasts that caused major damage, the following table gives the frequency (in cycles per second) and particle displacement (in inches).[†]

Frequency	Displacement	Frequency	Displacement
2.5	.390	11.0	.180
2.5	.250	11.0	.093
2.5	.200	11.0	.080
3.0	.360	16.0	.052
3.5	.250	25.0	.051
9.3	.077	25.0	.150
9.9	.140		

Find the empirical regression line of displacement on the logarithm of frequency. (If you do not have logs on your calculator, obtain the regression of displacement on frequency.)

Sections 11.4 and 11.5

*6. The empirical regression line calculated from the corrosion data in Example 11-1 is

$$y = 129.8 - 24.02x.$$

Give a 90 percent confidence interval for the slope, given that $r = -.9847$ and $S_x^2 = .439$.

7. Students in a statistics class recorded their ages and pulse rates,

[†]Nicholls, H. R., C. F. Johnson, and W. I. Duvall, "Blasting Vibrations and their Effects on Structures," U.S. Dept. of the Interior, Bureau of Mines Bulletin 656, p. 17, Fig. 3.3 (1971).

as follows (age in months, pulse rate in beats per minute):

Age	Pulse	Age	Pulse	Age	Pulse	Age	Pulse	Age	Pulse
268	78	269	68	376	86	225	80	268	80
355	90	221	80	294	64	243	72	327	72
254	65	379	84	417	60	251	68	238	80
383	62	308	68	245	74	274	70	234	68
262	70	285	62	256	70	460	56	242	72

(a) Calculate r and the least squares regression line.

(b) Given $r = -.2089$, $a = 79.97$, $b = -.0273$, and $S_Y = 8.364$, calculate the t-statistic that would be used for testing the hypothesis that the true regression line has slope $\beta = 0$ (assuming a random sample and a normal distribution about the regression line).

(c) Notice that the sample includes five students over 30 years of age. When these are deleted from the sample, the revised values are $n = 20$, $r = .0819$, $a = 68.15$, $b = .01653$, and $S_Y = 6.719$. Recompute T for this smaller sample.

*8. Referring to Example 11–5, find the mean value of second exam scores (Y) for individuals who scored 35 on the first exam (X). Calculate the proportion of these individuals who will score *less* than 35 on the second exam. [Given from Example 11–5: $r = .90$, m.v.(X) = m.v.(Y) = 70, and s.d.(X) = s.d.(Y) = 10.] Assume a normal distribution.

9. An educator finds a correlation of .90 between IQ (X) and college aptitude score (Y) based on 500 observations (X_i, Y_i).

(a) Given $S_X = 20$, $S_Y = 60$, determine the *slope* of the regression line of Y on X.

(b) Given $\bar{X} = 100$, $\bar{Y} = 400$, what college aptitude score would be predicted for a student with an IQ of 130?

(c) Taking the value .90 as pretty close to the actual population correlation coefficient, how much better a prediction is part (b), using the X-value given, than predicting Y (without knowing X) to be 400?

*10. Suppose you want to guess the weight of a 29-year-old female on the basis of her 64-inch height. If it is given that (1) the s.d. of the weights of 29-year-old females is 10 pounds, (2) the mean weight of 29-year-old females is $4h - 130$ (where h is height in inches), and (3) the correlation of height and weight is $\rho = .8$, what is a good guess, and what is the r.m.s. error?

11. The following are scores of 15 students on two midquarter examinations in an advanced statistics course.

1st	2nd	1st	2nd	1st	2nd
97	100	64	85	50	46
50	67	57	47	65	67
50	46	45	50	78	78
81	89	40	49	60	58
94	85	47	53	60	41

(a) Find the regression line of the score of the second examination on the score of the first examination.
(b) Given $S_1 = 17.0$ and $S_2 = 18.4$, determine the correlation coefficient from the slope in part (a).
(c) Predict the second examination score of a student with 97 on his first examination; with 30 on his first examination.

Review

12. A study reported in the *Annals of Surgery*[†] gives data on the increase in pancreatic intraductal pressure in dogs following the administration of a potent cholinesterase inhibitor. Successive doses of 5 mg/kg were administered 30 minutes apart. Data for Dog 577 are as follows:

Dose	Pressure
0	14.6
5	24.5
10	21.8
15	34.5
20	35.1
25	43

Find the least-squares regression line of pressure on dose.

13. The volume of wood in a mature tree is related to its more

[†]T. D. Dressel, et al., "Sensitivity of the Canine Pancreatic Intraductal Pressure to Subclinical Reduction in Cholinesterase Activity, *Annals of Surgery* **190**, No. 1, July 1979.

readily measurable dimensions: height and diameter. Knowing this relationship would permit estimating the amount of lumber in a stand of trees before cutting them down. Data on several black cherry tress are given in the book *Forest Mensurations* (State College, Pa.: Pennsylvania Valley Publishers, 1953), from which these results are computed:

Height × Diameter	Height × (Diameter)^2	(Diameter)^3	Volume
46.6	33.4	.368	10.3
72.2	64.4	.709	18.8
72.2	68.6	.857	21.0
79.6	85.5	1.24	22.2
95.3	105.64	1.36	27.4
91.0	106.2	1.588	34.5
96.0	128.0	2.370	38.3
116.8	168.4	2.966	55.4
119.3	178.0	3.319	58.3
149.4	256.4	5.059	77.0

Determine which of the first three quantities has the highest linear correlation with volume and give the empirical regression line in that case.

14. Scores of the verbal (V) and analytical (A) parts of the Graduate Record Examination are given for 26 applicants to a graduate program in statistics in 1 year:

V	A	V	A	V	A
470	670	790	650	790	620
290	420	420	550	710	680
380	660	450	600	290	420
640	730	420	670	780	730
240	540	450	470	370	560
390	520	490	550	670	670
330	560	780	710	470	670
270	500	600	690	610	710
430	580	540	690		

(a) Determine r and the least-squares line for regression of A on V.

(b) Given $r = .721$, $\bar{V} = 502.7$, $\bar{A} = 608.46$, $S_V = 169.5$, $S_A = 90.96$, predict the A-score for a student with a V-score of 400. What error is involved in this prediction?

15. A random sample of 500 individuals from the thousands taking the Graduate Record Examination has these characteristics, where A is the score on the "analytical" and V the score on the "verbal" portion of the examination:

$$\bar{A} = 490, \quad S_A = 170,$$
$$\bar{V} = 550, \quad S_V = 160.$$

The correlation between A and V is $r = .75$.

(a) Determine the empirical regression line for A on V.

(b) Assuming a bivariate normal distribution, give a 95 percent confidence interval for the slope of the regression line of A on V.

12

Analysis of Variance

IN CHAPTER 9 we gave some procedures for deciding whether an observed difference between a treatment group mean and a control group mean was evidence of a treatment effect or simply a manifestation of sampling variability. In this chapter we extend the analysis to cases in which there are several types or levels of treatment.

Each treatment type defines a population of possible responses, and we assume that each response consists of a treatment mean plus a random error. In Section 12.1 we consider the case of a single factor or controlled variable that operates at several levels or in several categories. In Section 12.2 we analyze data from an experimental design in which a "treatment" is a combination of levels or categories of two factors.

12.1
ONE-WAY ANOVA

It will help to keep in mind a specific application. Suppose that the "response" is the tensile strength of a spot-weld of an aluminum-clad material, and the "factor" is the machine used to make the weld. Each machine, in its use in the welding process, is a category or type of treatment.

Experience shows that different welds from the same machine vary in tensile strength. So it is usual to take a sample of more than one weld from each machine, a sample from the population or distribution of weld strengths that might be made by that machine. As a model for this process, we take the tensile strength to be composed of a mean for that machine plus a random error that has zero mean.

To illustrate the analysis, we shall act as "Nature" and make up some data according to the model we are assuming. Suppose that there are four machines and the mean tensile strengths (in some convenient units) are

$$\text{Machine 1:} \quad \mu_1 = 7$$
$$\text{Machine 2:} \quad \mu_2 = 2$$
$$\text{Machine 3:} \quad \mu_3 = 7$$
$$\text{Machine 4:} \quad \mu_4 = 4.$$

These are clearly different—there *is* a machine effect (or treatment effect); and if statisticians knew these population means, they would not need any data. However, in practice, there are random errors that would tend to obscure the differences.

Before adding in random errors, we rewrite each mean μ_i as an overall mean response μ plus a number τ_i, which we call the *treatment effect*. The four means μ_1, μ_2, μ_3, and μ_4 have an average value of $(7 + 2 + 7 + 4)/4 = 5$, so we write

$$\mu_1 = 7 = 5 + 2$$
$$\mu_2 = 2 = 5 - 3$$
$$\mu_3 = 7 = 5 + 2$$
$$\mu_4 = 4 = 5 - 1.$$

Thus,

$$\tau_1 = 2, \quad \tau_2 = -3, \quad \tau_3 = 2, \quad \tau_4 = -1.$$

These treatment effects will always add to 0 when defined as deviations about the overall mean.

The null hypothesis to be tested is that there is no treatment effect, or that all treatment means are the same—or that treatment effects τ_i are all 0:

$$H_0: \tau_1 = \tau_2 = \tau_3 = \tau_4 = 0 \quad \text{(or equivalently, } \mu_1 = \mu_2 = \mu_3 = \mu_4\text{).}$$

In the context of the welding machines, this says that on the average there is no difference in machines, as regards the tensile strength of the welds. Given the means we have assumed, this null hypothesis is false. It remains to be seen whether our statistical analysis will reach this conclusion when applied to data that include random errors.

A simple assumption for the analysis of the data is that the random-error components are independent and *normal* with mean 0 and a variance that is the same at each measurement and for each machine. Acting as Nature, we select numbers at random from a normal population with mean 0 and s.d. = 2 (to be specific) to be these errors. This we can do using Table IX of Appendix B, starting at an arbitrary point and multiplying entries by 2 (because the table entries come from a population with s.d. = 1) and rounding to the nearest hundreth for convenience:

	Errors (ϵ_{ij})		
For machine 1:	.26,	1.70,	-3.50
For machine 2:	-1.16,	-1.42,	1.46
For machine 3:	- .80,	1.82,	.82
For machine 4:	- .46,	- .22,	-3.70

Adding these to the means chosen earlier we obtain the data. We denote the *j*th observation from machine *i* by X_{ij}:

	From the Model (Unknown)			Data (Observed)
Machine Number	μ	τ_i	$\mu + \tau_i$	$X_{ij} = \mu + \tau_i + \epsilon_{ij}$
1	5	2	7	7.26, 8.70, 3.50
2	5	-3	2	0.84, 0.58, 3.46
3	5	2	7	6.20, 8.82, 7.82
4	5	-1	4	3.54, 3.78, 0.30

In any real-life situation, the statistician will see only the *data*. He may assume that the data have the structure $X_{ij} = \mu + \tau_i + \epsilon_{ij}$, as we have assumed, but he does not know what part of any observation X_{ij} is μ, what part is τ_i, or what part is error.

As in the case of two populations, a first step in comparing population means is to calculate the sample means—$\bar{X}_1$, the mean tensile strength for the welds from machine 1, $\bar{X}_2$ the mean for machine 2, and so on:

Machine (Factor Level)	Data—Tensile Strength (Responses, X_{ij})	Mean, $\bar{X}_i$
1	7.26, 8.70, 3.50	6.487
2	0.84, 0.58, 3.46	1.627
3	6.20, 8.82, 7.82	7.613
4	3.54, 3.78, 0.30	2.540

The sample means are not the same, but this would happen even when the population means *are* the same. What is important is not the fact that they differ, but the *amount* of their variation about their mean, $\bar{X} = 4.567$.

If the variation among sample means is much greater than would be expected to result from sampling fluctuations—from the error components of the data—it would be reasonable to conclude that there *are* treatment differences, or that the null hypothesis of no treatment effect is false. So, how much variation among sample means is "expected"?

We know that the mean $\bar{X}_i$ of the random sample of n_i measurements from machine i has variance

$$[\text{s.d.}(\bar{X}_i)]^2 = \frac{\sigma^2}{n_i},$$

where σ^2 is the "error variance"—the variance of the error component. This suggests dividing the quantity

$$\text{SSTr} = \sum n_i (\bar{X}_i - \bar{X})^2,$$

called the *treatment sum of squares*, by the number of degrees of freedom to obtain a quantity that should be close to σ^2 *if there is no treatment effect*. It will tend to be inflated by the presence of treatment effects, because these will make the $\bar{X}_i$ more variable. In the example at hand, with $k = 4$ levels,

$$\text{SSTr} = 3(6.487 - 4.567)^2 + 3(1.627 - 4.567)^2$$
$$+ 3(7.613 - 4.567)^2 + 3(2.540 - 4.567)^2 = 77.15,$$

and division by $k - 1 = 3$ yields the *treatment mean square:*

$$\text{MSTr} = \frac{\text{SSTr}}{k-1} = \frac{77.15}{3} = 25.7.$$

This is far from $\sigma^2 = 4$, the error variance we used in making up the data, because of the treatment components. But σ^2 is not known in practice, and we need another way of getting at σ^2.

The error variation shows up in the variation of the observations in each sample about that sample mean, variation that is unaffected by any nonzero treatment components τ. As we did in the case of two samples, to obtain an estimate of σ^2, we first pool the squared deviations about the sample means:

$$\text{SSE} = \sum \sum (X_{ij} - \bar{X}_i)^2.$$

This is the *error sum of squares.* (The symbol $\sum \sum$ calls for a "double summation"—first a summation within each sample, from $j = 1$ to n_i for sample i, and then a summation of these sample sums over the samples $i = 1$ to k.) For the data at hand,

$$\begin{aligned}
\text{SSE} = {} & (7.26 - 6.487)^2 + (8.70 - 6.487)^2 + (3.50 + 6.487)^2 \\
& + (.84 - 1.627)^2 + (.58 - 1.627)^2 + (3.46 - 1.627)^2 \\
& + (6.20 - 7.613)^2 + (8.82 - 7.613)^2 + (7.82 - 7.613)^2 \\
& + (3.54 - 2.54)^2 + (3.78 - 2.54)^2 + (.30 - 2.54)^2 \\
= {} & 30.54.
\end{aligned}$$

Next, we divide by the number of degrees of freedom—the sum of the $n_i - 1$ degrees of freedom in the samples:

$$\text{d.f.(error)} = (3 - 1) + (3 - 1) + (3 - 1) + (3 - 1) = 8$$

to obtain the *error mean square:*

$$\text{MSE} = \frac{\text{SSE}}{\text{d.f.(error)}} = \frac{30.54}{8} = 3.82.$$

This is quite reasonable as an estimate of $\sigma^2 = 4$.

The fact that MSTr is so much larger than MSE is taken as evidence of a treatment effect. When the errors are *normally* distributed, the ratio

$$F = \frac{\text{MSTr}}{\text{MSE}}$$

has an "F-distribution" with $\sum (n_i - 1)$ degrees of freedom in the numerator and k degrees of freedom in the denominator when there is no treatment effect. The 95th percentile of F is given in Table VIIa of Appendix B:

$$F_{.95}(3,8) = 4.07,$$

and because the observed ratio exceeds this:

$$F = \frac{25.7}{3.82} = 6.74 > 4.07,$$

we reject H_0. The F-value is "significant" at the 5 percent level. So the treatment effect that we knew to be present all along (since we put it there!) has been uncovered.

It can be shown algebraically that

$$\sum\sum(X_{ij} - \bar{X})^2 = \sum\sum(X_{ij} - \bar{X}_i)^2 + \sum n_i(\bar{X}_i - \bar{X})^2,$$

or that the sum of squared deviations of the observations about the overall mean, called the *total sum of squares* (SST), is decomposable into two terms:

$$\text{SST} = \text{SSE} + \text{SSTr},$$

one that is associated with treatment difference and one associated with the error component. The analysis is referred to as an *analysis of variance* (ANOVA).

In the present example, the total sum of squares is computed like this:

$$\sum\sum(X_{ij} - \bar{X})^2 = (7.26 - 4.567)^2 + (8.70 - 4.567)^2 + (3.50 - 4.567)^2$$
$$+ (.84 - 4.567)^2 + \cdots (12 \text{ terms in all}) = 107.70.$$

Observe (as a check on the computations) that this *is* the sum of SSE and SSTr. The computations are often summarized in an *ANOVA table:*

Source	Sum of Squares	d.f.	Mean Square	F
Machine	77.15	3	3.82	6.74
Error	30.54	8	25.7	—
Total	107.70	11	—	—

You will find that ANOVA tests based on the F-distribution are routinely carried out without much thought about the underlying distributions. At least, such thought is seldom mentioned in research reports. When the response variable is a measurement or a yield or a quality of performance, the assumption of normality is reasonable. And, whatever the exact distribution, if its "tails" are not too heavy, use of F will seldom lead one astray.

Example 12-1   *Pinball—More Than Luck?*

Five people played four games each of pinball with the following scores (in hundreds):

A: 1142, 865, 524, 1024   (mean = 888.75)

B: 565, 1040, 684, 749   (mean = 759.50)

C: 467, 880, 583, 621   (mean = 637.75)

D: 883, 628, 963, 1192   (mean = 916.50)

E: 609, 575, 708, 775   (mean = 666.75).

(The overall mean is 773.85.) A null hypothesis would be that one player is as good as another—there is no way to play to one's advantage, and only "luck" is a factor.

The "treatment" here is the player, and the sums of squares are as follows:

$$SSTr = 4(888.75 - 773.85)^2 + 4(759.5 - 773.85)^2$$

$$+ \cdots = 255{,}002.3$$

$$SSE = (1142 - 888.75)^2 + (865 - 888.75)^2$$

$$+ \cdots (20 \text{ terms}) = 617{,}064.25$$

$$SST = (1142 - 773.85)^2 + (865 - 773.85)^2$$

$$+ \cdots (20 \text{ terms}) = 872{,}066.55.$$

The ANOVA table summarizes the computations:

Source	SS	d.f.	MS	F
Player	255,002	4	63,750	1.55
Error	617,064	15	41,138	—
Total	872,067	19	—	—

The critical value of F at $\alpha = .05$ is (from Table VIIa—with 4 d.f. in the numerator and 15 in the denominator) 5.86; and because the observed F-value does not exceed this ($1.55 < 5.86$), the null hypothesis is not rejected. On the basis of the evidence in this sample, we do not conclude that the score is determined by more than luck. ▲

Example 12-2   *Close Encounters of the Bench Kind*

Using telephoto lenses, researchers[†] photographed pairs of unacquainted men on park benches in San Francisco, California, in Tangiers, Morocco, and in Seville, Spain. The men in a pair were not of different racial groups and were not tourists. Data on space between torsos were obtained by making measurements on enlargements, with reference to the known length of the benches:

	n (pairs)	$\bar{X}$ (in.)	s.d. (in.)
San Francisco	35	28	10
Tangiers	25	35	10
Seville	22	33	7

Although the raw data are not given in the article, an analysis of variance can be made from the given summary statistics. First, observe that the overall mean is computed as a weighted average of the sample averages:

$$\bar{X} = \frac{1}{82}\sum\sum X_{ij} = \frac{1}{82}(35 \times 28 + 25 \times 35 + 22 \times 33) = 31.476$$

(the sum of the X's in the San Francisco sample is 35 times the mean of that sample, and so on). And then

$$\text{SSTr} = 35(28 - 31.476)^2 + 25(35 - 31.476)^2 + 22(33 - 31.476)^2$$
$$= 784.45.$$

The degrees of freedom in this sum are d.f.(Tr) = 3 − 1 = 2.

The error sum of squares, based on within samples variation, can be computed from the given sample s.d.'s:

$$\text{SSE} = \sum\sum (X_{ij} - \bar{X}_i)^2 = 35 \times 10^2 + 25 \times 10^2 + 22 \times 7^2 = 7078.$$

[The sum of squared deviations over the San Francisco sample is 35 times the variance, (s.d.)^2, and so on.] The degrees of freedom in SSE is the sum of the d.f.'s in the three samples:

$$\text{d.f.(error)} = (35 - 1) + (25 - 1) + (22 - 1) = 79.$$

[†]A. Mazur, "Interpersonal Spacing on Public Benches in 'Contact' vs. 'Noncontact' Cultures," *J. Soc. Psych.* **101**, 53–58 (1977).

The total sum of squares cannot be computed directly as a check, but we now have the ingredients of the ANOVA table:

Source	SS	d.f.	MS	F
Location	784.45	2	392.2	4.38
Error	7078	79	89.59	—
Total	7792.25	81	—	—

The entries for denominator d.f. do not go beyond 60 in Table VII; but it is clear that the 95th percentiles decrease as denominator degrees of freedom increase, so the critical value is *less* than 3.15. The null hypothesis is rejected.

The study laments about the "severe time and budget constraints which precluded gathering data for inferences about women, acquaintances, cross-race and cross-sex pairs." ▲

12.2
TWO FACTORS

The production manager, when told by his industrial engineer about the plans to test for a welding machine effect (as described at the beginning of Section 12.1), might protest that there are other factors to take into account. The machine operator, for example, might affect the tensile strength of a weld. And if a given machine is always manned by the same operator, any operator effect would be part of— or "confounded" with—what would otherwise be interpreted as a machine effect.

If various operators are used on each machine, an analysis by the method of the preceding section would, in effect, consign any variation in their contributions to the random error component of the response. If there is an operator effect and if it could be removed from the error component, the residual error would be smaller, and the test for a machine effect would be more sensitive. The aim of this section is to show how to isolate the operator and machine effects.

When there are two categorical factors that can be controlled or set, a "treatment" consists of a combination of a level of one factor with a level of the other. If there are c levels or categories of factor A, and r levels or categories of factor B, then there are rc different treatment combinations. One might like to have several observations for each of these rc combinations; but this may be impractical, and (somewhat surprisingly) under certain conditions it is possible to sort out the factor effects with just one observation per combination. This is the case we consider here.

The treatment combining level i of factor A with level j of factor B will be assumed to result in a random response consisting of a mean plus a random deviation, where the mean is of the form

$$\mu + \alpha_i + \beta_j.$$

That is, there is an overall mean μ and a contribution α_i attributed to factor A which combines by addition to a contribution β_j of factor B. (This simple model is not always correct. In particular, there may be an "interaction" contribution to the mean that only comes into play when level i of A combines with level j of B. For example, an operator may be better on some machines than others.) The constant μ can always be defined to make $\sum \alpha$ and $\sum \beta$ equal to 0.

Again we shall play Nature and make up some data according to the assumed model. Suppose that factor A has four levels, with contributions to the mean

$$\alpha_1 = -2, \quad \alpha_2 = 3, \quad \alpha_3 = -5, \quad \alpha_4 = 4.$$

And let factor B have three levels, with contributions to the mean of

$$\beta_1 = 2, \quad \beta_2 = -1, \quad \beta_3 = -1.$$

To have a specific application in mind, suppose that factor A is fertilizer type and factor B is brand of seed. The response is the yield (in bushels/acre, or pounds, or whatever is appropriate).

Combining the contributions α and β, with an overall mean of $\mu = 20$, we obtain the following means for the 4×3 or 12 treatment combination, written in cells of a two-way array:

		β_j (Factor B)			
		2	-1	-1	
	-2	20	17	17	18
α_i	3	25	22	22	23
(Factor A)	-5	17	14	14	15
	4	26	23	23	24
		22	19	19	20

The mean of these cell means is 20, because we picked α's and β's that summed to zero.

Now if the experimenter knew these means, he or she would not need to experiment—it is clear that there *is* a factor A effect (because of the variation of 18, 23, 15, and 24 about 20) and there *is* a factor B effect (22, 19, and 19 about 20). But the experimenter will only see data, which are these means corrupted by random errors. As Nature, we go again to Table IX to get 12 random normal errors, starting at an arbitrary point, as follows (this time we take $\sigma = 1$):

$$\begin{array}{rrr} -.77 & -.15 & -1.31 \\ -.63 & 1.22 & -.22 \\ .83 & .67 & -.15 \\ 1.11 & -2.17 & .21 \end{array}$$

Combining these errors with the means, cell by cell, we finally obtain the *data*, which is where an actual analysis would start:

		Factor B			
		1	2	3	Means
	1	19.23	16.85	15.69	17.26
Factor	2	24.37	23.22	21.78	23.12
A	3	17.82	14.67	13.85	15.45
	4	27.11	21.83	23.21	24.05
Means		22.13	19.14	18.63	19.97

Row and column means have been written in, as estimates of the true row and column means (unknown to the experimenter) in the earlier table. However, in the presence of random errors it is not clear whether the variation among the means 22.13, 19.14, and 18.63 is attributable to a factor B effect or to random variation in the error. This will be judged by first calculating sums of squares giving the variation of row means about the overall mean, and of column means about the column means:

$$\text{SSA} = 3 \times [(17.26 - 19.97)^2 + (23.12 - 19.97)^2$$
$$+ (15.45 - 19.97)^2 + (24.05 - 19.97)^2] = 163.03$$
$$\text{SSB} = 4 \times [(28.13 - 19.97)^2 + (19.14 - 19.97)^2$$
$$+ (18.63 - 19.97)^2] = 28.60.$$

The multipliers 3 and 4 are included so that dividing by degrees of freedom would produce—in the absence of factor effects—estimates of the error variance. (Each row mean comes from a sample of 3, and each column mean from a sample of 4.)

To judge whether these sums of squares are inflated by factor effects, we need an estimate of error variance that is valid even in the presence of factor effects. Looking back at the table of cell means μ_j, we see that in each case the cell entry plus the overall mean in the corner is equal to the sum of the mean in that row and the mean in that column. Thus,

$$(20 + 20) - (22 + 18) = 0$$
$$(17 + 20) - (19 + 18) = 0$$
$$(25 + 20) - (23 + 22) = 0,$$

and so on. So in a table giving the data, deviations of this kind of combination from 0 are estimates of the error component:

$$(19.23 + 19.97) - (22.13 + 17.26) = -.19$$
$$(16.85 + 19.97) - (19.14 + 17.26) = .42$$

and so on. The results of these calculations are as follows:

$$
\begin{array}{rrr}
-.19 & .42 & -.23 \\
-.91 & .93 & 0 \\
.21 & .05 & -.26 \\
.90 & -1.39 & .50
\end{array}
$$

(Compare these with the actual errors that we, as Nature, know we put into the problem.) The sum of squares of the entries is

$$
\begin{aligned}
\text{SSE} &= (-.19)^2 + (.42)^2 + (-.23)^2 \\
&+ (-.91)^2 + (.93)^2 + 0^2 \\
&+ (.21)^2 + (.05)^2 + (-.26)^2 \\
&+ (.90)^2 + (-1.39)^2 + (.50)^2 \\
&= 5.06.
\end{aligned}
$$

When divided by degrees of freedom, this gives an estimate of the error variance.

The degrees of freedom are as follows:

$$\text{SSA:} \quad (\text{number of levels of } A) - 1 = r - 1$$

$$\text{SSB:}\quad (\text{number of levels of } B) - 1 = c - 1$$

$$\text{SSE:}\quad (r-1)(c-1)$$

$$\text{Total:}\quad rc - 1.$$

The results are summarized in an ANOVA table:

Source	SS	d.f.	MS	F
A	163.24	3	54.41	64.5
B	28.6	2	14.3	16.9
Error	5.06	6	.844	—
Total	196.9	11	—	—

The F-ratios shown are obtained by comparing the mean squares for treatment with the mean square for error:

$$F_A = \frac{\text{MSA}}{\text{MSE}} = \frac{54.41}{.844} = 64.5$$

$$F_B = \frac{\text{MSB}}{\text{MSE}} = \frac{14.3}{.844} = 16.9$$

Notice that MSE is not far from the $\sigma^2 = 1$ that Nature put into the data, but that MSA is very large, in comparison. If the random component in the data is normally distributed, the ratio F_A would exceed 4.76 with only a 5 percent probability (from Table VIIa, using 3 d.f. in the numerator and 6 in the denominator). At $\alpha = .05$, then, the hypothesis of no factor A effect (no fertilizer effect) would be rejected, because $64.0 > 4.76$. And this test has been carried out in the presence of a possible factor B effect, which has been taken into account but separated out from the factor A effect.

If there is interest in factor B itself, the F-ratio can be used to test for significance. In this case,

$$F_B = 16.9 > 5.14 = F_{.95}(2,6),$$

the latter obtained from Table VIIa, entering it with 2 d.f. in the numerator and 6 in the denominator. So at $\alpha = .05$ the hypothesis of no factor B effect is rejected. However, it should be pointed out that as an error size, the value $\alpha = .05$ applies to only one factor test at a time, not to a simultaneous test of both factors.

Example 12-3 *Comparing Detergents*

Whiteness readings on clothes washed with three detergents and three water temperatures were obtained as follows:

		Detergent			
		A	B	C	
Water Temperature	Cold	45	47	55	49
	Warm	36	41	55	44
	Hot	42	47	46	45
		41	45	52	46

The sum of squares for temperature is

$$\text{SSTemp} = 3[(49 - 49)^2 + (44 - 46)^2 + (45 - 46)^2] = 42,$$

and for detergent,

$$\text{SSDet} = 3[(41 - 46)^2 + (45 - 46)^2 + (52 - 46)^2] = 186.$$

The error sum of squares is computed as follows:

$$\text{SSE} = (45 + 46 - 41 - 49)^2 + (47 + 46 - 45 - 49)^2$$
$$+ \cdots \text{(nine terms in all)} = 78.$$

The total sum of squares provides a check:

$$\text{SST} = (45 - 46)^2 + (47 - 46)^2 + \cdots \text{(nine terms in all)} = 306.$$

The ANOVA table is then the following:

Source	SS	d.f.	MS	F
Detergent	186	2	93	4.77
Temperature	42	2	21	1.08
Error	78	4	19.5	—
Total	306	8	—	—

The F-ratio is not quite significant at $\alpha = .05$, since 4.77 does not exceed 6.94, the 95th percentile of F with 2 and 4 degrees of freedom. Our tables do not give P-values, but there seems to be some evidence of a detergent effect. The F-ratio for temperature, on the other hand, is so near 1 that one must accept the hypothesis of no temperature effect. ▲

Example 12-4 *Baking Meat Loaves*

Different methods of freezing meat loaf were to be compared, in a University of Wisconsin study.[†] Eight loaves could be baked at a time; but some parts of the oven perform differently than others, and it was realized that such differences might hide the differences among freezing methods. So a preliminary study was conducted to test for oven-position effect.

Three batches of eight loaves each were baked and analyzed for percentage of "drip loss." The results were as follows:

| | | Batch | | | |
		1	2	3	Means
	1	7.33	8.11	8.06	7.833
	2	3.22	3.72	4.28	3.740
	3	3.28	5.11	4.56	4.317
	4	6.44	5.78	8.61	6.943
Oven	5	3.83	6.50	7.72	6.017
Position	6	3.28	5.11	5.56	4.650
	7	5.06	5.11	7.83	6.000
	8	4.44	4.28	6.33	5.017
	Means	4.610	5.465	6.619	5.565

(The numbering of oven positions was arbitrary.) The components of each observation would be assumed to be an overall average, for the type of meat used (the same for all 24 loaves), an oven-position-effect term, a batch-effect term, and an error. The null hypothesis is that there is no oven position effect—that the oven-effect term is always zero.

The computation of the sums of squares proceeds as follows:

$$\text{SSBatch} = 8[(4.610 - 5.565)^2 + (5.465 - 5.565)^2$$
$$+ (6.619 - 5.565)^2] = 16.259$$
$$\text{SSOvenPos} = 3[(7.833 - 5.565)^2 + (3.740 - 5.565)^2$$
$$+ \cdots] = 40.396$$

[†]Study by Barbara Bobeng and Beatrice David, described (along with the data) by T. J. Ryan, Jr., et al., in *Minitab: A Student Handbook* (North Scituate, Mass.: Duxbury Press, 1976).

SSE $= (7.33 - 7.833 - 4.610 + 5.565)^2$

$$+ (8.11 - 7.833 - 5.465 + 5.565)^2 + \cdots = 9.290.$$

Degrees of freedom are

> d.f.(batch) $= 3 - 1 = 2$
>
> d.f.(oven pos.) $= 8 - 1 = 7$
>
> d.f.(error) $= (3 - 1)(8 - 1) = 14.$

Dividing sums of squares by the corresponding numbers of degrees of freedom yields the mean squares, as shown in the ANOVA table:

Source	SS	d.f.	MS	F
Oven position	40.396	7	5.771	8.70
Batch	16.259	2	8.1295	12.25
Error	9.290	14	.6636	—
Total	65.945	23	—	—

The F-ratio of 8.7 for oven position far exceeds even the 99th percentile of $F(7, 14)$ or about 4.3 (from Table VIIb of Appendix B), so there is strong evidence of an oven-position effect. There also seems to be a batch effect $[F = 12.25 > 6.51 = F_{.99}(2, 14)]$, but in spite of this, the oven-position effect could be sorted out and detected.

The oven-position effect would not be so pronounced if the batch factor had not been taken into account. Taking the data as eight samples of three, corresponding to the eight oven positions, the ANOVA table would be as follows, obtained by combining SSBatch with SSE (above) to get the new SSE:

Source	SS	d.f.	MS	F
Oven position	40.396	7	5.77	3.61
Error	25.549	16	1.60	—
Total	65.945	23	—	—

The critical F-value is about 2.66, so the calculated F is significant—but not as clear evidence of a position effect as when the batch factor was taken into account. ▲

KEY VOCABULARY

Treatment effect
Error component
Factor
Treatment sum of squares
Error sum of squares
F-ratio
Analysis of variance
Interaction

QUESTIONS

1. Why cannot one conclude that there is a treatment effect when the sample means corresponding to different levels of treatment are not all equal?
2. In a one-way ANOVA, why is the mean square for error used as the basis for judging whether or not the mean square for treatment is large enough to warrant the conclusion that there is a treatment effect?
3. What assumptions give rise to the F-distribution for the F-ratio used in an analysis of variance?
4. In a one-way ANOVA, the treatment sum of squares meaures the "between samples" variation of the mean. What quantity in a two-sample t-test plays the same role, and how is it different?
5. In a situation where there is actually an "interaction," so that the cell mean is not as simple as $\mu + \alpha_i + \beta_j$, what happens to that interaction if we carry out a two-way ANOVA that assumes no interaction?
6. If there are two factors influencing a response, and we take only one of them into account in a one-way ANOVA, what happens to the contribution of the other factor?

PROBLEMS

Section 12.1

*1. To help develop facility with ANOVA without excessive numerical difficulties, consider the following artificial data:

Treatment	Sample
A	2, 6, 4
B	12, 6, 6
C	4, 6, 8

Calculate the F-statistic for testing the null hypothesis of no treatment effect, and give its degrees of freedom.

2. Suppose that sample observations exhibit no variability, as in these data:

Treatment	Sample
A	4, 4, 4
B	8, 8, 8
C	6, 6, 6

What is the error sum of squares for these data? What does this imply for testing the hypothesis of no treatment effect? The sample means here are the same as in Problem 1; why is the conclusion now apt to be different?

*3. Independent samples were used for testing a treatment by means of the two-sample t-test (Chapter 9), assuming nearly normal populations. The ANOVA approach developed for comparing several samples works just as well for two samples. Consider these (artificial) data:

$$\text{Sample 1: } 8, 10, 16, 14.$$
$$\text{Sample 2: } 9, \ 7, \ \ 7, 13.$$

(a) Calculate the two sample T-statistic of Chapter 9, and determine the critical value of T in a two-sided test at $\alpha = .05$.

(b) Calculate the F-statistic from an ANOVA and determine the critical value of F at $\alpha = .05$. Compare these with the value of T^2 and the square of the critical level in part (a).

4. Five varieties of clover were to be compared as to yield, each variety being planted on three plots (at the Rosemount Experiment Station of the University of Minnesota). Yields in tons per acre were as follows:

Variety	Yield	Mean
Spanish	2.79, 2.26, 3.09	2.713
Evergreen	1.93, 2.07, 2.45	2.150
Commercial yellow	2.76, 2.34, 1.87	2.323
Madrid	2.31, 2.30, 2.49	2.367
Wis. A46	2.39, 2.05, 2.68	2.373

Test the hypothesis of no difference among varieties. (Any effects due to plot variability, being ignored, are included in the random error.)

*5. Using a computer-TV pinball simulator, 20 observations were made in each of three paddle positions with these results:

Paddle Position	Mean	$(s.d.)^2$
None	40.78	32.30
Far left	48.53	33.25
Medium left	50.50	119.05

Test the null hypothesis of no paddle effect.

*6. To compare the effectiveness of three different types of phosphorescent coating of airplane instrument dials, eight dials each are coated with the three types. Then the dials are illuminated by an ultraviolet light, and the following are the number of minutes each glowed after the light source was shut off:

Type 1	Type 2	Type 3
52.9	58.4	71.3
62.1	55.0	66.6
57.4	59.8	63.4
50.0	62.5	64.7
59.3	64.7	75.8
61.2	59.9	65.6
60.8	54.7	72.9
53.1	58.4	67.3

Test the null hypothesis that there is no difference in the effectiveness of the three coatings at the 1 percent significance level.

7. The experiment described in Example 12–2 included data on spacing between skulls:

Location	Mean	s.d.	n
San Francisco	38	10	34
Tangiers	44	11	25
Seville	42	9	24

Use an F-test for the null hypothesis of no difference among locations.

8. About 20,000 small holes are drilled in a certain experimental printed-circuit board. A substantial clearance between each hole and a layer of copper plating is desirable to minimize shorts. Six holes are selected from each of five locations on a board and the clearances measured. These are as follows (in thousandths of an inch):

		Location		
A	B	C	D	E
6.9	10.1	5.0	8.8	7.8
6.9	10.1	8.0	8.0	8.8
7.5	7.3	6.4	7.3	8.9
	7.3	6.8	8.5	7.2
	9.1	7.2	7.8	8.1
	9.9	8.0	8.9	7.1

(Three of the holes at location A were ground through accidentally in the measuring process.) It is possible that certain layers of the board have expanded, or perhaps shifted, during the manufacturing process, so that there may be systematic differences in clearance from one location to another. Test the hypothesis of no difference.

Section 12.2

*9. Here are some artificial data—made simple to give practice in the method without numerical complications:

	1	Factor B 1	2	3
	1	3	5	4
Factor A	2	11	10	12
	3	16	21	17

Obtain the ANOVA table and the relevant F-statistics.

10. In a two-factor experiment, one observation is taken under each treatment combination. There are three levels of factor A and two levels of factor B, making $3 \cdot 2 = 6$ treatments in all. The data are as follows:

		Factor A 1	2	3
Factor B	1	14	18	19
	2	16	20	21

(a) Compute the "error sum of squares." Find F_A—what conclusion is drawn from this?

(b) If factor B were not taken into account in the experimental design, the data above could be thought of as comprising three samples of size 2, corresponding to the three categories of factor A:

> Sample 1: 14, 16.
>
> Sample 2: 18, 20.
>
> Sample 3: 19, 21.

Now what conclusion is drawn? Does the two-factor design have any advantage?

*11. Four drivers participate in a test of the hypothesis that "all gasolines are alike," in which four brands are compared as to miles per gallon. Each driver will calculate mpg after each of four trials of 100 miles of normal driving.

(a) How would it be to assign one brand to each driver for all three of his trials? Is there a better way?

(b) Suppose that each driver uses a different brand at each trial, with these results:

| | | Brand | | | |
		A	B	C	D
Driver	1	12.6	10.9	11.3	13.2
	2	16.4	16.1	17.6	17.9
	3	13.0	10.6	11.8	12.6
	4	19.2	18.0	18.9	19.9

Test the hypothesis of no brand difference.

(c) Using the data in part (b), test the hypothesis of no driver difference.

(d) Realizing that "driver" includes the car used as well as the driver's driving habits and driving patterns, can you think of a better experimental design? (Do not worry about its analysis.)

12. A wear-testing machine, with three weighted brushes under which samples of fabrics are fixed, is used to determine resistance to abrasion, measured as loss of weight after a given number of cycles. The data are as follows:

| | | Brush Position | | |
		1	2	3
Fabric	A	1.93	2.38	2.20
	B	2.55	2.72	2.75
	C	2.40	2.68	2.31
	D	2.33	2.40	2.28

Calculate F-statistics for testing fabric differences and for testing for effect of brush position. Is either significant?

*13. Six students in a statistics class measured their pulse rates, and gave information about smoking habits and whether or not they wore corrective lenses:

	No Glasses	Glasses
Smoke Now	78	84
Did Smoke but Quit	64	80
Never Smoked	68	76

(a) Construct an ANOVA table and obtain the F-statistics for testing the null hypothesis of no effect on pulse rate because of smoking and no effect because of wearing glasses.

(b) Does the computation in part (a) have meaning other than as an exercise in calculating F? (If you should happen to ask, "how were the subjects obtained?" we answer that they were taken from a class survey, starting at the top of a list in random order and going down the list until one subject was found for each category.)

(c) Suppose that the six subjects could be thought of as chosen at random from the six populations corresponding to the six cells. And suppose that the F-statistic for eyeglasses had turned out to be significant. Would this mean that wearing glasses causes a high pulse rate?

(d) Suppose that you wanted to know whether smoking causes a high pulse rate. Apart from the question of whether wearing glasses (or any other factor) should be considered, how would you design the experiment?

14. Two different methods were used in determining the fat content of meat, with these results:

Meat Sample	Fat (Method 1)	Fat (Method 2)
1	23.1	22.7
2	23.2	23.6
3	26.5	27.1
4	26.6	27.8
5	27.1	27.4
6	48.3	46.8
7	40.5	40.4
8	23.0	24.9

These data can be analyzed by a two-way ANOVA, with measurement method as one factor and meat sample as another factor.

(a) Compute the F-statistic for testing no difference between methods, and compare with the appropriate 95th percentile of F.

(b) Compute the one-sample t-statistic using the eight differences between measurements in a pair and compare with the appropriate 97.5 percentile of t. Square each of these numbers and compare with the observed and critical values of F in part (a).

APPENDIX A
Glossary of Abbreviations and Symbols

Abbreviations

ANOVA Analysis of variance
MS Mean square (in ANOVA)
m.v.( ) Mean value of the random quantity in parentheses
r.m.s. Root mean square (square root of the average square)
s.d. Standard deviation
s.d.( ) Standard deviation of the random quantity in parentheses
s.e. Standard error (in estimation)
s.e.( ) Standard error of the estimator in parentheses
SS Sum of squares (in ANOVA)

Greek Letters

α (alpha) Significance level of a test (type I error size)
β (beta) Size of the type II error
δ (delta) Difference of population means
μ (mu) Population mean (a subscript would indicate the population variable)
ρ (rho) Population correlation coefficient
σ (sigma) Population standard deviation (a subscript would indicate the population variable)
Σ (sigma) Instruction to sum quantities of the type that follow it
χ (chi) χ^2 is the goodness-of-fit statistic

Latin Letters

f Frequency (of a category or a class interval)

F Ratio of mean squares (in ANOVA)

k Number of categories of a classification or number of distinct values in a frequency distribution

n Sample size

p Population proportion of individuals of a given type, or probability of "success"

P As in P-value: level at which an observed result is just significant

$P(\ )$ Probability of the event in parentheses

q $1 - p$ (probability of "failure")

r Sample correlation coefficient

R_+, R_- Signed-rank statistics

R_x, R_y Rank-sum statistics for comparisons

R Range of values in a set of numbers or sample

S Standard deviation of a set of numbers or of a sample

S_{xy} Covariance of a sample of pairs (X, Y)

T Statistic for a one-sample or two-sample t-test

$x, y, \ldots$ Generic names for specific values of random variables

$X, Y, \ldots$ Generic names for random variables

$\bar{X}$ Mean of a set of numbers or of a sample (from X)

Z Standardized score:

$$\frac{\text{observation} - \text{its mean}}{\text{its s.d.}}$$

APPENDIX B
Tables

Table I. Binomial Probabilities $f(k) = P(k$ successes in n trials)

n	k	.01	.05	.10	.15	1/6	.20	.25	.30	1/3	.35	.40	.45	.50
5	0	.9510	.7738	.5905	.4437	.4019	.3277	.2373	.1681	.1317	.1160	.0778	.0503	.0312
	1	.0480	.2036	.3280	.3915	.4019	.4096	.3955	.3601	.3292	.3124	.2592	.2059	.1562
	2	.0010	.0214	.0729	.1382	.1608	.2048	.2637	.3087	.3292	.3364	.3456	.3369	.3125
	3	.0000	.0011	.0081	.0244	.0322	.0512	.0879	.1323	.1646	.1811	.2304	.2757	.3125
	4	.0000	.0000	.0004	.0022	.0032	.0064	.0146	.0283	.0412	.0488	.0768	.1128	.1562
	5	.0000	.0000	.0000	.0001	.0001	.0003	.0010	.0024	.0041	.0053	.0102	.0185	.0312
6	0	.9415	.7351	.5314	.3771	.3349	.2621	.1780	.1176	.0878	.0754	.0467	.0277	.0156
	1	.0571	.2321	.3543	.3993	.4019	.3932	.3560	.3025	.2634	.2437	.1866	.1359	.0938
	2	.0014	.0305	.0984	.1762	.2009	.2458	.2966	.3241	.3293	.3280	.3110	.2780	.2344
	3	.0000	.0021	.0146	.0415	.0536	.0819	.1318	.1852	.2195	.2355	.2765	.3032	.3125
	4	.0000	.0001	.0012	.0055	.0080	.0154	.0330	.0595	.0823	.0951	.1382	.1861	.2344
	5	.0000	.0000	.0001	.0004	.0006	.0015	.0044	.0102	.0165	.0205	.0369	.0609	.0938
	6	.0000	.0000	.0000	.0000	.0000	.0001	.0002	.0007	.0014	.0018	.0041	.0083	.0156
7	0	.9321	.6983	.4783	.3206	.2791	.2097	.1335	.0824	.0585	.0490	.0280	.0152	.0078
	1	.0659	.2573	.3720	.3960	.3907	.3670	.3115	.2471	.2049	.1848	.1306	.0872	.0547
	2	.0020	.0406	.1240	.2097	.2344	.2753	.3115	.3177	.3073	.2985	.2613	.2140	.1641
	3	.0000	.0036	.0230	.0617	.0781	.1147	.1730	.2269	.2561	.2679	.2903	.2918	.2734
	4	.0000	.0002	.0026	.0109	.0156	.0287	.0577	.0972	.1280	.1442	.1935	.2388	.2734
	5	.0000	.0000	.0002	.0012	.0018	.0043	.0115	.0250	.0384	.0466	.0774	.1172	.1641
	6	.0000	.0000	.0000	.0001	.0001	.0004	.0013	.0036	.0064	.0084	.0172	.0320	.0547
	7	.0000	.0000	.0000	.0000	.0000	.0000	.0001	.0002	.0005	.0006	.0016	.0037	.0078
8	0	.9227	.6634	.4305	.2725	.2326	.1678	.1001	.0576	.0390	.0319	.0168	.0084	.0039
	1	.0746	.2793	.3826	.3847	.3721	.3355	.2670	.1977	.1561	.1373	.0896	.0548	.0312
	2	.0026	.0515	.1488	.2376	.2605	.2936	.3115	.2965	.2731	.2587	.2090	.1569	.1094
	3	.0001	.0054	.0331	.0839	.1042	.1468	.2076	.2541	.2731	.2786	.2787	.2568	.2187
	4	.0000	.0004	.0046	.0185	.0260	.0459	.0865	.1361	.1707	.1875	.2322	.2627	.2734
	5	.0000	.0000	.0004	.0026	.0042	.0092	.0231	.0467	.0683	.0808	.1239	.1719	.2187
	6	.0000	.0000	.0000	.0002	.0004	.0011	.0038	.0100	.0171	.0217	.0413	.0703	.1094
	7	.0000	.0000	.0000	.0000	.0000	.0001	.0004	.0012	.0024	.0033	.0079	.0164	.0312
	8	.0000	.0000	.0000	.0000	.0000	.0000	.0000	.0001	.0002	.0002	.0007	.0017	.0039
9	0	.9135	.6302	.3874	.2316	.1938	.1342	.0751	.0404	.0260	.0207	.0101	.0046	.0020
	1	.0830	.2985	.3874	.3679	.3489	.3020	.2253	.1556	.1171	.1004	.0605	.0339	.0176
	2	.0034	.0629	.1722	.2597	.2791	.3020	.3003	.2668	.2341	.2162	.1612	.1110	.0703
	3	.0001	.0077	.0446	.1069	.1302	.1762	.2336	.2668	.2731	.2716	.2508	.2119	.1641
	4	.0000	.0006	.0074	.0283	.0391	.0661	.1168	.1715	.2048	.2194	.2508	.2600	.2461
	5	.0000	.0000	.0008	.0050	.0078	.0165	.0389	.0735	.1024	.1181	.1672	.2128	.2461
	6	.0000	.0000	.0001	.0006	.0013	.0028	.0087	.0210	.0341	.0424	.0743	.1160	.1641
	7	.0000	.0000	.0000	.0000	.0001	.0003	.0012	.0039	.0073	.0098	.0212	.0407	.0703
	8	.0000	.0000	.0000	.0000	.0000	.0000	.0001	.0004	.0009	.0013	.0035	.0083	.0176
	9	.0000	.0000	.0000	.0000	.0000	.0000	.0000	.0000	.0001	.0001	.0003	.0008	.0020
10	0	.9044	.5987	.3487	.1969	.1615	.1074	.0563	.0282	.0173	.0135	.0060	.0025	.0010
	1	.0914	.3151	.3874	.3474	.3230	.2684	.1877	.1211	.0867	.0725	.0403	.0207	.0098
	2	.0042	.0746	.1937	.2759	.2907	.3020	.2816	.2335	.1951	.1757	.1209	.0763	.0439
	3	.0001	.0105	.0574	.1298	.1550	.2013	.2503	.2668	.2601	.2522	.2150	.1665	.1172
	4	.0000	.0010	.0112	.0401	.0543	.0881	.1460	.2001	.2276	.2377	.2508	.2384	.2051
	5	.0000	.0001	.0015	.0085	.0130	.0264	.0584	.1029	.1366	.1536	.2007	.2340	.2461
	6	.0000	.0000	.0001	.0012	.0022	.0055	.0162	.0368	.0569	.0689	.1115	.1596	.2051
	7	.0000	.0000	.0000	.0001	.0002	.0008	.0031	.0090	.0163	.0212	.0425	.0746	.1172
	8	.0000	.0000	.0000	.0000	.0000	.0001	.0004	.0014	.0030	.0043	.0106	.0229	.0439
	9	.0000	.0000	.0000	.0000	.0000	.0000	.0000	.0001	.0003	.0005	.0016	.0042	.0098
	10	.0000	.0000	.0000	.0000	.0000	.0000	.0000	.0000	.0000	.0000	.0001	.0003	.0010

Note: For $p > .5$, reverse the roles of p and $q = 1 - p$. (For example, the probability of k successes when $p = .7$ is found as the entry for $n - k$ under $p = .3$.)

Table II.   Cumulative Probabilities for the Standard Normal Distribution

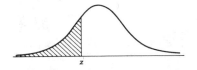

z	0	1	2	3	4	5	6	7	8	9
-3.	.0013	.0010	.0007	.0005	.0003	.0002	.0002	.0001	.0001	.0000
-2.9	.0019	.0018	.0017	.0017	.0016	.0016	.0015	.0015	.0014	.0014
-2.8	.0026	.0025	.0024	.0023	.0023	.0022	.0021	.0021	.0020	.0019
-2.7	.0035	.0034	.0033	.0032	.0031	.0030	.0029	.0028	.0027	.0026
-2.6	.0047	.0045	.0044	.0043	.0041	.0040	.0039	.0038	.0037	.0036
-2.5	.0062	.0060	.0059	.0057	.0055	.0054	.0052	.0051	.0049	.0048
-2.4	.0082	.0080	.0078	.0075	.0073	.0071	.0069	.0068	.0066	.0064
-2.3	.0107	.0104	.0102	.0099	.0096	.0094	.0091	.0089	.0087	.0084
-2.2	.0139	.0136	.0132	.0129	.0126	.0122	.0119	.0116	.0113	.0110
-2.1	.0179	.0174	.0170	.0166	.0162	.0158	.0154	.0150	.0146	.0143
-2.0	.0228	.0222	.0217	.0212	.0207	.0202	.0197	.0192	.0188	.0183
-1.9	.0287	.0281	.0274	.0268	.0262	.0256	.0250	.0244	.0238	.0233
-1.8	.0359	.0352	.0344	.0336	.0329	.0322	.0314	.0307	.0300	.0294
-1.7	.0446	.0436	.0427	.0418	.0409	.0401	.0392	.0384	.0375	.0367
-1.6	.0548	.0537	.0526	.0516	.0505	.0495	.0485	.0475	.0465	.0455
-1.5	.0668	.0655	.0643	.0630	.0618	.0606	.0594	.0582	.0570	.0559
-1.4	.0808	.0793	.0778	.0764	.0749	.0735	.0722	.0708	.0694	.0681
-1.3	.0968	.0951	.0934	.0918	.0901	.0885	.0869	.0853	.0838	.0823
-1.2	.1151	.1131	.1112	.1093	.1075	.1056	.1038	.1020	.1003	.0985
-1.1	.1357	.1335	.1314	.1292	.1271	.1251	.1230	.1210	.1190	.1170
-1.0	.1587	.1562	.1539	.1515	.1492	.1469	.1446	.1423	.1401	.1379
- .9	.1841	.1814	.1788	.1762	.1736	.1711	.1685	.1660	.1635	.1611
- .8	.2119	.2090	.2061	.2033	.2005	.1977	.1949	.1922	.1894	.1867
- .7	.2420	.2389	.2358	.2327	.2297	.2266	.2236	.2206	.2177	.2148
- .6	.2743	.2709	.2676	.2643	.2611	.2578	.2546	.2514	.2483	.2451
- .5	.3085	.3050	.3015	.2981	.2946	.2912	.2877	.2843	.2810	.2776
- .4	.3446	.3409	.3372	.3336	.3300	.3264	.3228	.3192	.3156	.3121
- .3	.3821	.3783	.3745	.3707	.3669	.3632	.3594	.3557	.3520	.3483
- .2	.4207	.4168	.4129	.4090	.4052	.4013	.3974	.3936	.3897	.3859
- .1	.4602	.4562	.4522	.4483	.4443	.4404	.4364	.4325	.4286	.4247
- .0	.5000	.4960	.4920	.4880	.4840	.4801	.4761	.4721	.4681	.4641

Table II (cont.)

z	0	1	2	3	4	5	6	7	8	9
.0	.5000	.5040	.5080	.5120	.5160	.5199	.5239	.5279	.5319	.5359
.1	.5398	.5438	.5478	.5517	.5557	.5596	.5636	.5675	.5714	.5753
.2	.5793	.5832	.5871	.5910	.5948	.5987	.6026	.6064	.6103	.6141
.3	.6179	.6217	.6255	.6293	.6331	.6368	.6406	.6443	.6480	.6517
.4	.6554	.6591	.6628	.6664	.6700	.6736	.6772	.6808	.6844	.6879
.5	.6915	.6950	.6985	.7019	.7054	.7088	.7123	.7157	.7190	.7224
.6	.7257	.7291	.7324	.7357	.7389	.7422	.7454	.7486	.7517	.7549
.7	.7580	.7611	.7642	.7673	.7703	.7734	.7764	.7794	.7823	.7852
.8	.7881	.7910	.7939	.7967	.7995	.8023	.8051	.8078	.8106	.8133
.9	.8159	.8186	.8212	.8238	.8264	.8289	.8315	.8340	.8365	.8389
1.0	.8413	.8438	.8461	.8485	.8508	.8531	.8554	.8577	.8599	.8621
1.1	.8643	.8665	.8686	.8708	.8729	.8749	.8770	.8790	.8810	.8830
1.2	.8849	.8869	.8888	.8907	.8925	.8944	.8962	.8980	.8997	.9015
1.3	.9032	.9049	.9066	.9082	.9099	.9115	.9131	.9147	.9162	.9177
1.4	.9192	.9207	.9222	.9236	.9251	.9265	.9278	.9292	.9306	.9319
1.5	.9332	.9345	.9357	.9370	.9382	.9394	.9406	.9418	.9430	.9441
1.6	.9452	.9463	.9474	.9484	.9495	.9505	.9515	.9525	.9535	.9545
1.7	.9554	.9564	.9573	.9582	.9591	.9599	.9608	.9616	.9625	.9633
1.8	.9641	.9648	.9656	.9664	.9671	.9678	.9686	.9693	.9700	.9706
1.9	.9713	.9719	.9726	.9732	.9738	.9744	.9750	.9756	.9762	.9767
2.0	.9772	.9778	.9783	.9788	.9793	.9798	.9803	.9808	.9812	.9817
2.1	.9821	.9826	.9830	.9834	.9838	.9842	.9846	.9850	.9854	.9857
2.2	.9861	.9864	.9868	.9871	.9874	.9878	.9881	.9884	.9887	.9890
2.3	.9893	.9896	.9898	.9901	.9904	.9906	.9909	.9911	.9913	.9916
2.4	.9918	.9920	.9922	.9925	.9927	.9929	.9931	.9932	.9934	.9936
2.5	.9938	.9940	.9941	.9943	.9945	.9946	.9948	.9949	.9951	.9952
2.6	.9953	.9955	.9956	.9957	.9959	.9960	.9961	.9962	.9963	.9964
2.7	.9965	.9966	.9967	.9968	.9969	.9970	.9971	.9972	.9973	.9974
2.8	.9974	.9975	.9976	.9977	.9977	.9978	.9979	.9979	.9980	.9981
2.9	.9981	.9982	.9982	.9983	.9984	.9984	.9985	.9985	.9986	.9986
3.	.9987	.9990	.9993	.9995	.9997	.9998	.9998	.9999	.9999	1.0000

Notes: 1. Enter table at Z, read out $P(Z \leqslant z)$, the shaded area.
2. For a general normal X, enter table at $z = (x - \mu/\sigma$ to read $P(X \leqslant x)$.
3. Entries opposite 3 are for 3.0, 3.1, 3.2, ... , 3.9.
4. For $z \geqslant 4$, probabilities are 1.0000; for $z \leqslant -4$, probabilities are 0.0000.

Table IIa. Percentiles of the Standard Normal Distribution

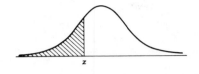

$P(Z \leqslant z)$	z
.001	-3.0902
.005	-2.5758
.01	-2.3263
.02	-2.0537
.03	-1.8808
.04	-1.7507
.05	-1.6449
.10	-1.2816
.15	-1.0364
.20	-.8416
.25	-.6745
.30	-.5244
.35	-.3853
.40	-.2533
.45	-.1257
.50	0
.55	.1257
.60	.2533
.65	.3853
.70	.5244
.75	.6745
.80	.8416
.85	1.0364
.90	1.2816
.95	1.6449
.96	1.7507
.97	1.8808
.98	2.0537
.99	2.3263
.995	2.5758
.999	3.0902

Table IIb. Two-Tailed Probabilities for the Standard Normal Distribution

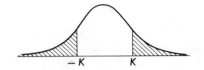

$P(|Z| > K)$	K
.001	3.2905
.002	3.0902
.005	2.8070
.01	2.5758
.02	2.3263
.03	2.1701
.04	2.0537
.05	1.9600
.06	1.8808
.08	1.7507
.10	1.6449
.15	1.4395
.20	1.2816
.30	1.0364

Table IIc. Multipliers of the Standard Error of the Mean for Confidence Interval Construction (Large n)

Confidence Coefficient	Multiplier
.68	1.000
.80	1.282
.90	1.645
.95	1.960
.99	2.576

Table III. Percentiles of the *t*-Distribution

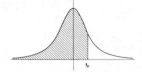

Degrees of Freedom	p								
	.60	.70	.80	.85	.90	.95	.975	.99	.995
1	.325	.727	1.38	1.96	3.08	6.31	12.7	31.8	63.7
2	.289	.617	1.06	1.39	1.89	2.92	4.30	6.96	9.92
3	.277	.584	.978	1.25	1.64	2.35	3.18	4.54	5.84
4	.271	.569	.941	1.19	1.53	2.13	2.78	3.75	4.60
5	.267	.559	.920	1.16	1.48	2.01	2.57	3.36	4.03
6	.265	.553	.906	1.13	1.44	1.94	2.45	3.14	3.71
7	.263	.549	.896	1.12	1.42	1.90	2.36	3.00	3.50
8	.262	.546	.889	1.11	1.40	1.86	2.31	2.90	3.36
9	.261	.543	.883	1.10	1.38	1.83	2.26	2.82	3.25
10	.260	.542	.879	1.09	1.37	1.81	2.23	2.76	3.17
11	.260	.540	.876	1.09	1.36	1.80	2.20	2.72	3.11
12	.259	.539	.873	1.08	1.36	1.78	2.18	2.68	3.06
13	.259	.538	.870	1.08	1.35	1.77	2.16	2.65	3.01
14	.258	.537	.868	1.08	1.34	1.76	2.14	2.62	2.98
15	.258	.536	.866	1.07	1.34	1.75	2.13	2.60	2.95
16	.258	.535	.865	1.07	1.34	1.75	2.12	2.58	2.92
17	.257	.534	.863	1.07	1.33	1.74	2.11	2.57	2.90
18	.257	.534	.862	1.07	1.33	1.73	2.10	2.55	2.88
19	.257	.533	.861	1.07	1.33	1.73	2.09	2.54	2.86
20	.257	.533	.860	1.07	1.32	1.72	2.09	2.53	2.84
21	.257	.532	.859	1.06	1.32	1.72	2.08	2.52	2.83
22	.256	.532	.858	1.06	1.32	1.72	2.07	2.51	2.82
23	.256	.532	.858	1.06	1.32	1.71	2.07	2.50	2.81
24	.256	.531	.857	1.06	1.32	1.71	2.06	2.49	2.80
25	.256	.531	.856	1.06	1.32	1.71	2.06	2.48	2.79
26	.256	.531	.856	1.06	1.32	1.71	2.06	2.48	2.78
27	.256	.531	.855	1.06	1.31	1.70	2.05	2.47	2.77
28	.256	.530	.855	1.06	1.31	1.70	2.05	2.47	2.76
29	.256	.530	.854	1.06	1.31	1.70	2.04	2.46	2.76
30	.256	.530	.854	1.05	1.31	1.70	2.04	2.46	2.75
40	.255	.529	.851	1.05	1.30	1.68	2.02	2.42	2.70
∞	.253	.524	.842	1.04	1.28	1.64	1.96	2.33	2.58

Notes: 1. The area to the right of the table entry is $1 - p$.
2. The distribution is symmetric. For example, for 10 degrees of freedom, $P(-1.37 < t < 1.37) = .9 - .1 = .8$.

Table IV. Tail Probabilities for the Wilcoxon Signed-Rank Statistic

c	n=4	5	6	7	8	9	10	11	12	13	14	15	c
0	.062	.031	.016	.008									0
1	.125	.062	.031	.016	.008								1
2	.188	.094	.047	.023	.012	.006							2
3	.312	.156	.078	.039	.020	.010							3
4		.219	.109	.055	.027	.014	.007						4
5			.156	.078	.039	.020	.010						5
6			.219	.109	.055	.027	.014						6
7				.148	.074	.037	.019	.009					7
8				.188	.098	.049	.024	.012					8
9				.234	.125	.064	.032	.016	.008				9
10					.156	.082	.042	.021	.010				10
11					.191	.102	.053	.027	.013				11
12					.230	.125	.065	.034	.017	.009			12
13						.150	.080	.042	.021	.011			13
14						.180	.097	.051	.026	.013			14
15						.213	.116	.062	.032	.016	.008		15
16							.138	.074	.039	.020	.010		16
17							.161	.087	.046	.024	.012		17
18							.188	.103	.055	.029	.015		18
19							.216	.120	.065	.034	.018	.009	19
20								.139	.076	.040	.021	.011	20
21								.160	.088	.047	.025	.013	21
22								.183	.102	.055	.029	.015	22
23								.207	.117	.064	.034	.018	23
24									.113	.073	.039	.021	24
25									.151	.084	.045	.024	25
26									.170	.095	.052	.028	26
27									.190	.108	.059	.032	27
28									.212	.122	.068	.036	28
29										.137	.077	.042	29
30										.153	.086	.047	30
31										.170	.097	.053	31
32										.188	.108	.060	32
33										.207	.121	.068	33
34											.134	.076	34
35											.148	.084	35
36											.163	.094	36
37											.179	.104	37
38											.196	.115	38
$\frac{n(n+1)}{2}$	10	15	21	28	36	45	55	66	78	91	105	120	

Notes: 1. Table entries are $P(R_+ \leq c) = P[R_+ \geq n(n+1)/2 - c]$.
2. $R_- = n(n+1)/2 - R_+$.
3. For $n > 15$, use a normal approximation (Section 8.7).

Table V. Percentiles of the Chi-Square Distribution

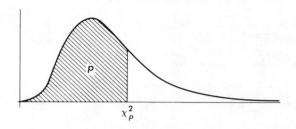

$$\chi^2_p$$

Degrees of Freedom	p												
	.01	.025	.05	.10	.20	.30	.50	.70	.80	.90	.95	.975	.99
1	.000	.001	.004	.016	.064	.148	.455	1.07	1.64	2.71	3.84	5.02	6.63
2	.020	.051	.103	.211	.446	.713	1.39	2.41	3.22	4.61	5.99	7.38	9.21
3	.115	.216	.352	.584	1.01	1.42	2.37	3.66	4.64	6.25	7.81	9.35	11.3
4	.297	.484	.711	1.06	1.65	2.19	3.36	4.88	5.99	7.78	9.49	11.1	13.3
5	.554	.831	1.15	1.61	2.34	3.00	4.35	6.06	7.29	9.24	11.1	12.8	15.1
6	.872	1.24	1.64	2.20	3.07	3.83	5.35	7.23	8.56	10.6	12.6	14.4	16.8
7	1.24	1.69	2.17	2.83	3.82	4.67	6.35	8.38	9.80	12.0	14.1	16.0	18.5
8	1.65	2.18	2.73	3.49	4.59	5.53	7.34	9.52	11.0	13.4	15.5	17.5	20.1
9	2.09	2.70	3.33	4.17	5.38	6.39	8.34	10.7	12.2	14.7	16.9	19.0	21.7
10	2.56	3.25	3.94	4.87	6.18	7.27	9.34	11.8	13.4	16.0	18.3	20.5	23.2
11	3.05	3.82	4.57	5.58	6.99	8.15	10.3	12.9	14.6	17.3	19.7	21.9	24.7
12	3.57	4.40	5.23	6.30	7.81	9.03	11.3	14.0	15.8	18.5	21.0	23.3	26.2
13	4.11	5.01	5.89	7.04	8.63	9.93	12.3	15.1	17.0	19.8	22.4	24.7	27.7
14	4.66	5.63	6.57	7.79	9.47	10.8	13.3	16.2	18.2	21.1	23.7	26.1	29.1
15	5.23	6.26	7.26	8.55	10.3	11.7	14.3	17.3	19.3	22.3	25.0	27.5	30.6
16	5.81	6.91	7.96	9.31	11.2	12.6	15.3	18.4	20.5	23.5	26.3	28.8	32.0
17	6.41	7.56	8.67	10.1	12.0	13.5	16.3	19.5	21.6	24.8	27.6	30.2	33.4
18	7.01	8.23	9.39	10.9	12.9	14.4	17.3	20.6	22.8	26.0	28.9	31.5	34.8
19	7.63	8.91	10.1	11.7	13.7	15.4	18.3	21.7	23.9	27.2	30.1	32.9	36.2
20	8.26	9.59	10.9	12.4	14.6	16.3	19.3	22.8	25.0	28.4	31.4	34.o	37.6
21	8.90	10.3	11.6	13.2	15.4	17.2	20.3	23.9	26.2	29.6	32.7	35.5	38.9
22	9.54	11.0	12.3	14.0	16.3	18.1	21.3	24.9	27.3	30.8	33.9	36.8	40.3
23	10.2	11.7	13.1	14.8	17.2	19.0	22.3	26.0	28.4	32.0	35.2	38.1	41.6
24	10.9	12.4	13.8	15.7	18.1	19.9	23.3	27.1	29.6	33.2	36.4	39.4	43.0
25	11.5	13.1	14.6	16.5	18.9	20.9	24.3	28.2	30.7	34.4	37.7	40.6	44.3
26	12.2	13.8	15.4	17.3	19.8	21.8	25.3	29.2	31.8	35.6	38.9	41.9	45.6
27	12.9	14.6	16.2	18.1	20.7	22.7	26.3	30.3	32.9	36.7	40.1	43.2	47.0
28	13.6	15.3	16.9	18.9	21.6	23.6	27.3	31.4	34.0	37.9	41.3	44.5	48.3
29	14.3	16.0	17.7	19.8	22.5	24.6	28.3	32.5	35.1	39.1	42.6	45.7	49.6
30	15.0	16.8	18.5	20.6	23.4	25.5	29.3	33.5	36.2	40.3	43.8	47.0	50.9
40	22.1	24.4	26.5	29.0	32.4	35.0	39.3	44.2	47.3	51.8	55.8	59.3	63.7
50	29.7	32.3	34.8	37.7	41.5	44.4	49.3	54.7	58.2	63.2	67.5	71.4	76.2
60	37.5	40.5	43.2	46.5	50.7	53.9	59.3	65.2	69.0	74.4	79.1	83.3	88.4

Note: For degrees of freedom $k > 30$, use $\chi^2_p = \frac{1}{2}(z_p + \sqrt{2k - 1})^2$, where z_p is the corresponding percentile of the standard normal distribution.

Table VI. Tail Probabilities for the Rank-Sum Statistic

m	4	4	4	4	4	4	4
n	4	5	6	7	8	9	10
c							
10	.014	.008	.005				
11	.029	.016	.010	.006	.004		
12	.057	.032	.019	.012	.008	.006	
13	.100	.056	.033	.021	.014	.010	.007
14	.171	.095	.057	.036	.024	.017	.012
15	.243	.143	.086	.055	.036	.025	.018
16		.206	.129	.082	.055	.038	.027
17			.176	.115	.077	.053	.038
18				.158	.107	.074	.053
19				.206	.141	.099	.071
20					.184	.130	.094
21						.165	.120
M	36	40	44	48	52	56	60

m	5	5	5	5	5	5
n	5	6	7	8	9	10
c						
15	.004					
16	.008	.004				
17	.016	.009	.005			
18	.028	.015	.009			
19	.048	.026	.015	.009		
20	.075	.041	.024	.015	.009	
21	.111	.063	.037	.023	.014	.010
22	.155	.089	.053	.033	.021	.014
23	.210	.123	.074	.047	.030	.020
24		.165	.101	.064	.041	.028
25		.214	.134	.085	.056	.038
26			.172	.111	.073	.050
27			.216	.142	.095	.065
28				.177	.120	.082
29					.149	.103
M	55	60	65	70	75	80

Notes: 1. m is the size of the smaller sample.

2. The entry opposite c is the cumulative tail probability:

$$P(R \leqslant c) = P(R \geqslant M - c),$$

where R is the rank sum for the smaller sample, and M is the sum of the minimum and maximum values of R. [Note that $M/2 = $ m.v.(R).]

3. For m or $n > 10$, use a normal approximation (see Section 9.4).

Table VI *(cont.)*

m	6	6	6	6	6		m	7	7	7	7
n	6	7	8	9	10		n	7	8	9	10
c							c				
24	.008						34	.009			
25	.013	.007					35	.013	.007		
26	.021	.011	.006				36	.019	.010		
27	.032	.017	.010	.006			37	.027	.014	.008	
28	.047	.026	.015	.009			38	.036	.020	.011	
29	.066	.037	.021	.013	.008		39	.049	.027	.016	.009
30	.090	.051	.030	.018	.011		40	.064	.036.	.001	.012
31	.120	.069	.041	.025	.016		41	.082	.047	.027	.017
32	.155	.090	.054	.033	.021		42	.104	.060	.036	.022
33	.197	.117	.071	.044	.028		43	.130	.076	.045	.028
34		.147	.091	.057	.036		44	.159	.095	.057	.035
35		.183	.114	.072	.047		45	.191	.116	.071	.044
36			.141	.091	.059		46		.140	.087	.054
37			.172	.112	.074		47		.168	.105	.067
38				.136	.090		48		.198	.126	.081
39				.164	.110		49			.150	.097
							50			.176	.115
M	78	84	90	96	102		M	105	112	119	126

m	8	8	8		m	9	9		m	10
n	8	9	10		n	9	10		n	10
c					c				c	
46	.010				59	.009			74	.009
47	.014	.008			60	.012			75	.012
48	.019	.010			61	.016	.009		76	.014
49	.025	.014	.008		62	.020	.011		77	.018
50	.032	.018	.010		63	.025	.014		78	.022
51	.041	.023	.013		64	.031	.017		79	.026
52	.052	.030	.017		65	.039	.022		80	.032
53	.065	.037	.022		66	.047	.027		81	.038
54	.080	.046	.027		67	.057	.033		82	.045
55	.097	.057	.034		68	.068	.039		83	.053
56	.117	.069	.042		69	.081	.047		84	.062
57	.139	.084	.051		70	.095	.056		85	.072
58	.164	.100	.061		71	.111	.067		86	.083
59	.191	.118	.073		72	.129	.078		87	.095
60		.138	.086		73	.149	.091		88	.109
61		.161	.102		74	.170	.106		89	.124
M	136	144	152		M	171	180		M	210

Table VIIa. Ninety-fifth Percentiles of the F-Distribution

						Numerator Degrees of Freedom								
	1	**2**	**3**	**4**	**5**	**6**	**8**	**10**	**12**	**15**	**20**	**24**	**30**
1	161	200	216	225	230	234	239	242	244	246	248	249	250
2	18.5	19.0	19.2	19.2	19.3	19.3	19.4	19.4	19.4	19.4	19.4	19.5	19.5
3	10.1	9.55	9.28	9.12	9.01	8.94	8.85	8.79	8.74	8.70	8.66	8.64	8.62
4	7.71	6.94	6.59	6.39	6.26	6.16	6.04	5.96	5.91	5.86	5.80	5.77	5.75
5	6.61	5.79	5.41	5.19	5.05	4.95	4.82	4.74	4.68	4.62	4.56	4.53	4.50
6	5.99	5.14	4.76	4.53	4.39	4.28	4.15	4.06	4.00	3.94	3.87	3.84	3.81
7	5.59	4.74	4.35	4.12	3.97	3.87	3.73	3.64	3.57	3.51	3.44	3.41	3.38
8	5.32	4.46	4.07	3.84	3.69	3.58	3.44	3.35	3.28	3.22	3.15	3.12	3.08
9	5.12	4.26	3.86	3.63	3.48	3.37	3.23	3.14	3.07	3.01	2.94	2.90	2.86
10	4.96	4.10	3.71	3.48	3.33	3.22	3.07	2.98	2.91	2.85	2.77	2.74	2.70
11	4.84	3.98	3.59	3.36	3.20	3.09	2.95	2.85	2.79	2.72	2.65	2.61	2.57
12	4.75	3.89	3.49	3.26	3.11	3.00	2.85	2.75	2.69	2.62	2.54	2.51	2.47
13	4.67	3.81	3.41	3.18	3.03	2.92	2.77	2.67	2.60	2.53	2.46	2.42	2.38
14	4.60	3.74	3.34	3.11	2.96	2.85	2.70	2.60	2.53	2.46	2.39	2.35	2.31
15	4.54	3.68	3.29	3.06	2.90	2.79	2.64	2.54	2.48	2.40	2.33	2.29	2.25
16	4.49	3.63	3.24	3.01	2.85	2.74	2.59	2.49	2.42	2.35	2.28	2.24	2.19
17	4.45	3.59	3.20	2.96	2.81	2.70	2.55	2.45	2.38	2.31	2.23	2.19	2.15
18	4.41	3.55	3.16	2.93	2.77	2.66	2.51	2.41	2.34	2.27	2.19	2.15	2.11
19	4.38	3.52	3.13	2.90	2.74	2.63	2.48	2.38	2.31	2.23	2.16	2.11	2.07
20	4.35	3.49	3.10	2.87	2.71	2.60	2.45	2.35	2.28	2.20	2.12	2.08	2.04
21	4.32	3.47	3.07	2.84	2.68	2.57	2.42	2.32	2.25	2.18	2.10	2.05	2.01
22	4.30	3.44	3.05	2.82	2.66	2.55	2.40	2.30	2.23	2.15	2.07	2.03	1.98
23	4.28	3.42	3.03	2.80	2.64	2.53	2.37	2.27	2.20	2.13	2.05	2.01	1.96
24	4.26	3.40	3.01	2.78	2.62	2.51	2.36	2.25	2.18	2.11	2.03	1.98	1.94
25	4.24	3.39	2.99	2.76	2.60	2.49	2.34	2.24	2.16	2.09	2.01	1.96	1.92
30	4.17	3.32	2.92	2.69	2.53	2.42	2.27	2.16	2.09	2.01	1.93	1.89	1.84
40	4.08	3.23	2.84	2.61	2.45	2.34	2.18	2.08	2.00	1.92	1.84	1.79	1.74
60	4.00	3.15	2.76	2.53	2.37	2.25	2.10	1.99	1.92	1.84	1.75	1.70	1.65

Denominator Degrees of Freedom

Note: The fifth percentiles are obtainable as follows:

$$F_{.05}(r, s) = \frac{1}{F_{.95}(s, r)}.$$

[For example, $F_{.05}(3, 6) = 1/F_{.95}(6, 3) = 1/8.94$.]

Table VIIb. Ninety-ninth Percentiles of the F-Distribution

		Numerator Degrees of Freedom												
		1	2	3	4	5	6	8	10	12	15	20	24	30
Denominator Degrees of Freedom	1	4050	5000	5400	5620	5760	5860	5980	6060	6110	6160	6210	6235	6260
	2	98.5	99.0	99.2	99.2	99.3	99.3	99.4	99.4	99.4	99.4	99.4	99.5	99.5
	3	34.1	30.8	29.5	28.7	28.2	27.9	27.5	27.3	27.1	26.9	26.7	26.6	26.5
	4	21.2	18.0	16.7	16.0	15.5	15.2	14.8	14.5	14.4	14.2	14.0	13.9	13.8
	5	16.3	13.3	12.1	11.4	11.0	10.7	10.3	10.1	9.89	9.72	9.55	9.47	9.38
	6	13.7	10.9	9.78	9.15	8.75	8.47	8.10	7.87	7.72	7.56	7.40	7.31	7.23
	7	12.2	9.55	8.45	7.85	7.46	7.19	6.84	6.62	6.47	6.31	6.16	6.07	5.99
	8	11.3	8.65	7.59	7.01	6.63	6.37	6.03	5.81	5.67	5.52	5.36	5.28	5.20
	9	10.6	8.02	6.99	6.42	6.06	5.80	5.47	5.26	5.11	4.96	4.81	4.73	4.65
	10	10.0	7.56	6.55	5.99	5.64	5.39	5.06	4.85	4.71	4.56	4.41	4.33	4.25
	11	9.65	7.21	6.22	5.67	5.32	5.07	4.74	4.54	4.40	4.25	4.10	4.02	3.94
	12	9.33	6.93	5.95	5.41	5.06	4.82	4.50	4.30	4.16	4.01	3.86	3.78	3.70
	13	9.07	6.70	5.74	5.21	4.86	4.62	4.30	4.10	3.96	3.82	3.66	3.59	3.51
	14	8.86	6.51	5.56	5.04	4.69	4.46	4.14	3.94	3.80	3.66	3.51	3.43	3.35
	15	8.68	6.36	5.42	4.89	4.56	4.32	4.00	3.80	3.67	3.52	3.37	3.29	3.21
	16	8.53	6.23	5.29	4.77	4.44	4.20	3.89	3.69	3.55	3.41	3.26	3.18	3.10
	17	8.40	6.11	5.18	4.67	4.34	4.10	3.79	3.59	3.46	3.31	3.16	3.08	3.00
	18	8.29	6.01	5.09	4.58	4.25	4.01	3.71	3.51	3.37	3.23	3.08	3.00	2.92
	19	8.18	5.93	5.01	4.50	4.17	3.94	3.63	3.43	3.30	3.15	3.00	2.92	2.84
	20	8.10	5.85	4.94	4.43	4.10	3.87	3.56	3.37	3.23	3.09	2.94	2.86	2.78
	21	8.02	5.78	4.87	4.37	4.04	3.81	3.51	3.31	3.17	3.03	2.88	2.80	2.72
	22	7.95	5.72	4.82	4.31	3.99	3.76	3.45	3.26	3.12	2.98	2.83	2.75	2.67
	23	7.88	5.66	4.76	4.26	3.94	3.71	3.41	3.21	3.07	2.93	2.78	2.70	2.62
	24	7.82	5.61	4.72	4.22	3.90	3.67	3.36	3.17	3.03	2.89	2.74	2.66	2.58
	25	7.77	5.57	4.68	4.18	3.86	3.63	3.32	3.13	2.99	2.85	2.70	2.62	2.54
	30	7.56	5.39	4.51	4.02	3.70	3.47	3.17	2.98	2.84	2.70	2.55	2.47	2.39
	40	7.31	5.18	4.31	3.83	3.51	3.29	2.99	2.80	2.66	2.52	2.37	2.29	2.20
	60	7.08	4.98	4.13	3.65	3.34	3.12	2.82	2.63	2.50	2.35	2.20	2.12	2.03

Table VIII. Random Digits

42916	50199	26435	97117	77100	62919	74498	14252	11052	70038
49019	02101	14580	14421	58592	30885	60248	29783	39125	97534
04421	62261	52644	36493	53146	31906	00208	98915	27613	58180
74606	07765	21788	03093	69158	44498	51540	61267	70550	90599
76288	24031	13826	61989	54283	95614	20378	35853	86644	68259
42866	46273	43621	93636	23582	59351	29828	53006	06004	00427
99017	74447	14581	32223	89571	38437	43037	17654	32705	02726
67245	80759	07378	06307	51311	52458	57898	15213	72105	18792
29317	02377	60654	51918	97109	38972	71750	81431	69776	00892
56457	56692	88071	93055	31559	77054	33921	24189	47537	18470
66908	96815	00106	47915	72072	34460	04085	74036	99640	88672
83966	92418	68500	70046	30009	99166	49224	68804	34733	69265
53196	82252	58476	40657	09612	15380	70717	33052	93954	14642
63291	73919	67613	81329	27561	97499	79346	28385	20829	73829
62205	91166	04127	19669	17699	31072	16918	81168	72908	00561
83502	34546	70327	79999	26659	68085	43541	69983	09041	05677
20293	65765	45954	12799	49028	44691	19957	40928	81503	07030
72932	94622	89404	69024	73518	29828	35482	83798	92363	13918
69803	06247	23872	32055	36776	77634	01444	88377	50827	83716
01155	81380	11691	18090	13236	34313	13390	31223	64796	40116
44290	82296	81987	09423	44272	24414	43248	50536	52161	18884
16980	43552	32970	87214	99340	79058	70912	03514	87351	05102
73249	52463	51467	18602	28336	41484	49543	74121	04575	78007
06050	29975	60715	02040	12974	02831	52032	69726	67679	13772
34216	50564	74588	70102	62585	25511	38134	13802	98334	76947
04607	52269	21767	98347	69224	44987	31255	00344	60841	53970
92738	66714	58465	83216	95109	31032	99817	18844	31514	44004
76152	98002	84257	47518	53932	46337	96349	17004	81135	26247
24405	52117	41434	82281	02756	40000	26893	71507	55783	78195
99046	61444	59911	58255	45299	60971	72833	61883	52645	60945
26312	73154	21070	90104	42013	27302	55283	13166	14051	81929
36315	59502	91215	86654	44578	04159	63389	43516	48971	40922
52467	19775	71391	63601	84377	63350	59557	74397	06289	74426
66790	72193	63999	20307	47423	55164	93870	43783	06851	90065
16427	71681	64661	59249	74118	46257	69308	31035	64498	19592
63988	01319	15012	95770	82029	99778	81793	73836	11528	81863
67468	22553	71756	30281	28244	58696	72161	46240	63452	56485
60477	14463	49722	95808	73193	37865	84147	46004	43753	92444
95384	28822	12047	59393	14588	22723	64262	93653	00284	05594
51396	45671	08283	96848	27039	20852	38008	65531	65322	51775
70321	26394	01403	77390	52111	27816	33570	28064	41906	81867
98710	50639	43559	34442	25514	32178	83688	31018	11232	70459
61664	16238	04228	33224	18550	02255	34597	64773	97872	28450
12906	19628	77265	38578	00958	67476	92199	70519	32591	80452
07633	02489	78236	70986	74294	29591	31175	20817	64727	70957
35933	31203	16796	66581	55006	90733	07198	65126	54346	42214
57652	46065	59420	33920	44589	70899	41795	86683	27317	74817
86860	69306	49382	48964	92022	98252	47414	05190	66648	35104
54447	02332	11406	27021	60064	70307	42155	15810	08324	36194
69865	39302	09057	46982	14177	94534	90536	44442	43337	16371

Table VIII *(cont.)*

40500	21406	00571	87320	81683	42788	86367	44686	22159	67015
35892	49668	83991	72088	30210	74009	86370	97956	02132	93512
54819	26094	51409	21485	94764	85806	13393	48543	07042	76538
64224	47909	09994	23750	17351	52141	30486	60380	86546	66606
36913	58173	45709	83679	82617	23381	09603	61107	00566	06572
64745	10614	86371	43244	97154	10397	50975	68006	20045	16942
25536	74031	31807	70133	78790	40341	68730	39635	39013	66841
44043	96215	21270	59427	25034	40645	84741	52083	54503	36861
27659	95463	53847	40921	70116	61536	56756	08967	31079	20097
76014	99818	16606	19713	66904	27106	24874	96701	73287	76772
06073	57343	51428	91171	28299	17520	64903	04177	36071	94952
59008	28543	11576	74547	13260	20688	41261	02780	06633	37536
08844	95774	49323	30448	14154	83379	71259	23302	68402	43750
88505	15575	44927	06584	29867	21541	65763	12154	86616	79877
73259	68626	98962	68548	86576	48046	51755	64995	03661	64585
81550	46798	49319	50206	22024	05175	12923	23427	55915	91723
55831	83784	81034	86779	34622	84570	18960	48798	42970	95789
39465	82353	68905	44234	18244	54345	05592	89361	14644	67924
66415	89349	88530	72096	44459	05258	48317	48866	56886	90458
75889	04514	37227	11302	04667	02129	80414	86289	15887	87380
50749	83220	50529	20619	11606	36531	23409	78122	19566	76564
33045	66703	30017	35347	35038	12952	13971	03922	98702	11786
38388	69556	76728	60535	59961	23634	42211	98387	34880	27755
93182	99040	96390	65989	38375	03652	59657	57431	24666	11061
64713	85185	72849	58611	31220	26657	77056	24553	24993	05210
89024	32054	46997	92652	28363	98992	22593	97710	47766	37646
93573	95502	33790	92973	27766	62671	89698	10877	73893	41004
96035	18795	48080	59666	30241	35233	87353	43647	13404	41982
19264	29229	61369	08309	39383	42305	25944	13577	51545	68990
69801	37145	79189	55897	57793	66816	21930	56771	79296	73793
21632	42301	23696	72641	56310	85576	03004	25669	69221	32996
23040	65782	23712	13414	10758	15590	97298	74246	51511	46900
36795	38292	03852	06384	84421	03446	91670	45312	27609	87034
06683	83891	88991	16533	09197	34427	60384	48525	90978	46107
21693	12956	21804	46558	37682	81207	85840	53238	35026	04835
53264	41376	17783	64756	39278	25403	33042	20954	31193	24247
45911	92453	25370	86602	48574	57865	26436	16122	76614	17028
21262	59718	77821	14036	31033	90563	45410	15158	90209	84089
38053	60780	54166	14255	33120	27171	71798	91214	80040	56699
12475	40193	59415	04769	75920	01036	02692	75862	16612	73670
61182	03305	90334	00187	91659	28063	75684	50017	82643	09282
77376	85469	08164	05584	36623	82597	83859	03435	98460	70095
80257	04381	06501	08924	35514	14297	54373	71369	05172	15955
82441	04636	48215	06821	03385	17663	40107	55679	30366	42390
95895	16083	58499	17176	55993	51034	49296	04010	78974	35930
02019	96226	27167	68245	53109	59037	37843	79243	10262	58797
61490	82590	52411	54783	29447	94551	30026	97959	93939	73217
82573	62154	78291	33728	39102	11484	86210	43794	73553	87435
01110	77108	56521	78610	08254	01842	43068	70415	79195	26136
49786	47279	38471	20379	54704	86614	91138	51595	50818	80186

Table VIII *(cont.)*

```
95809 54837   55978 10534   46194 00273   03659 57186   73342 05949
76742 70505   64773 48334   00869 80439   69374 35279   99952 85860
03880 30798   40515 66819   40691 72678   17590 76085   62741 93844
81045 58617   37788 64693   50968 31853   95733 08068   21988 60613
54802 96997   52909 14310   08726 09630   49081 66952   05603 08950

63119 68055   76641 87635   64835 06121   86006 76257   51695 44571
85397 39692   66765 50318   33763 45429   30943 20128   14439 68279
06618 14101   87706 77153   54866 00025   34092 92939   12528 24763
19525 93122   11658 06188   43735 43104   18115 28815   21863 22218
49494 16854   95248 00045   40357 73893   50732 11319   38804 15121

51694 84242   33341 77153   71970 66070   80879 10293   70875 61168
01838 72046   40042 59287   78115 74332   36858 17687   14357 58846
32697 10332   92643 06454   39300 20099   61461 15730   29333 95548
75128 04855   96418 26636   53328 69758   17597 56658   81043 40374
15675 20566   45945 10123   21679 38139   95843 76372   78669 14598

78142 72095   47327 43718   48286 88374   65046 27199   50484 03834
72869 28546   22578 94059   56817 88443   65557 75239   38101 17180
13896 62838   09470 31133   65941 84219   23017 34539   77391 52502
95808 40466   39870 79974   71187 02420   23124 15714   91874 86307
52320 60822   38657 81962   32388 50425   53231 62797   95490 87063

03564 46703   30528 28041   86108 99297   31593 21021   12451 90445
89432 52921   28068 71091   12944 06524   42605 02606   69417 81733
88573 55150   01443 97336   79910 49014   02237 85000   32344 45649
32920 01391   01105 15435   10918 92181   03839 92364   84229 83989
52704 39386   81791 35616   97616 64947   60456 16196   79527 43770

02696 86377   34209 38850   43712 58088   58490 42162   16423 79089
83961 43893   81108 79331   27601 30995   25447 05835   26029 01069
21914 31443   85624 29878   97401 66466   88421 76385   65526 93134
60215 43656   42638 13774   87380 32166   44914 57637   95151 08573
17644 23867   35765 75634   22484 86921   29597 94523   39661 15403

69499 38275   93129 99455   06429 10947   62748 09375   53925 65096
96761 14313   79554 48204   56142 39889   98293 07233   25422 43510
42115 20104   10771 93968   76480 82630   09458 50774   00461 17435
96628 79221   70360 47978   68880 91249   42500 92943   89942 94929
98056 88721   38743 37395   23774 29013   39877 56221   08293 74795

62011 34646   99276 37811   30494 51236   30385 22514   77077 68381
97562 82265   27078 02950   45701 53691   27376 17196   53122 40779
81485 72983   95838 93212   68260 21176   33964 32478   98334 81713
53465 74671   88519 84254   65937 56020   45728 52449   17785 75868
14640 29533   35425 50917   85742 38691   31928 68477   70081 31907

58410 98236   76474 16076   17250 91650   83632 54718   16705 22827
26780 38018   96714 32836   11929 56912   71592 29622   71248 49260
74439 29381   62148 13205   68606 03817   35829 21987   40162 35558
51015 49183   15384 28173   53705 96163   72306 10015   63078 95319
28516 56674   30562 96465   17886 00360   11265 05653   51383 85153

79501 69898   55076 54853   66742 70410   44434 15140   88331 75362
35468 52850   17797 78112   42126 33055   99776 40129   70370 27342
29763 07791   20976 69285   32965 95201   96582 71055   16511 13122
76537 32386   84442 97095   31922 39406   56418 76857   51158 43193
74519 89378   58353 83848   02802 06046   74264 76358   08642 31973
```

Table VIII *(cont.)*

94988 12022	77021 60277	39048 03087	18920 98682	26756 05107
72363 40974	09594 10276	09631 43203	13227 90021	35899 21515
74967 66480	83894 82989	24784 42757	24447 44970	60048 26514
26236 32399	81419 47377	93952 89101	79748 03446	23212 10489
05632 68465	67842 85597	02094 42059	86912 23145	43060 94694
67352 41392	17545 30949	87565 83820	19827 34043	00575 23260
92727 35027	03117 80848	74559 96797	08118 72948	91838 00281
18223 91136	39695 39943	77413 48937	32672 94704	99738 50907
80723 91394	02992 11530	67845 05881	76173 18594	91937 74215
75007 85671	88211 55080	15581 02685	07889 26594	47083 76723
60050 80463	30926 74970	38951 14928	81875 61424	62060 96004
69715 28522	73974 99491	50647 20252	44455 66593	23255 94807
92096 43555	48882 60717	07963 39375	21441 66090	80430 68380
35482 63353	08086 66635	71009 95777	70335 65808	69105 42800
24879 78061	38949 21123	28430 72627	04565 14741	85781 20795
68985 60486	58133 07709	25899 68531	10370 46536	46506 86675
96601 96785	20850 70389	74637 34020	61780 33461	32496 51247
66706 67664	93292 05934	71050 68192	51898 18872	01371 95558
39273 41912	40198 36441	89472 38835	41709 85397	57429 33822
42539 21771	58672 71421	92528 67229	57837 89729	97494 40943
78209 77315	08393 95809	15832 31381	83170 32933	90911 49431
31279 02627	93411 96192	88570 88861	79463 64823	93331 97785
76915 11168	58452 30237	43211 88094	40120 44502	89799 09877
31714 15972	03620 07957	84828 01328	66806 45975	01001 72953
23131 71925	82240 57451	86216 08900	42868 39434	74956 93714
95186 13811	57341 15008	70542 51583	01563 50348	01403 32881
69801 54360	53265 70858	65549 02535	43657 75928	00109 59990
22293 14758	37440 49589	96421 74696	46442 27647	63950 60872
91834 04476	12776 34486	64027 15943	12307 36791	13600 32570
18615 53505	27034 34479	81642 20618	19992 12006	37023 48116
97224 65695	75788 47328	86654 52342	31619 97675	40129 57606
26686 27899	76013 33828	00554 35016	56081 74251	51761 20546
71559 57727	59162 88726	75150 37162	90733 79311	48085 67398
46858 08197	42531 34583	46155 77480	13712 15607	61265 81522
68979 78824	23553 16776	72208 12765	33682 68422	53269 99473
82578 98058	70623 49929	23105 14041	96639 89657	53337 31527
26830 21649	96350 76934	95421 23398	24309 97283	55317 06715
66780 42434	36053 11518	10114 29978	35558 73063	47919 41275
26850 83856	07550 60184	71509 33950	19410 57673	82122 98268
83552 46929	39407 19738	08959 95800	66562 70100	16777 17829
69664 34188	44958 12585	14608 48201	21973 64338	56105 01633
91846 59409	44903 39135	09758 79483	93450 39965	19354 55190
94766 30727	36436 87104	45035 27194	40273 71694	57000 40477
06482 16026	77039 23656	00584 01539	43904 13910	79229 91014
38098 51630	27394 80495	28942 12426	84937 10365	44686 73746
42624 08957	40485 75135	35223 18951	86245 86562	52403 37080
75639 11741	34530 96298	61180 04574	41471 76100	49195 68552
58681 84924	85898 94144	40948 88720	92349 75081	72752 03225
49384 67921	07641 03287	85245 06555	59403 71346	25280 03515
02176 23783	96594 05593	94006 41335	81326 28049	25784 22043

Table VIII *(cont.)*

04427 91941	28411 61926	05091 80116	27678 57044	07800 84220
40340 65310	96210 55328	41531 94918	57002 35599	03358 53340
98328 84687	92431 84619	81785 23800	37648 54398	33192 25522
35222 17971	82925 55038	61876 18784	29955 54903	77482 22822
02215 78263	70394 48269	39084 02420	87574 70394	77197 48801
00663 32518	21620 14766	00208 61501	81367 55121	98950 13374
95940 15988	63010 65689	82645 63299	71458 53879	96490 60874
72150 00691	26260 77788	39202 26373	06606 90187	02699 74662
26750 91609	78698 73651	60910 61747	23542 83251	67920 99495
79393 73431	90813 88577	60588 33597	45109 10244	18562 17903
23051 57058	54998 19079	29084 75546	20645 56793	75394 03138
37296 68692	28556 46842	31524 39581	48003 42089	16177 53839
96823 10760	28105 62075	43405 78449	42253 87637	50780 38439
70952 23508	01322 23974	78382 28359	69756 78049	88541 54049
70351 97343	34034 74893	20255 59081	26857 12532	93557 56388
34636 19478	74001 48594	39996 69571	91386 12571	22537 86232
70312 19527	35476 21474	59539 53327	02519 92874	64401 16908
55353 72710	70156 07101	84714 05185	12678 56232	46255 16716
90388 13730	16120 25122	90522 19876	63652 34123	23606 31510
61177 86244	68832 87827	83552 71569	08380 57632	39787 63944
40268 67249	82293 82287	47654 37093	79714 12981	41227 15768
90113 03927	77870 13652	24204 97211	50410 90813	38829 10786
06149 12136	98972 87973	00081 07982	34577 75853	10540 20318
51255 69854	09777 61282	81833 92904	66731 05620	55181 35443
79042 75404	50913 29839	13155 54721	91971 53744	32816 63961
13404 54241	08323 33434	63532 40574	60968 07680	61478 17184
36505 14866	30781 10772	43777 81058	75365 26010	94141 98158
52712 91306	32234 85403	66208 36267	47291 38749	90806 20855
47704 39463	43277 00476	51957 91849	62326 89333	47574 60731
07579 47308	68495 37428	47790 94349	41404 55339	34073 37117
71216 16914	07734 83356	87540 63554	19247 49986	74703 37502
41364 69095	27803 28120	46961 80172	32796 46654	88473 92257
28730 88643	76114 78189	59028 23382	60273 54234	99130 71974
34993 17394	21127 59043	70356 33220	55236 92612	60809 05685
28067 64186	55481 64655	08239 76084	87795 73725	75869 77508
03385 15665	26855 59615	38405 66863	01186 79189	20461 08922
78180 00999	45633 64132	12140 22229	16982 89195	87149 41013
94233 25325	11755 93689	14826 56758	19796 78285	31041 16965
88659 89137	09396 52029	61765 48534	22548 90526	33807 11337
17585 85278	77482 77410	68468 92470	04936 38517	19410 37980
94742 51432	29572 62015	36880 94423	03179 59708	07823 20340
86059 07982	76029 73154	59123 12899	18899 84852	30109 46997
40788 59424	45541 07501	49539 29967	37247 15966	18005 34461
59564 89198	21965 90705	45538 55894	82002 96641	44273 88581
29392 53058	29439 39922	82990 67171	40537 57251	53456 83807
85532 87232	26852 06072	20552 37337	43185 87156	02774 81224
96818 98016	85405 71729	86948 43048	10382 12016	11606 16368
45410 27000	35418 64141	08731 20285	97055 74760	85335 95673
09848 60615	11085 64537	71296 71800	09895 69411	93490 04683
91228 87269	43082 28147	04175 54874	03737 93493	97159 07515

Table VIII *(cont.)*

37370	12924	09734	11945	16699	63695	97271	10085	94018	88720
81449	01495	88354	95326	98842	05351	79822	45913	85930	89116
70653	06205	24049	98497	25305	97719	75952	61813	65910	79846
07226	50068	41313	88697	14318	81930	39809	67058	48867	31782
24797	34171	44550	32971	93170	39723	67902	04470	92985	17001
10778	79910	10816	42511	07048	15281	43390	44768	25771	68355
29066	34322	98209	59788	21244	24795	80587	00667	86758	05876
39944	74378	82952	39684	13674	18636	85356	68279	97352	53596
44392	14599	02972	93981	82686	11495	17849	27417	08396	77429
60420	68938	23660	07710	41404	07310	75862	13754	29307	05521
43791	03773	87643	47727	37051	86406	59421	95703	51136	73002
10652	47844	19624	02108	59309	52024	84080	81176	15985	74684
78083	23436	74701	78473	43159	41323	71985	48152	98475	10700
11416	23157	51335	01227	83042	94421	31688	37373	58196	05831
19189	90702	43779	97629	78531	07331	56303	92912	53303	66085
56997	53578	55166	58132	30342	15437	34661	29324	87946	24518
82001	98953	83951	61438	32323	75493	25037	89827	39983	50100
98333	67613	65743	86072	23445	84413	90068	90305	51225	51863
62873	74023	43718	21969	13153	75996	70130	29277	18871	03337
42994	70048	65656	15626	66899	82126	11273	50734	97160	18360
80944	39683	50779	12806	89213	83556	33795	68541	54390	11743
04019	41952	80355	71294	23315	69325	98008	63158	02668	59855
16460	29092	96267	14125	49265	76020	56135	18432	92043	70545
88408	33301	73868	18007	69469	25690	64594	99549	19800	14225
71946	00476	19799	90651	32590	38707	80649	77913	11278	55362
92382	87283	38693	24996	22024	19199	28140	42381	59086	00912
49805	23508	08680	74161	33254	71091	72112	18652	19549	05536
60276	95349	85591	24301	49100	39057	99048	39533	40440	77864
59877	45032	05836	59119	89279	00908	25294	54092	96517	85432
34742	15755	39939	39600	43080	21606	71859	05232	70913	71807
84318	78069	40638	11503	11461	95654	82140	22218	87042	29199
78215	35856	32463	09807	68070	13808	44544	51398	23208	20482
35695	93257	42934	30877	65341	26631	99044	94223	64625	07516
01620	94471	68238	55012	42963	14265	06625	48744	16181	02287
15979	84434	62898	71805	08792	11495	80934	62840	19700	38011
73511	08079	45387	90085	99065	79351	51671	15827	97297	64778
17432	55144	03105	26000	92767	69812	42500	82656	36932	63869
60024	97910	75276	65894	06248	43761	35194	43920	88209	35043
66107	50506	70616	90063	29481	79706	63795	58500	93602	59115
53353	83680	41106	43683	95163	25136	98942	05560	05280	49947
89033	68608	55675	31344	06176	77204	58338	63812	59547	68362
38867	59860	11216	69161	55322	71409	38689	38108	79384	09768
08759	37061	90185	17551	43601	59100	59726	90265	54295	13348
90845	71945	85849	76992	27306	79650	31692	70867	66585	55744
63766	07888	95949	27659	51419	11439	53558	04274	09768	60006
12568	46270	85421	25095	30734	96358	05254	73037	26042	48201
27998	10931	17068	48275	57896	22709	97116	58484	22385	97098
34808	05053	65848	18600	87278	37045	17445	62410	79014	79944
06984	94805	99744	00348	76221	04210	43575	32569	63591	08300
98579	74989	07382	81862	96687	56820	32313	23786	06981	41270

Table VIII *(cont.)*

```
76087 08080   41582 68035   96551 05820   32568 94865   73340 87465
05745 90566   12461 65594   21917 64503   13054 43525   26514 22473
12676 97987   86298 28636   11370 44242   42498 16150   48322 74248
90971 87090   32283 04856   02763 11155   39742 83483   90232 95340
65052 74517   05905 99738   21882 79821   72303 08983   84020 34220

74392 92688   57180 35038   19183 40836   83963 66277   85838 39183
61374 44371   58369 95300   84932 71824   91309 48997   58208 09138
09243 93675   10530 34015   35820 39996   06411 37090   46595 63275
52675 43070   40623 33070   70155 74851   93564 95149   17366 91546
21595 36963   21644 19724   42805 83605   06232 91720   54320 19808

11346 38862   33998 16965   92460 52099   91390 92293   44092 60679
44334 29601   47168 77779   37281 40838   48002 00621   96864 46220
02469 93747   25931 01218   08945 82668   14044 37688   48891 96651
95661 99743   56230 06783   07317 28417   84408 25012   68342 43917
65777 59608   02888 92821   24570 06377   39023 21602   51650 25601

35526 38430   81663 68824   39540 21996   02921 94762   62214 93871
27818 55178   07803 93776   01524 66503   71858 09702   27317 62723
10647 91105   03556 85620   15528 34738   72662 91130   48471 41541
39820 73347   04577 16800   11485 75384   00715 37915   14057 65396
49342 18040   37249 26714   33865 33595   62540 05651   65062 67557

92035 56857   54034 34770   66045 08651   80846 56459   21225 23558
14212 17484   93674 95566   75855 11758   58285 60774   86185 93984
12278 56602   44263 14657   46626 28331   76057 17543   86502 70996
77243 42205   34033 10504   94087 94971   84060 44230   83514 41108
68839 93401   14557 63275   37514 22665   22878 84697   06983 38142

76454 84743   03305 10878   34973 08948   09965 23712   44164 13418
63221 45006   58082 98829   26354 41483   55479 30330   80354 02731
83618 36126   78502 86020   52690 29290   06658 04507   29418 13306
17060 82744   06045 45432   81993 48272   44720 17362   77797 90805
17942 21429   67588 54367   19686 52490   12331 57933   37688 87089

15695 05135   44576 04961   27836 71940   56596 68780   72511 15667
41191 36920   03496 28796   30489 67742   04424 15603   82903 66631
66871 00255   66182 21278   81377 48820   00116 49367   20202 60406
88620 81661   63152 33852   13899 17813   31636 53397   07201 66823
93668 89860   66652 71739   13202 82472   32102 53809   67088 16504

87553 99144   38065 91174   48963 97650   04559 02220   39587 82118
88473 96135   36371 50362   02588 85421   97691 88402   40217 80682
58399 09848   93846 26288   31680 12762   77424 83070   26029 64236
64740 71267   08520 72224   35799 50335   00663 76778   15911 76777
94470 35800   17936 62191   60792 42553   41820 61828   68558 57204

96729 58046   63765 51927   15191 37516   51055 88292   04822 28171
58744 90144   90033 93669   60905 58675   92684 30909   76625 72145
91069 91333   16730 16843   24633 90623   21725 48401   32282 45624
16895 84817   67159 14764   67615 78674   95583 40342   45251 56026
83845 92008   75960 91934   00508 63683   26691 20335   20171 56435

29214 73033   34942 52750   89947 17560   27683 80763   34511 15849
78357 04329   78278 16218   54540 62535   89822 93254   32630 72449
08639 08761   96537 66833   92335 55928   71242 22356   44429 58534
63405 20883   47156 83312   66901 06727   98913 29652   90484 98845
11652 84507   34402 48389   60768 57283   85683 31918   26890 18989
```

Table VIII *(cont.)*

35721	16570	41666	75332	21224	42085	31418	57377	96829	97911
61079	36493	81024	75877	77677	42301	19248	41286	22754	42846
17905	63119	76009	17778	76788	64848	68179	78998	61269	82285
45235	82838	00121	20378	28411	89772	70904	52037	67828	89812
44041	04631	31637	67566	69429	25945	95544	15618	59577	56155
69733	07671	05760	78242	95175	65369	13705	32407	36883	46519
67857	20130	16284	80291	91534	88495	43780	64467	65106	42929
37242	53919	29386	94380	97947	63637	98757	20831	47458	05089
41857	01241	21379	01301	68143	76586	52380	77003	89353	52572
87401	05533	45155	98928	68387	32717	67058	11749	22846	98983
67784	79023	61325	60557	22310	30409	64819	02087	86283	42646
76880	74318	40728	32521	91857	76883	71385	72508	94470	88380
02151	69381	64990	54511	47393	90504	35634	92843	29702	31220
99798	46782	66080	75657	30586	61272	67332	58261	01230	86738
12909	75373	29163	57038	66890	70639	04131	76317	86186	47461
66844	25788	19009	64535	13697	79728	85815	43684	16247	37760
91596	20152	32083	46805	30953	64181	27897	12859	96409	92802
58182	83954	59492	53415	69665	99414	22884	99524	50244	06054
84995	70403	53064	59331	54145	09229	49110	03102	59430	58455
38324	96649	92746	06856	63714	39771	45536	41678	13514	12087
03080	59503	84461	58905	83436	53680	27900	06818	70936	96202
39044	36957	21093	22709	00124	04854	67543	36566	31157	36996
52250	69590	18653	55580	51853	19510	69654	14246	79890	53483
78628	16374	02831	56973	85636	63325	02431	60298	12751	21754
16713	17673	80815	42320	20279	28187	59046	78723	01816	66795
05608	32418	75113	04505	50291	88245	88721	83957	77352	61300
75382	30316	38463	22627	66525	96774	35250	32465	52724	67262
24858	25083	74999	85805	20839	25231	14628	33698	54507	40131
11878	07670	96685	37782	16194	74410	74589	40017	51580	30312
79129	81328	34051	14354	30252	53890	01550	56755	07711	63032
46671	68850	37077	78799	88479	72640	45271	79842	17806	17833
91188	30393	41584	65195	22830	34659	85898	47518	10263	78557
24698	93005	84570	09093	06252	11339	64708	48373	32750	02524
98567	76501	06137	73464	99822	44783	53221	89661	31041	77218
89522	04581	57879	79336	03066	33079	62058	17217	61522	69314
22600	96884	25595	45871	20707	36044	43796	60881	74744	29333
21231	14767	56140	92059	64726	14328	46742	43336	08729	06938
52426	85990	49775	31859	22784	77690	06769	22266	57210	19614
62915	42464	36760	88273	32325	45688	41200	36136	91339	75673
21371	27256	12321	77812	36440	12974	43561	43292	00191	16423
73494	97853	70628	26173	99311	98334	83958	04911	02664	22267
28023	31765	56433	45912	83834	04224	57776	02432	14422	87720
32027	89340	06725	59769	87449	96581	82910	35379	12531	12464
46635	24133	64952	85641	26425	75873	69942	38380	45169	87282
39780	86720	24985	23675	54215	09342	01291	90784	29026	59083
10419	27375	93180	03610	55268	88413	98176	60411	79754	28406
20330	99946	03609	07812	11912	92837	61331	59754	41341	11518
17274	46628	99420	63494	72116	54872	90957	20382	47123	26002
87653	48905	29551	00000	99755	37360	44654	28780	16738	51069
37567	21732	61611	77332	72388	40899	07618	48938	07723	29153

Table IX. Normal Random Numbers

.638	.158	1.136	2.145-	.602-	.474	.185-	1.007	.049	1.181
.398	1.779	.632-	.372-	.849	.410-	.117-	1.389	.100-	.821-
2.310	1.964-	.929	.478-	.856	.322	.614	.298-	.862-	1.367
.391-	1.466-	.583	.073-	1.875	1.580-	.902-	.256	1.656	2.209
1.320	.080-	1.675-	.610-	.971-	.039-	1.672-	.056	.372-	.867
.489-	.737	.303-	.495	.563-	.075-	.036	.585-	.051-	.915
.124-	1.409	.502-	.651-	.301	1.621-	1.700-	1.605	1.583-	.452-
1.342-	1.696-	2.305-	.158	.222	.334	.380	.219-	.477-	.551-
.321	.239-	.851-	1.322-	.535	1.233-	.812-	.887-	.581-	2.939-
3.402-	.013-	.628	1.286-	.540	.354-	.254	.543-	.330	1.573
.069	.150	.095	.269	.430-	.253	1.514	1.467	.196-	1.790-
2.029	.204-	2.024	.419	.718	.231	.940-	1.030-	.815	.453-
2.507	.662-	1.370	.175-	1.938	.508-	.131-	1.441	.783	.799-
.388	.894	1.628	1.216-	.826	.124-	2.307-	.595-	.490-	.214-
.195-	.404-	1.770	2.368	.360-	.836	.522-	1.241-	1.111-	1.621
1.172-	.049-	.215	.469-	.614	.059	1.031-	1.488	.649	1.509-
1.380	.642-	.814-	.086-	1.419-	1.164	.316-	.139	.542-	.411-
1.128	.437	.272-	1.296-	1.947-	1.477-	.425-	.508	.752	.143-
.551	1.116-	.383	1.194-	.847	.188	1.355	.594	.096-	.198
.359-	.301-	1.725	1.369-	.179	.558-	.255	1.673	1.604	1.904-
1.288	1.262-	.753-	1.467-	.784-	.621-	.732	1.054-	.320-	1.041
1.683	1.997	.319	1.129	1.094	1.457-	.060-	.363-	.480-	1.270-
.723-	.833-	.579	.042-	.147	1.370	.234	.404	1.522	1.406
.413	1.672	.843-	2.219	.525	.957-	.054-	.111-	.763-	1.406
1.390-	.975	1.151	.702	.017	.585-	1.277	1.270-	.679-	.876
.856-	.506	.419-	.234	.112-	.065-	.369-	.503-	.927-	.516
.078-	.401-	1.334	1.799-	2.176-	.693	1.630	.652-	.258-	.062
.611	1.061	.686	.291	.057-	1.084-	1.513-	.048-	1.136	.523-
.135	.543	1.130-	.012	.161	.085	.864	2.715	.081	1.582
.333-	.317-	2.097-	.782	.868-	.462-	.820-	.147-	.405	.616
.450-	.756-	.096-	.247-	.741	.311	.670-	1.028-	1.749-	1.502
.216-	.134-	.260-	2.319-	.734-	.471-	.164-	1.856-	.524-	.098-
.003-	.736-	.572	2.223	1.561-	2.532	.490	.339	.807-	.276
1.327-	.355-	.495	.333	.559	.386	.339	.380-	2.039-	1.089-
1.214-	.864-	.048	.645	.046	.198	1.003-	.173-	.988	.063-
.395-	1.049-	1.060-	.440-	.641-	.405	.345-	.735-	1.656	.458
.309-	.541-	.181	1.161-	.226	.837-	1.347	2.651	.389-	.875-
.453	.746	.286	1.387	1.878	1.165	2.725-	.010	1.589	2.842-
1.333-	1.053-	.096-	.802	1.078-	.714	.027	.242-	.123	.758-
1.023-	.580-	.341-	.081-	.713-	.247	.069	.317-	.122	.889-
1.293	.910-	.952	.271-	1.912	1.902	.470	.087-	2.270-	.226
1.480-	.738-	1.939-	1.407-	2.615-	2.257-	.878-	.318-	1.788	1.812-
.382	.229-	.199	.311-	.154-	1.743-	.412-	.249	.370-	1.838
.715	.379	.623	1.520-	.772	.164	.075-	.038-	.294	.173-
.079-	2.231	.669-	.924-	.365	.707	.203	.879	.813	.674-
1.522	.210-	1.111	.669-	.160	.648-	.931	1.122	.492-	.894
.938-	.133-	.600-	.808	1.063-	1.022-	2.292	.705-	.284-	.915
1.705	.669-	2.209	1.280-	2.380-	.568	.635-	1.524-	.382-	1.096
.573-	.796	.680-	1.393	1.196	1.677-	1.409	.789	.529	.106-
1.511-	2.557-	.242-	.800	.483-	.167-	.458-	.773-	.847-	1.401-

Reprinted with permission from *A Million Random Digits with 100,000 Normal Deviates*, Rand Corporation, Santa Monica, Calif.

Table IX *(cont.)*

```
 .670-   .518    .387    .523    .641   1.243    .322   2.607- 1.097-   .012-
2.912-  1.448   1.343    .122-   .726    .617-   .609   2.319   .450-  1.197-
 .028-   .790-   .057   1.425   1.940   1.161    .878-   .716-   .244- 1.151-
1.257-   .774    .003    .388   1.060   1.028    .236-  1.172   .442    .157-
2.372   1.376-  1.318-  1.236    .738    .337    .534-   .090   .886    .676

 .970-   .438    .672-   .180-   .667   1.370    .481-   .329   .842    .449
1.228-   .129    .426-   .165-   .028   2.696   1.201   1.351-  .724   1.017-
 .369-   .310    .432    .237    .884   1.224-   .539    .852   .497    .283-
1.161   1.219   1.615    .336   1.100    .528-   .161    .278   .675   1.143-
 .284-  2.609    .792   1.825    .249-  1.654    .621    .979  1.472-  1.173-


 .578-   .789-   .106    .832    .597-   .496    .561-  1.033-  .578-   .378-
 .074    .261    .766-  1.046-   .361    .043-  1.927-  1.527   .605   1.475
 .230    .046    .978   1.901-  1.162    .545-   .697   1.151  2.033    .080
2.162    .562-  1.190    .925   1.057-   .015   1.371-  1.067  1.080-  1.129-
1.020-  1.130-   .315-   .628    .140-  2.050    .030-   .629-  .128   1.221-

1.323    .836-   .284-   .249-   .768-  1.242    .879-   .417-  .013    .502-
2.329   1.884    .033    .598    .217-   .260    .431   1.914-  .205   1.155
2.761   1.800    .562-   .714    .407-   .009    .724-  1.168-  .247   1.166
 .232-   .605    .023-   .531-   .542    .155-   .697   1.037   .316-   .003-
 .742-   .210    .741-  1.099-   .158   2.112    .765-   .319-  .247-   .345


1.410-   .413    .705   1.444   1.057    .843-   .043    .571-  .001-   .203
2.272    .719-   .679   2.007    .180-   .698   1.137-   .688   .571-   .100-
2.832    .925   1.350-  1.529    .260-  1.007-  2.350-  1.501-  .289   1.522
1.086-   .558-   .973-  1.285-   .021-   .077    .915    .241-  .249-   .529-
 .134   1.815    .313   1.571    .216-  2.261    .696    .130-  .393    .017

 .783    .600    .745-  1.127    .684-   .519-   .125    .499- 1.543    .082-
 .174-   .897-   .575    .751-   .694   2.959-   .529   1.587   .339    .813-
1.319-   .556   2.963   1.218   1.199   1.746-  1.611    .467   .490-   .202
1.298    .940-  1.143-  1.136-  1.516-   .548    .629    .250  1.087-   .322
 .676-  1.107-  1.483-   .278    .493    .442-  1.078    .336-  .177-   .057-


1.287-   .775   1.095-  1.161   1.877-  1.874   1.703   1.619-  .725-  1.407-
 .260    .028-  1.982-   .811    .999   1.662    .908   1.476  1.137-   .945-
 .481   1.060   1.441    .163    .720   1.490    .026-   .502-  .427    .351-
 .794    .725   1.971    .384    .579-  1.079-  1.440-   .859-  .346-   .077
 .584    .554-  1.460    .791    .426-   .682-   .430   1.922  2.099-   .221

 .114-   .379    .698-  1.570    .511-   .725-   .680    .591- 1.091-   .357
1.128-  1.707-   .921    .859-  1.566-  1.523    .900-   .988-  .264    .282
 .691    .153    .076   1.691    .553    .457   1.107-   .322   .633    .007
1.115    .777    .738-   .868   1.484   1.792-   .950    .842-  .192-   .620
 .389-   .559    .670    .315-  1.234    .475   1.117   1.286   .649-  1.880-


 .330    .750    .642-   .148    .608-   .866   1.720-   .653   .210-   .959-
 .333-   .084-  1.239    .049-   .095-   .197-   .213-  1.420-  .491-   .102
1.718   1.111    .548-   .653-  1.534    .456-   .395-  1.614   .531-   .785-
 .182-   .620   1.178   1.071-   .444    .072-  1.001-  1.325   .302-  1.119-
1.260   1.192-   .182    .397-   .705-  1.085-  1.492-  1.642   .673    .707-

1.204-  1.725-  1.695   1.473    .665    .489-   .020    .267  1.230    .865
 .619-   .307    .226-   .096-   .987   1.195-  1.412-   .433  2.052    .022
 .272-   .096-   .137    .361-   .653    .156-  1.309    .480-  .397-  1.302
 .245    .690-   .493   1.123-  1.465    .132    .582    .429-  .225    .125
 .101    .855-   .782   1.040-  2.113   1.423-  1.010-   .158   .106   1.232-
```

Table IX *(cont.)*

1.172	.574	.231	1.011	.616-	1.386	.044-	.207	.724	.452
.034	.365-	2.277	2.116-	.746	.418	1.239-	.606-	.133	.644
.793-	.821	1.331	1.178-	.508-	1.267-	.640	.099	2.054	.251
.118	.169-	1.705	.312	1.039-	.364-	.190	.379-	.263-	1.008
.092	.604-	.616	.053	.584-	1.652-	.189	.117-	.817-	.521
1.633	.637	.852	1.101-	.035	1.623	.287	.959-	.295-	2.341
.789	1.855	1.581	.004	1.134	2.749	.580-	1.681-	.061-	.319
.192	.992-	.398	.195	1.359	.919	.040	.012	.770-	.206-
2.178-	.611-	1.218	.673-	.754-	1.136	.345	1.033-	1.286	2.784-
.167	.037-	.471-	.818-	1.551-	1.321	.369	.228	1.097-	1.295
1.226-	.228	.348-	1.113	1.383-	.410	.701	.215	.597-	.597-
1.089-	.304-	.185-	.290	.250-	.267	.498	.138-	.265	1.054-
1.832-	.468-	1.506	.007-	1.101-	1.272-	1.491-	.745	.148	.604-
2.127-	1.142-	1.225-	1.355	1.414-	1.215	.739-	.241	.877	.009-
1.844-	.872-	.642	.543-	.162	1.494	.442	1.029	.317	1.198-
1.656	.784-	.163	.589-	1.074	.046	1.435	.292-	.682	1.809-
.096	.028	.430-	.722-	.042-	.368	.636	.171	.356	.919-
.260-	1.756	.150-	.123-	.074-	.323-	.174-	.494-	.393	.244-
.101-	.063	1.444-	.604	.430-	1.147-	.116	1.309-	.982	.806-
.518-	.073	1.490-	.119-	.953-	1.947-	.820	.852	1.223	1.217-
.033-	.915	.054-	1.339	.393	.103	1.208-	1.442	1.408	2.080
2.057	.715	.496	1.224-	.763	2.011-	.577	.372-	.033-	.991
.026	.562-	.934-	.028	.677	1.273	1.363	1.952-	.369	.064
1.342-	.673	.659	.234	.552-	.656	2.019	.523-	1.599	.307-
.789-	.671-	1.704	1.095	.670	.247	.377-	.137	.008	.152
.233	2.316-	1.339-	.834	1.742	.314-	.136	.961-	.699-	.574
1.474-	.719	1.471-	.674-	1.207	.246	.050-	.426	1.004	.216-
.619	.927	1.590-	.301-	.630	.637	.325-	1.119-	.869-	.800-
.082-	1.100	1.196	.156	.877-	1.294	.966-	2.253	1.825-	1.435
1.616-	.518-	.620-	.321-	.238-	1.892-	1.107-	.113-	1.899-	1.385
.354-	.627-	.386	.073-	.219-	.531-	.398	2.333-	1.018	.144-
.424-	1.794-	.080-	.089	1.739	1.204-	.400-	1.711	.682	.611
.778	.482	.146-	.506	1.029	.086-	.414-	1.287-	1.579	.876-
1.560	.673-	.294-	.379-	2.073-	1.046	.361	1.500-	.514	.097
.840-	.042	.626-	2.200	1.625-	.043-	1.075	.809-	.525	1.889-
2.587-	.242	.192	1.341-	.360	.241-	.402-	.770	.047-	1.920
.754-	2.069-	1.738-	1.176-	.479	.181-	.708	1.303-	.044	.543
1.062-	.617	.692	.132-	1.285-	.283	1.260	.168-	.740-	.049-
.352	2.274	.797-	.295	.016-	1.434-	2.298	.311-	.635-	.845
.905-	1.272-	.571	.300-	1.025	.904-	1.317-	1.098	.048-	.337-
2.248	.494-	.011	.613	.279	.069	.727	.005-	.305	.604-
1.359	.097-	.587-	1.344	.483-	1.876-	.226	.569-	.649	.252-
.860-	.291-	.168	1.902-	.916	.990-	1.576-	.661-	2.239-	.301-
.696	1.320	.043	1.510-	.490-	1.634-	.954-	1.123-	.381-	1.061
.610-	.362	.984-	.584-	1.419-	.080-	.907	1.429	.478	.051
.861	.527	.678-	.203	1.091	.406-	1.232-	.347-	.298	.309
.282-	1.780	1.487	.750-	.255-	.763-	.249-	.965-	.105	.507-
.782-	1.261	.513-	2.559	.795	.408	.641-	1.149	.315-	1.755-
.420-	.172	.674	.143	.601-	.400	.751	.232	.689-	2.226-
.919-	.006	.534-	1.707-	.071	1.050-	.747-	1.151-	.446-	.706-

Table IX *(cont.)*

```
.011    1.059   1.062-   .897    2.152    .428-    .780    .720-    .092-    .099-
.842-   1.278    .515-   .833-    .999    .252   1.351-  2.191   1.109     .693
.019     .130   1.707-   .892-  1.049    .699-    .236   1.490-   .679-   1.624-
.128-   1.357    .306-   .626     .842    .378-    .066-   .470    .621     .685-
1.099    .615-  1.464  1.036-   1.103    .401-    .174   1.373    .501-    .342-

.044-    .749-   .033-  1.895    .102-    .244-  2.016   1.580    .474     .575-
.044     .718    .182-  1.561-   .480-  1.699-    .477    .554    .297     .317-
.453-   1.353    .364    .207     .143   1.121    .379    .188-   .604-    .994-
.726     .973    .154-   .225     .088   2.092-   .564-   .008    .565    1.512
.850     .362    .498-   .364-    .359-   .576-   .016-   .279    .319     .716-

1.414    .723    .306   1.036   2.172-    .014-  2.228   1.169-   .649-   1.494-
.524-    .132-   .056    .508     .881-    .820    .512    .650-  2.338-    .624-
.182     .069    .687-  2.523     .915-    .842-   .642-   .667-   .253     .077-
.854-    .221   1.763   1.172-    .447-    .880    .913-   .985    .649    1.169-
1.213-   .709-  1.253    .430     .653   1.291    .116-  1.434   1.371-    .079-

.013-   2.301    .875    .184-  1.339-  1.808    .640-    .176    .887-    .294
.155-    .838-   .526    .934     .511-   .101   1.485    .372    .476     .286
.192-   1.043-   .827   1.346     .222-   .011-   .970-   .599    .667-    .958
.288-    .177-   .250   1.373-  1.186-  1.576    .088-  1.447    .408-    .609-
.638-   2.137-  1.171- 1.340-    .008   1.733    .958    .575    .676-    .858

.468-   1.199    .635-  1.066   1.385-  1.025    .250-  1.202    .228-    .533
1.164-   .794    .090-   .027     .464   1.494   1.393-  1.362   1.067-    .799-
.081     .882    .689    .400   1.735   2.005   1.966-   .389    .652-    .620
1.498    .334-   .104-  2.035   2.680-   .255    .049-  1.641    .368-    .547
.550     .047-   .906-   .106-    .046-   .754    .667-   .781    .306    1.031-

1.789   1.010   1.089    .412-    .678    .166    .261-   .102   1.252-    .551-
.349-   1.089-  1.026    .640-    .671    .857    .531   1.539    .134-    .633
1.054   1.106-  1.005    .817-    .555    .481-  2.082-  2.458-  1.214     .611-
.442-    .971    .492-   .046     .121    .211-   .472-   .083    .443     .360
.370-   1.419-  1.568-  1.329     .442-  1.013   1.752-  1.310-   .236-    .123-

.218     .601-   .977    .206-    .798    .525    .544-   .323   1.902     .454-
.411-    .353-   .588    .016     .298-  1.341-  2.433   1.015-   .974-    .956
.680     .037-  1.099-   .721    1.670   2.079    .372    .356-  1.426-    .858
.297     .421-   .513    .265-   1.160   1.050   1.410    .008-   .775     .974
.963-    .903    .331    .687    1.854   1.743-   .285-   .598-   .772-    .762-

2.069   1.251    .862-  1.086    .251-   .252-   .599-   .061   1.040    1.505-
.672    1.554    .050    .035-    .548-   .613-   .356-  1.453-   .525-   1.247
.526-    .595    .136-   .338     .187   2.680   1.219-   .089    .186-    .059
.345-   2.634    .224-  1.152-    .843-   .095-   .503-   .347    .601     .560-
.817-   1.097-   .139-  1.406     .249   1.501-  1.606    .260-   .581-   1.597

.784-    .419    .599    .588     .794   1.237-   .394    .192    .273     .526
.590     .786    .843-   .706-    .161-   .891    .626-   .692-   .655     .828-
.692     .672   1.560-  2.389   1.137-   .181-   .247    .693-   .615-    .316
.919-   1.147-   .903-  1.301    1.174    .478-   .487    .220    .084    1.692
.230    1.316   1.812-  1.104     .099    .167   2.354-   .904-   .165-    .448

2.977    .168   2.353    .639-  1.999    .884   1.381    .345    .455     .657
.685-    .092-   .984   1.155-    .391-   .955   1.313-  2.778   1.367     .080-
1.964   1.025-   .617-  1.070-  1.133-   .584-   .090    .382    .495     .001-
1.248-  1.612    .895    .004     .081-   .349    .696-  1.745    .618     .603-
.820-    .467    .149-   .239-    .229-  1.302    .021-   .692    .977-    .634-
```

APPENDIX C

DATA SET A

Vital signs for 142 student nurses are listed below. These were, typically (but not exclusively) female, between 20 and 30, and healthy. Measurements were made by the students on each other and checked by the instructor. In the case of blood pressure, where there seemed to be more discrepancies, the measurement by the instructor is given where there was a difference.

Pulse and respiration are given in beats and breaths per minute, temperature in degrees Fahrenheit, and pressure in millimeters of mercury. (In a reading A/B, A is the systolic and B the diastolic pressure.)

| | | | | | Blood Pressure | |
Subject	Radial Pulse	Apical Pulse	Respiration	Temperature	By Student	By Instructor
1	84	86	20	98.2	122/78	
2	88	90	18	98.6	130/98	
3	84	98	12	98.8	132/74	
4	80	79	12	98.2	102/68	
5	100	96	16	100.0	108/74	110/78

Subject	Radial Pulse	Apical Pulse	Respiration	Temperature	Blood Pressure By Student	By Instructor
6	76	71	12	98.6	130/76	130/72
7	76	75	20	99.4	100/76	102/76
8	104	105	16	98.6	110/68	110/70
9	104	102	12	98.8	104/76	
10	84	86	18	98.0	108/70	
11	68	70	24	98.6	110/70	110/74
12	84	76	20	98.2	100/60	102/60
13	72	88	16	99.0	90/60	
14	96	85	16	98.0	102/76	
15	80	78	16	98.8	100/78	102/74
16	80	106	20	100.0	118/74	120/60
17	76	81	12	99.6	114/82	
18	68	72	20	99.4	110/62	110/52
19	104	97	12	99.2	120/74	122/72
20	80	81	16	98.6	116/72	
21	100	99	16	98.2	112/62	
22	72	67	16	98.4	132/72	
23	100	105	24	99.2	118/80	118/78
24	80	90	16	99.0	142/92	
25	60	60	20	98.0	100/78	
26	92	96	16	98.8	124/88	126/88
27	120	100	12	100.2	104/76	104/74
28	68	65	18	98.8	104/68	
29	76	75	20	99.2	120/78	
30	96	98	16	98.2	130/96	
31	100	102	20	99.0	112/78	
32	104	102	12	98.6	128/76	
33	100	98	20	99.8	126/78	
34	88	70	16	100.0	126/90	124/88
35	100	95	16	99.2	150/108	154/110
36	68	70	16	99.0	112/70	
37	68	69	12	98.6	120/78	
38	136	118	16	100.0	140/82	140/80
39	92	76	12	98.8	130/90	130/80
40	56	65	16	98.2	108/72	
41	88	76	20	96.8	122/88	124/86
42	96	94	20	100.0	110/70	
43	76	70	18	99.0	112/70	
44	80	92	20	98.8	132/92	
45	84	96	20	99.4	94/62	

Subject	Radial Pulse	Apical Pulse	Respiration	Temperature	Blood Pressure	
					By Student	By Instructor
46	96	100	16	98.6	112/80	112/84
47	60	63	20	98.8	98/54	102/78
48	72	77	20	98.8	102/78	
49	92	92	16	99.0	96/74	
50	80	85	28	97.2	132/78	
51	68	68	16	98.6	116/70	
52	88	93	12	99.2	108/84	110/84
53	100	100	16	98.6	110/70	108/68
54	84	76	16	98.6	136/68	134/64
55	80	73	20	98.6	98/70	98/74
56	96	88	20	99.2	124/68	
57	64	76	20	98.6	118/88	
58	92	86	12	99.0	98/74	98/76
59	80	86	16	98.6	108/76	106/74
60	72	92	24	99.8	112/34	
61	84	80	16	99.2	134/96	130/98
62	76	71	28	99.6	120/78	
63	104	100	16	99.0	110/72	114/70
64	92	98	24	99.2	106/70	
65	100	105	20	99.6	104/78	106/80
66	78	75	16	98.6	110/82	114/82
67	88	102	16	100.0	110/72	
68	76	82	20	99.4	108/78	106/76
69	112	98	20	98.6	110/70	112/72
70	80	80	20	99.2	104/60	
71	88	87	12	99.0	110/72	
72	80	73	16	97.8	106/70	
73	80	73	20	98.6	120/88	122/84
74	68	66	18	99.4	110/70	
75	92	100	20	98.8	118/76	
76	64	77	12	99.2	118/78	118/76
77	84	89	20	98.2	102/64	104/64
78	84	82	12	99.6	112/76	114/76
79	76	80	20	98.2	120/60	
80	80	90	12	99.6	110/76	120/60
81	96	85	12	99.6	108/84	
82	100	104	12	99.6	128/90	
83	100	104	16	98.6	124/74	
84	96	87	16	99.0	130/88	142/84
85	72	77	14	98.6	128/92	

Subject	Radial Pulse	Apical Pulse	Respiration	Temperature	Blood Pressure By Student	Blood Pressure By Instructor
86	92	99	16	99.4	108/86	110/86
87	80	82	20	98.6	116/86	
88	80	79	16	99.0	120/80	124/78
89	84	90	20	98.6	104/58	102/56
90	100	100	16	99.2	118/72	
91	112	100	16	99.6	134/90	138/90
92	108	100	16	99.2	118/72	
93	112	100	16	99.6	134/90	138/96
94	92	94	24	98.6	120/80	116/82
95	88	87	16	99.4	136/90	
96	80	81	20	100.4	122/78	120/70
97	68	106	16	99.4	122/80	
.98	84	92	16	99.6	112/72	114/78
99	76	76	12	98.6	116/80	114/84
100	88	94	20	99.2	118/90	
101	96	100	16	98.6	100/78	
102	84	89	16	99.8	110/72	
103	72	69	12	98.8	126/84	
104	104	102	12	98.6	110/68	
105	100	85	20	100.2	116/74	120/78
106	64	68	28	99.6	110/76	112/74
107	84	80	24	99.4	108/82	110/70
108	72	75	20	98.2	118/64	
109	88	91	20	98.6	158/100	
110	96	96	16	99.0	118/74	
111	92	96	12	99.4	130/72	
112	96	92	12	100.0	130/84	134/72
113	88	94	20	99.2	118/90	
114	84	94	16	99.2	110/86	100/84
115	76	69	16	97.8	110/72	110/70
116	104	90	20	99.4	118/80	120/80
117	76	86	16	98.4	118/84	118/80
118	96	95	20	99.0	102/72	
119	90	102	20	98.6	100/68	
120	68	67	16	98.8	122/74	
121	76	71	16	99.0	132/92	134/94
122	80	90	20	99.2	140/104	138/100
123	92	100	12	99.0	116/78	
124	60	84	16	98.6	132/80	
125	80	82	20	98.2	112/80	

Subject	Radial Pulse	Apical Pulse	Respiration	Temperature	Blood Pressure	
					By Student	By Instructor
126	104	99	16	99.2	140/70	
127	80	82	20	98.6	116/86	
128	80	79	16	99.0	120/80	124/78
129	84	90	20	98.6	104/58	102/56
130	100	100	16	99.2	118/72	
131	102	100	16	99.6	134/90	138/90
132	92	94	24	98.6	120/80	116/82
133	88	87	16	99.4	136/90	
134	80	81	20	100.4	122/78	120/68
135	68	106	16	99.4	122/80	
136	84	92	16	99.6	112/72	114/78
137	76	76	12	98.6	116/80	114/84
138	80	119	20	99.6	130/88	
139	88	80	16	98.4	110/78	110/74
140	96	92	20	98.6	106/78	
141	116	86	16	100.0	138/74	140/76
142	72	73	12	99.4	122/82	126/80

DATA SET B

Students in an elementary statistics course were surveyed, and the results are listed below. There were two sections, one on the main campus and the other on an "Ag" campus (a couple of miles away). The variables were recorded as follows:

1. Class: 1 = freshman
            2 = sophomore
            3 = junior
            4 = senior
            5 = adult special

2. Major: 0 = education   5 = forestry
            1 = agriculture   6 = home economics
            2 = liberal arts   7 = nursing
            3 = biological sciences 8 = i.t.
            4 = business   9 = undecided

3. Number of math and other statistics courses previously taken or currently enrolled in.
4. Percent confidence of doing well in a statistics course.
5. Grade-point average (to nearest tenth): A = 4, B = 3, etc.
6. Height in inches.
7. Weight in pounds.
8. Life satisfaction index; range 10-70.

Main Campus: Males

No.	Class	Major	No. Math.	Conf.	GPA	Ht.	Wt.	Life
1	2	4	1	90	3.0	68	155	—
2	4	5	3	80	2.7	75	182	30
3	2	4	3	60	2.4	70.5	156	54
4	2	4	2	50	2.5	71	160	65
5	2	9	3	99	2.7	72	146	50
6	2	2	3	90	2.1	73	180	61
7	2	4	3	90	2.7	71	150	51
8	2	4	2	78	2.7	67	145	38
9	2	4	2	90	2.7	71	175	42
10	2	4	3	100	3.6	68	135	44
11	2	4	2	80	3.1	68	165	54
12	2	4	3	85	3.2	67	145	59
13	2	4	3	85	3.8	70	150	51
14	2	4	2	80	2.9	72	160	54
15	4	1	1	95	2.2	66	155	56
16	1	4	1	65	2.5	69	155	47
17	2	4	1	90	2.8	73	195	59
18	1	8	4	100	3.6	66	105	59
19	2	8	5	80	3.0	74	165	64
20	4	4	3	86	2.3	66	138	54
21	2	4	3	88	3.0	66	162	51
22	1	4	2	75	3.0	69	150	58
23	2	4	2	90	2.5	75	240	50
24	2	4	2	90	2.0	74	195	46
25	2	4	2	85	2.9	73	170	49
26	2	4	2	85	2.7	72	155	48
27	1	9	0	85	2.5	73	165	60
28	5	9	1	100	3.6	74	—	65
29	2	4	2	70	2.7	70	150	54
30	2	4	2	65	2.3	69	150	57
31	2	2	1	100	3.5	71	190	45
32	2	4	3	75	2.5	70	185	49
33	3	2	3	65	3.0	66	150	47
34	1	9	2	75	2.1	74	170	46
35	3	4	2	90	3.7	67	140	59
36	3	2	3	65	—	72	195	38
37	2	4	0	80	2.8	70	155	39
38	2	4	2	90	2.4	70.5	145	57
39	2	4	4	100	2.7	73	146	70
40	1	4	3	90	2.7	72	185	42

No.	Class	Major	No. Math.	Conf.	GPA	Ht.	Wt.	Life
41	1	4	1	95	2.8	75	175	61
42	2	4	3	65	2.2	69	145	33
43	2	4	3	100	3.0	74	160	62
44	2	4	3	85	2.5	70	160	62
45	2	4	1	75	2.5	65	120	64
46	2	4	1	88	3.0	71	165	59
47	2	4	2	70	2.6	73	175	44
48	1	9	0	99	2.8	70	150	38
49	2	4	3	88	3.0	69	133	53
50	1	3	0	100	3.9	77.5	190	56
51	4	4	2	100	3.5	73	175	66
52	2	8	3	99	3.2	69	150	54
53	2	4	3	99	3.0	71	170	51
54	1	2	0	60	3.0	69	140	50
55	2	4	3	100	3.6	71	155	58
56	2	4	2	88	2.9	67	140	50
57	2	4	3	50	2.8	70	165	64
58	4	1	2	85	2.1	70	160	55
59	2	4	1	85	2.6	71	165	54
60	3	4	2	80	2.7	68	170	46
61	2	4	2	90	2.6	69	145	56
62	2	4	3	90	2.2	68	150	57
63	5	2	0	85	2.6	70.5	160	56
64	2	4	3	95	3.3	70	145	45
65	4	2	0	75	3.3	71	180	54
66	4	2	3	75	3.0	74	185	56
67	2	4	3	90	3.2	70	158	16
68	1	4	3	90	3.0	71	145	34
69	1	2	0	95	2.7	69	140	59
70	3	4	0	85	3.5	72	165	20
71	2	4	2	95	3.0	74	135	90
72	2	4	2	75	3.1	70	160	18

Main Campus: Females

No.	Class	Major	No. Math.	Conf.	GPA	Ht.	Wt.	Life
73	2	4	2	65	2.6	63	145	55
74	1	4	1	90	3.0	66	150	53
75	4	6	0	75	2.9	67	132	53
76	1	8	3	80	3.0	66	140	57
77	2	8	1	95	2.2	66	158	64
78	2	4	2	80	2.8	61	100	59
79	3	2	3	89	3.3	63	150	47
80	2	4	2	100	3.7	65.5	111	67
81	2	4	2	75	2.9	62	110	51
82	2	0	2	85	—	69	118	58
83	4	1	2	95	3.3	71	165	65
84	4	2	3	75	2.6	62	105	63
85	5	7	3	90	3.5	63	115	57
86	2	4	2	95	2.5	63	115	63
87	3	5	1	30	2.5	66	125	29
88	2	4	2	60	2.5	67	120	52
89	2	4	2	90	3.0	62	110	48
90	2	4	2	90	3.0	63	135	48
91	3	4	3	80	2.5	62	115	32
92	2	4	3	88	3.1	64	106	49
93	2	4	3	80	3.0	64	115	55
94	4	2	4	90	2.5	64	125	43
95	4	2	7	90	3.8	64	125	60
96	2	4	2	90	2.8	72	150	57
97	1	3	4	95	3.0	68	125	51
98	3	—	0	96	2.8	66	140	50
99	3	4	2	85	3.2	63	135	54
100	2	4	2	90	3.7	65	110	52
101	2	1	1	85	2.7	63	145	49
102	2	4	3	100	3.4	70.5	140	55

Ag Campus: Males

No.	Class	Major	No. Math.	GPA	Ht.	Wt.	Life
103	2	4	1	2.4	69	145	42
104	2	5	2	2.3	75.5	165	34
105	3	4	2	2.5	73	180	42
106	1	1	1	2.7	70	160	57
107	4	1	1	2.9	83	150	36
108	3	4	3	2.4	69	161	55
109	2	4	0	2.5	74	170	57
110	3	5	3	2.8	70	155	48
111	2	4	2	2.3	72	151	42
112	3	1	2	2.0	68	160	45
113	2	4	3	2.7	71	157	53
114	3	1	2	2.9	72	157	60
115	3	1	2	2.0	72	195	60
116	2	4	3	2.7	70	155	51
117	3	1	3	2.6	71	180	51
118	2	2	1	2.8	70	170	53
119	2	1	2	2.9	72.5	210	49
120	4	6	1	3.0	72	170	63
121	2	4	2	3.2	70	155	47
122	2	4	2	2.3	81	170	54
123	2	1	1	2.2	71	160	51
124	2	4	5	2.8	72	206	52
125	3	1	2	2.5	73	155	86
126	3	2	2	2.2	75	193	49
127	1	1	1	2.9	70	150	57
128	3	1	1	2.8	71	180	56
129	2	4	3	2.8	73	165	52
130	2	5	3	3.1	66	145	58
131	3	1	2	3.0	70.5	138	—
132	4	1	1	3.0	68	120	68
133	3	1	2	2.5	70	167	56
134	3	5	3	3.2	56	150	55
135	2	5	1	2.6	74	200	51
136	4	1	1	2.0	74	195	51
137	2	4	3	3.0	74	175	59
138	2	4	2	2.9	72	150	59
139	2	4	2	2.7	72	174	44
140	4	4	2	2.9	72	155	49
141	2	1	2	2.2	69	148	42
142	2	4	3	2.7	70	170	50
143	2	2	2	2.8	74	175	54
144	2	5	2	3.4	71.5	170	45
145	2	2	3	2.8	72	155	54

AG Campus: Males (continued)

No.	Class	Major	No. Math.	GPA	Ht.	Wt.	Life
146	2	1	1	2.6	72	190	27
147	5	5	4	3.8	70	145	58
148	2	5	4	2.7	67	130	52
149	2	4	2	2.5	67.5	150	68
150	2	4	3	2.7	69	148	49
151	2	1	2	2.5	70	155	37
152	4	1	1	2.5	74	210	37
153	2	5	1	2.4	75	190	59
154	4	1	3	2.6	78	210	57
155	1	9	0	2.5	72	145	55

Ag Campus: Females

No.	Class	Major	No. Math.	GPA	Ht.	Wt.	Life
156	2	3	3	2.8	66.5	150	51
157	2	4	4	3.0	63	114	55
158	3	6	1	2.8	62	130	57
159	2	1	2	2.5	68	150	49
160	2	1	1	3.1	70	150	45
161	2	1	1	2.5	65	120	52
162	4	6	2	3.5	69.5	115	64
163	2	1	1	2.0	63	175	45
164	2	9	1	3.0	64	124	58
165	1	1	2	2.9	66	115	56
166	2	1	1	3.7	65	140	43
167	3	6	2	2.8	65	135	57
168	2	6	1	3.5	70	130	55
169	2	1	2	2.8	70	170	53
170	1	1	2	2.8	65.5	107	43
171	3	1	2	3.1	67	136	47
172	2	4	3	2.8	63	118	54
173	4	1	2	2.8	62	145	54
174	2	1	1	3.0	68	210	35
175	3	6	2	2.1	59	110	58
176	2	6	1	3.0	63	110	39
177	3	5	0	3.4	55	130	49
178	2	4	3	3.2	66	130	52
179	3	7	1	3.6	66	113	53
180	2	1	1	3.0	66	115	53

Ag Campus: Females (continued)

No.	Class	Major	No. Math.	GPA	Ht.	Wt.	Life
181	2	4	2	3.7	66	125	43
182	2	4	4	3.6	68	114	55
183	1	1	0	2.8	64	200	53
184	3	1	1	2.8	65	120	51
185	2	9	0	3.8	67	136	59
186	4	6	2	2.9	58	163	63
187	4	1	2	2.5	64	125	58
188	5	1	1	3.7	64	105	63
189	2	1	0	2.5	64.5	150	56
190	3	4	2	2.7	64	103	47
191	2	4	3	3.7	66	117	55
192	5	1	1	2.5	68	140	67
193	3	6	1	3.0	62	130	61
194	3	1	1	3.3	66	127	58
195	3	6	2	3.7	66	127	59
196	4	6	2	3.5	62	105	67
197	3	1	1	3.2	56	120	57
198	3	6	1	3.6	66	130	48
199	3	4	3	3.0	63	120	60
200	4	1	1	3.2	65	115	52

Answers to Problems

Chapter 1

1. The sample is apt to be biased, in an unknown amount and direction. (One could argue that responders on a tax issue tend to be opponents of a tax, but this might not be true if, say, the taxes were for schools; for school supporters might be even more vigorous in their support of a tax.) One thing that might be done is to obtain a personal interview with, say, 100 of the 1808 nonresponders (drawn at random) to see if the support percentage is appreciably different from that of the responders.

3. Although the 354 is better than the initial 313, those not reached would surely tend to be in households with few persons. Bias is assured (and unknown).

5. Neither method yields a random sample as defined here, although they both fall in the category of "probability sampling." [Method (a) is "cluster sampling" and (b) is a "two-stage sample," both of which can be analyzed. Neither is the same as random sampling, but they are often used to increase the amount of information per unit cost.]

7. (1) Two sources of bias: Leaving the choice up to the interviewer, and using a building apt to be frequented by a nontypical subpopulation. (2) This could be good if math students at a certain level are representative—but you would have no way of

knowing. Better than (1) only in that the interviewer has no choice. (3) This is the best of the methods listed—and the most expensive. (4) This is not a random sample, but if last names are not correlated with opinions on abortion, it could give reasonable results. A bit risky. (5) Bad method. Lolling students are apt to be unrepresentative, and interviewers would tend to select on the basis of personal attractiveness.

9. (a) It depends on your sample, but it is not likely that they will be.

(b) As in part (a), this is unlikely, but not as unlikely as in (a).

11. Starting at an arbitrary point of Table VIII, read (across or down) groups of two digits. The first digit will be the outcome of the first die, and the second, the outcome of the second die. (Ignore groups in which either digit is outside the range 1 to 6.)

13. (a), (b) Most sequences will have *some* peculiarity of pattern.

(c) *Any* given sequence of 20 would be hard to find. The two sequences given are equally likely to occur. (Chances are about 1 in a million.)

14. (a) He could assign patients to drug A if their prognoses are good, to drug B if their prognoses are bad, and divide the middle group to achieve the desired balance.

(b) He can devote more attention to those who received drug A, so any advantage experienced by that group might come from this extra treatment.

(c) He can choose to admit a patient with a favorable prognosis only if it is drug A's turn and to admit one with an unfavorable prognosis only if it is drug B's turn. [Here, as in parts (a) and (b), there are other, less extreme ways to bias the trial.]

15. If "popularity" is measured according to the proportion of the population, then it might be less in 1965 than in 1960—because the total population increased. (One would need to know the ratio of the population size in 1960 to that of 1965.)

16. Yes, in one sense—the *number* decrease is largest for CLA. But CLA is the largest college, and the *percentage* decrease is smallest.

18. No. It may be, for example that the hospital with the 49 percent rate specializes in the most difficult cases. Or, hospitals may have quite different policies regarding this surgery, using it routinely (10 percent) or as a last resort (49 percent). It would be important to know the rates for various categories of heart disease and other patient characteristics. It is actually possible

for the hospital with the 49 percent rate to have a lower rate than the other hospitals in *every* such category. In particular, you would want to know the rates for patients with characteristics and disease category similar to yours.

Chapter 2

1. In the picture for 1974, the circles' areas are not proportional to the frequencies. In the 1966 representation, the areas are proportional to frequencies except for the one on top, which has been rounded to give the appearance of a tombstone.

3.
(a)

Fr	So	Jr	Sr	AS
15	62	10	12	3

(b)

Ed	Ag	Arts	Bio	Bus	For	HE	Nur	IT	Undec
1	4	13	2	67	2	1	1	5	5

5.

	Class—Males					College	Class—Females				
1	2	3	4	5			1	2	3	4	5
2	5	8	5			Ag	3	9	4	3	2
	3	1				Arts					
						Bio		1			
	15	2	1			Bus		6	2		
	6	2		1		For			1		
			1			HE		2	6	3	
						Nur			1		
1						Und		2			

(a) 15/29   (b) 3/11

8.

		Husband										
		1			2			3			4		
		S	N	Q	S	N	Q	S	N	Q	S	N	Q
	S	2				1	2	2			1		
Wife	N	2			3			1	1	2	1	6	
	Q	1							1	1		1	2

11.

Counts/min:	12	14	16	18	20	24	28
Frequency:	26	1	56	5	44	7	3

(Since numbers divisible by 4 predominate, it seems clear that breaths were usually counted for 15 seconds, the number counted then multiplied by 4.)

13. The answer depends on your choice of class intervals. One choice is this:

Height	Freq.—Main	Freq.—Ag
53–55		1
56–58		2
59–61	1	1
62–64	15	15
65–67	9	18
68–70	3	8
71–73	2	

15. (a) 504   (b) 9 percent   (c) 16 percent   (d) 450

17. Main: 64 inches; Ag: 65 inches

18.

(a)

Depth	Stem	Leaves
1	9	1
5	10	7917
12	11	0434463
18	12	793104
—		
12	13	562555818673
—		
20	14	8622280826
10	15	6100407
3	16	393

(b)

Class Interval	Class Mark	Frequency
9.0–9.9	9.45	1
10.0–10.9	10.45	4
11.0–11.9	11.45	7
12.0–12.9	12.45	6
13.0–13.9	13.45	12
14.0–14.9	14.45	10
15.0–15.9	15.45	7
16.0–16.9	16.45	3

20. Median = 4.38, range = .55.

22.

Stem	Leaves
96	8
97	2
97	88
98	2202024022222424
98	686686868866886886666666668666666686668666666
99	40422020200402202024204204022442404224002020202444
99	68866666666686666
100	002000042040

Median = 99.0, Q_1 = 98.6, Q_3 = 99.4.

Note: Since temperatures, at the nursing instructor's direction, were rounded to the nearest even number of tenths, stems are unbalanced in this division into 0-4 (which includes 0, 2, 4) and 5-9 (which includes 6, 8). The implied histogram would be distorted because the class intervals would be unequal. (The bar heights could be adjusted to take this into account.) On the other hand, using only one stem for 97, one for 98, and one for 99 gives a less informative display.

24. Overall there seems to be a strong relationship, although in the first 20 days the relationship is weak. (Plotting the points on a scatter diagram loses track of the order of observation.)

26. The points will all lie on a line—the line whose equation is the relation between Fahrenheit and Celsius temperatures.

28. (a) There seems to be a moderate relationship.

(b)

Pressure	Frequency
146–150	1
140–144	2
134–138	0
128–132	9
122–126	5
116–120	6
110–114	10
104–108	6
98–102	8
92–96	2
86–90	1

Chapter 3

1. 14/3

3. (a) Av. hi = 38.53, av. lo = 22.5.

(b) May not be too bad. It would not take into account, for instance, how long the temperature was near the high and how long near the low; however, in Minneapolis, at least, it works out pretty well.

(c) Yes—but it is not necessarily the same as the overall average temperature for the month.

5. (a) No—this would not take into account the different numbers of sales in the various areas.

(b) Divide 694,407,759 by 12,615. Either someone made a mistake, or they mean another kind of average.

(c) The median would be better as a "typical" price. The data may not have been available for determining the median. (It cannot be calculated from what is given here.)

7. 5

9. (a) 16   (b) ————   (c) $\bar{X} = 4$

x	f
1	3
3	5
5	6
7	1
9	1

12. 22.9

14. 4.5 yards

16. $\bar{X} = 14.5, S = 5.45$

17. 2.46

19. (a) Mean = 26.43, median = 30.
    (b) About mean: 16.94; about median: 16.43.

20. (a) 17   (b) 6.6

22. 2.12

24. (a) Main: 2.88, Ag: 2.85
    (b) Main: .429, Ag: .435
    (c) Mean = 2.874, s.d. = .449

26. (a) Number 5   (b) x-values   (c) 3   (d) Positive   (e) Equal

28. $r = .844$

30. -.617

Chapter 4

1. (a) 8/13   (b) 8/13   (c) 3/13

3. (a) 8/20   (b) 12/20   (c) 6/20   (d) 4/00
   (e) 4/20   (f) 4/20

5. (a) 5/7   (b) No (you should get $8 to your $2 in a fair bet).

6. (a) 5/12, 5/13, 1/3, 1/6   (b) The bookie is permitted a profit.

8. (a) $8800 (15.4/35 of $20,000)
   (b) No—he'd "expect" $3,080,000 for his $7,000,000 investment, but he *could* win it all!

10. Mean = 7, s.d. = $\sqrt{35/6}$

12. (a) 36/52, 4/52, 4/52, 4/52, 4/52   (b) 10/13   (c) 10

14. (a) Black, 18/45; brown, 18/45; blue, 9/45 (same as for *first* sock drawn).

(b) No, the columns in the table are not proportional.

(c) 6/6 + 8 + 4 = 1/3 (consistent—if one black is gone, the remaining three are equally likely).

(d)

	Black	Brown	Blue
Black	4/25	4/25	2/25
Brown	4/25	4/25	2/25
Blue	2/25	2/25	1/25

15. About 1 chance in 10. [Exact: $(116/366) \cdot (115/365) = .09986$; approximate: $(116/366)^2 = .10045$.]

17. (a) $(1/2)^{3+1} = 1/16$ (b) $(1/2)^{k+1}$

(c) $P(0 \text{ or } 1 \text{ or } 2 \text{ or } 3) = (1/2)^1 + (1/2)^2 + (1/2)^3 + (1/2)^4$ $= 15/16$ [Alternatively: $1 - P(\text{first four are tails}) = 1 - (1/2)^4$]

19. 94.7 cents on each bet except the 5-split, for which it is 92.1 cents (including your dollar bet).

Chapter 5

1. (a) About .23   (b) About .22   (c) Just over 70

3. (a) 1/4   (b) .19   (c) .293

5. (a) .99 – .90 = .09   (b) .90 – .10 = .80

(c) .89   (d). 78   (e) About .804

6. (a) 19 percent   (b) 24 percent   (c) 31 percent

(d) 77 percent   (e) 41 percent

8. (a) 70 (may be a little less)   (b) 12

9. 1/4

11. (a) 1/8   (b) 1   (c) .41

12. (a) 678   (b) .31   (c) 68 percent   (d) 4.56 percent

14. Examples:

Score	Percent	Z	Percent, Normal
680	91	1.192	90.2
540	51	.154	56.1
420	21	– .821	20.6
300	5	–1.797	3.6

Chapter 6

1. (a) Mean = 20 percent, s.d. = .08 (or 8 percentage points).
    (b) Mean number = 5, s.d. of the number selected is 2.
3. (a) You might, but it would be extremely unusual.
    (b) $1/2; 2.5$ (d) s.d.$(\hat{p})$ = .224
5. .010
7. 625 (obtained by setting $p = 1/2$ in the formula for the s.d. of $\hat{p}$)
9. (a) The figures 9 out of 10 and 2 out of 3 are doubtless rounded, and in that sense they are inaccurate. If the sample were a random sample from some population of interest, the actual sample proportions would be very accurate estimates of the populations proportions. But there is not much point to computing a standard error, since the sampling is clearly not random. (We would expect a huge response bias.)
    (b) Doubts of a revival are neither dispelled nor supported by such a sample.
10. .06 [= $1 - \Phi(.11/.071)$]
12. (a) .1445   (b) .1444
14. Limits: .46 ± .031 (s.e. = .0158)
16. (a) .714 (s.e. = .0092)   (b) Limits: .714 ± .0237
18. 921,000 (Here it is assumed that p will be near $1/6$, and we use that value in estimating the s.d. of $\hat{p}$.)

Chapter 7

1. m.v.$(\overline{X})$ = 10, s.d.$(\overline{X})$ = 0.20
3. $(\mu = 116.80, \sigma/\sqrt{n} = 4.89)$
5. 0.84
7. (a) Increases it by a factor $\sqrt{2} = 1.414$.   (b) Multiply it by 4.
8. (a) .0062   (b) .0124
10. (a) .80   (b) .45
11. $246 < \mu < 274$
13. (a) $70.48 < \mu < 71.42$ (b) $70.22 < \mu < 71.68$
15. Limits: 10.88 ± .727
16. Limits: 38 ± 5.2 (assumes a population distribution not far from normal)
18. (a) 71.36   (b) Limits: 71.36 ± 2.39
    (c) Limits: .7136 ± .0239
    (d) Because the proportion of U's among 100 would be nearly

normal if these 100 are independent, and the *number* of U's is just 100 times the proportion (so only the scale factor changes). However, the 100 tacks may not fall independently when dumped from the box.

Chapter 8

1. (a) .114   (b) .0000 [= 1 - $\Phi(4.3)$]   (c) 19
   (d) 5 percent of 200, or 10; 1 percent of 200, or 2.
3. P = .0007 (Z = 3.2), "highly significant"
5. P = .0002 [This can be calculated as $(.9)^{80}$—the probability of getting a white 80 times in a row. Using a normal approximation does not work too well in this case.]
7. T = 1.18 (6 d.f.); P-value is just under 15 percent.
9. T = -.28 (7 d.f.), not significant. [*Note:* If you had used all of the data (n = 25), you would have got $\bar{X}$ = -1.36, S = 22.9, and T = -.29. But using a t-test when the population is known to be nonnormal (with a considerable probability that the error is 0) would be hard to defend. The hypothesis that the mean of the nonzero errors is 0 is actually equivalent to the hypothesis that the mean of all errors is 0.]
11. R_- = 14.5, which is not significant, although the table does not give entries for a P-value when R_- is this close to its mean (18).
13. (a) Z = 2.24; the evidence suggest a shift.
    (b) T = 1.67 (.05 $<$ P $<$ .10); evidence of shift is not as strong as in part (a).
    (c) R_- = 2, P = .094—some evidence of a shift, but not as clear as with the t-test in (b).
15. (a) χ^2 = 2.56 (1 d.f.), .10 $<$ P $<$ .20
    (b) Z = 1.6, P = .055.
    (c) Neither is significant. Although the P-values seem to be different, the tests are equivalent in a sense: Z^2 = χ^2 and $P(|Z| > 1.6) = P(\chi^2 > 2.56)$. That is, the right tail of the distribution of χ^2 corresponds to the union of the two tails of Z. So Z discriminates between $p > 1/2$ and $p < 1/2$, whereas χ^2 will be large if $\hat{p}$ is much different from p = 1/2 in *either* direction. (This illustrates some ambiguity in a "P-value" unless one indicates clearly which statistic is being used.)
17. χ^2 = 2.6 (3 d.f.), $P >$ .40
19. χ^2 = 5.6 (3 d.f.), .10 $<$ P $<$ .20 (not significant)
21. The group of 25 *is* better, on the average. (The P-value just

signifies that if this were a random sample from some population, there is strong evidence that the population mean is greater
than 521.) Whether or not a difference of 79 points is "significant" with regard to analytical ability is not a statistical question. (It no doubt *is* significant; an applicant with a score of 600
is at the 70th percentile, according to Table 5.4, page 226.)

23. (a) Test 1, whose critical region is included within that of test 2.
    (b) Test 1, for the same reason as in part (a).

25. (a) 2/36   (b) 2/36
    (c) Test (b) will reject fairness with probability 1/3 when the
    altered dice are used, whereas test (a) never will.
    (d) Reject fairness if the point total is 7 or 11.

27. (a) .0026
    (b) The test with the larger α has a better chance of detecting a
    shift in mean, and this is what the state would be concerned
    with. (The larger α just means unnecessary expense or delay on
    the part of the lab.)

29. To test $p \leq .05$ against $p > .05$, calculate Z for testing $p = .05$
    to be 2.05, which exceeds 1.645. Reject $p \leq .05$ at the 5 per-
    cent significance level.

Chapter 9

1. s.d.$(X_1 - X_2) = \sqrt{8.25 + 8.25} = 4.06$ (The s.d. of your sample
   of 20 differences will depend on the sample, but it should not
   be too different from this.)

3. 1.0, s.e. = .26

5. $Z = 2.82, P = .0024$

7. This is not a statistical question. (Data given are for the whole
   population of interest—there is no question of inference.)

9. $T = 1.82, P \doteq .04$; accept H_0 at the 10 percent significance level.
   However, this is a sample of convenience, and there may be
   hidden biases. Moreover, care regarding any cause-effect type of
   inference is especially needed here. (For example, smoking is
   disdained by many physical exercise enthusiasts, and exercise
   has a marked effect on pulse rate.) The assumption of equal
   population variance is also questionable.

11. Reducing errors: $Z = .95$, reducing time: $Z = .46$. Neither seems
    to be grounds for rejection, but a firm conclusion would require

knowing more of the nature of the sampling. (And even if the difference had turned out to be statistically significant, this would not necessarily mean that meditation caused the effect. This was not a controlled experiment, and any treatment effect is confounded with factors that cause people to take up meditation.)

12. (a) $R_x = 105$ (b) $|Z| = 2.59 > 1.96$ (significant)
13. $R_T = 63.5, P \doteq .015$
15. (a) $Z = -.30$ ($R_{sm} = 167.5$), not significant
    (b) $T \doteq .63$, not significant
17. No: $Z = -3.48$
19. (a) $Z = 1.59, P = .056$ (almost significant)
    (b) $Z = 2.53, P = .0045$ (highly significant)
21. $\chi^2 = 6.89$ (2 d.f.), $P \doteq .04$; one could agree, but the tendency is not marked. (One could also argue that for "pro-abortion" one need only consider the two categories "for" and "not for" and compare the proportions 320/760 and 360/760, using Z or a 2 X 2 contingency table. This way, $P \doteq .02$, but again the tendency is slight.)
23. $\chi^2 = 5.95$ (3 d.f.), $P \doteq .12$
25. (a) $R_+ = 3, P = .039$—reject H_0
    (b) $T = -2.46, P = .025$—again reject H_0
27. $R = 9, P = .234$—if the samples are random.
29. (a) $P = .23$ (b) $P = .23$ (c) $P = .36$

Chapter 10

1. Despite the terminology of "control group," this is not a controlled experiment. The investigator had no control over which students underwent the "treatment," and the factors that determined a student's choice may also determine his sleeping problems and jitters. So, although there is a relationship, it is not necessarily of the "cause-effect" type.
3. (a) 75/200 or 3/8
   (b) Yes (proportion of glasses wearers is the same in both subpopulations).
   (c) 120/200, or 60 percent
   (d) 45/75, or 60 percent
   (e) 30/200

(f) In general, it is not very likely to happen (one chance in six), but it could.

(g) Again, it is generally not very likely that they would—but they could be:

2	3	5
8	12	20
10	15	25

5. (a) 6 percent   (b) 25 percent

(c)

	Black	White	
Unemployed	.015	.045	.06
Employed	.085	.855	.94
	.10	.90	

(d) No (e.g., among unemployed, odds on black are 1:3, as opposed to 1:9 in the population).

7. $\chi^2 = 6.25$ (1 d.f.). If this can be considered a random sample from some population, the evidence is strongly against the hypothesis of independence in that population ($P \doteq .02$).

9. $\chi^2 = 9.08$ (2 d.f.), $P \doteq .01$. Evidence in the sample is that there is a relationship. (But then, there is almost certain to be *some* relationship, and a large enough sample would detect it. This sample was apparently large enough.)

11. One might expect that the longer a student was in school, the more classes he would have taken: $\chi^2 = 30.0$ (12 d.f.), $P < .01$.

12. $T = .187$, not at all significant.

14. $T = 1.21$, $P = .115$. (With regard to female students on campus, there is no obvious reason that those in a statistics class would not represent all students, as regards the relationship in question. Whether you would expect a relationship is a personal matter. Influenced by popular "standards," one might expect that an unusually heavy student might be unhappy with life on account of this; if so, the correlation would be negative. The data show some evidence of an association, but it is not compelling. If one variable does "cause" the other, it is not clear which would be the cause and which the effect. Other factors may account for the relationship.)

16. Using students nos. 2 to 21, $C = 1$. The data do not give evidence

of a relationship. (The remarkable aspect of the data is that students are willing and able to give a numerical measure of their satisfaction—to two significant figures!)

Chapter 11

1. (a) $-1, 4; 0/5, -2; 1/4, -7/4$
    (b) $x - 3y = -18, 2x + y = 3, 4x - y = 12$
2. (a) $y = 2x$ (b) 1 (c) r measures linearity $(r = \sqrt{5/6})$.
4. $y = 3.01x - 93.2$ [Moral: It is not necessary to believe something just because it appears in print.]
6. 24.02 ± 2.11 (11 d.f.)
8. $38.5; 21$ percent
10. 126 pounds, r.m.s.e. = 6 pounds

Chapter 12

1. $F = 1.8; (2, 6)$ d.f.
3. (a) $T = 1.30$, critical value = 2.45
    (b) $F = 1.69 = T^2$, critical value = $5.99 = 2.45^2$
5. SSTr = 1056, SSE = 3692; $F = 8.15 > 3.16 = F_{.95}(2, 57)$. Reject H_0.
6. $F = 17.0 > 5.78 = F_{.99}(2, 21)$. Reject H_0.
9.

Source	SS	d.f.	MS	F
A	294	2	149	49
B	6	2	3	1
Error	12	4	3	—
Total	312	8	—	—

11. (a) Driver and brand effects could not be separated. It is better to have each driver use each brand.
    (b) $F_{brand} = 9.86$, (3, 9) d.f. [Reject the hypothesis of no "brand effect" at $\alpha = .05$, and also at $\alpha = .01$: $F_{.99}(3, 9) = 6.99$.]
    (c) $F_{driver} = 176.7$—strong evidence of a driver effect.
    (d) One way, using the same number of runs, would be to have the drivers use car makes F, C, P, and V (say), according to this scheme:

		Brand			
		A	B	C	D
Driver	1	F	C	P	V
	2	V	F	C	P
	3	P	V	F	C
	4	C	P	V	F

Data obtained in this way can be analyzed. The pattern shown is called a *Latin square*. In it, each car make appears once with each driver and once with each brand; and each brand is used once by each driver and once in each make of car. (See more advanced texts under the subject "Design of Experiments.")

13. (a)

Source	SS	d.f.	MS	F	d.f. for F
A	108	2	54	54/14	(2, 2)
B	150	1	150	150/14	(1, 2)
Error	28	2	14	—	

(b) It depends on the way the list was compiled. Strictly speaking, the sample was not obtained as a random selection. But unless there is some connection between the order on the list and the characteristics tabulated, the sample might serve to give information about some population. (Of course, with the small-sample size here, the information would be limited.)

(c) No—the experiment is not controlled as to who gets what treatment combination.

(d) It is not feasible to assign the treatment (smoking) to subjects chosen at random for the experiment, so it is difficult to show statistically that smoking does or does not cause a high pulse rate. One might try taking a random sample of smokers and giving the "treatment" of making them give up smoking, to see if giving it up lowered the pulse rate—either as a "before-after" experiment, or by a comparison with a control group of nonsmokers. However, there are problems with this; those who give up smoking tend to compensate for it in other ways, such as running or eating, that would be apt to affect pulse rate. Another approach, sometimes used, would be to apply the treatment of smoking to animals; sometimes animals react to treatments in approximately the same way humans do.

Index